AF835719

Author
Dr. Christian Friedrich Steinwachs
School of Mathematical Sciences
University of Nottingham
Nottingham
UK

Supervisor
Prof. Dr. Claus Kiefer
Institut für Theoretische Physik
Universität zu Köln
Köln
Germany

ISSN 2190-5053 ISSN 2190-5061 (electronic)
ISBN 978-3-319-01841-6 ISBN 978-3-319-01842-3 (eBook)
DOI 10.1007/978-3-319-01842-3
Springer Cham Heidelberg New York Dordrecht London

Library of Congress Control Number: 2013947361

© Springer International Publishing Switzerland 2014
This work is subject to copyright. All rights are reserved by the Publisher, whether the whole or part of the material is concerned, specifically the rights of translation, reprinting, reuse of illustrations, recitation, broadcasting, reproduction on microfilms or in any other physical way, and transmission or information storage and retrieval, electronic adaptation, computer software, or by similar or dissimilar methodology now known or hereafter developed. Exempted from this legal reservation are brief excerpts in connection with reviews or scholarly analysis or material supplied specifically for the purpose of being entered and executed on a computer system, for exclusive use by the purchaser of the work. Duplication of this publication or parts thereof is permitted only under the provisions of the Copyright Law of the Publisher's location, in its current version, and permission for use must always be obtained from Springer. Permissions for use may be obtained through RightsLink at the Copyright Clearance Center. Violations are liable to prosecution under the respective Copyright Law.
The use of general descriptive names, registered names, trademarks, service marks, etc. in this publication does not imply, even in the absence of a specific statement, that such names are exempt from the relevant protective laws and regulations and therefore free for general use.
While the advice and information in this book are believed to be true and accurate at the date of publication, neither the authors nor the editors nor the publisher can accept any legal responsibility for any errors or omissions that may be made. The publisher makes no warranty, express or implied, with respect to the material contained herein.

Printed on acid-free paper

Springer is part of Springer Science+Business Media (www.springer.com)

Publications Based on this Work

1) Barvinsky, A.O., Kamenshchik, A.Yu., Kiefer, C., Starobinsky, A.A., Steinwachs, C.: Asymptotic freedom in inflationary cosmology with a nonminimally coupled Higgs field. J. Cosmol. Astropart. Phys. **12**, 003 (2009)

2) Barvinsky, A.O., Kamenshchik, A.Yu., Kiefer, C., Starobinsky, A.A., Steinwachs, C.: *The Higgs Field as an Inflaton.* In *Proceedings of the 12th Marcel Grossmann Meeting On General Relativity* (ed. T. Damour, R.T. Jantzen, R. Ruffini), pp. 1244–1246 (2009)

3) Barvinsky, A.O., Kamenshchik, A.Yu., Kiefer, C., Steinwachs, C.F.: Tunneling cosmological state revisited: Origin of inflation with a nonminimally coupled Standard Model Higgs inflaton. Phys. Rev. D **81**, 043530 (2010)

4) Steinwachs, C.F., Kamenshchik, A.Yu.: One-loop divergences for gravity nonminimally coupled to a multiplet of scalar fields: Calculation in the Jordan frame. I. The main results. Phys. Rev. D **84**, 024026 (2011)

5) Barvinsky, A.O., Kamenshchik, A.Yu., Kiefer, C., Starobinsky, A.A., Steinwachs, C.: Higgs boson, renormalization group, and naturalness in cosmology. Eur. Phys. J. C**72**, 2219 (2012)

6) Steinwachs, C.F., Kamenshchik, A.Yu.: One-loop divergences for gravity nonminimally coupled to a multiplet of scalar fields: Calculation in the Jordan frame. II. Comparison with the Einstein frame results. (in preparation)

Supervisor's Foreword

The field of cosmology is undergoing an exciting phase of development. Observations have reached an amount of precision that can compete with the precision known from other fields of physics. The most recent manifestation of this state is the release of data from the PLANCK mission in March 2013. An important aspect of current cosmology is its deep connection with particle physics and quantum theory.

The present book gives an impressive testimony of this connection. It is based on the author's Ph.D. thesis, which he completed at the University of Cologne in 2012. Two of the main features of current cosmology have to do with an accelerated expansion of the Universe. As in particular the observations of type Ia supernovae indicate, our Universe is presently accelerating. The reason for this may be a positive cosmological constant or a so far unknown form of dark energy. The second type of acceleration is more speculative and has probably happened when the Universe was younger than 10^{-30} s. This phase is called inflation. Although there is strong support for inflation for both conceptual and empirical reasons, the exact mechanism leading to inflation has not yet been revealed. Most models make use of one or more scalar fields, whose dynamics is more or less introduced in an ad hoc way.

In this book, the idea is entertained that the scalar field responsible for inflation is the one scalar field that has so far been observed: the Higgs field of the Standard Model of the strong and electroweak interaction. The corresponding Higgs particle was detected at the LHC in Geneva in 2012 and possesses a mass of around 126 GeV. As has been clear for some time, *Higgs inflation* will only work if an additional coupling of the Higgs field to gravity is included—a strong non-minimal coupling of the field to the Ricci scalar. The consequences of this coupling are the main topic of this monograph.

In the first half of his book, Christian Steinwachs presents a detailed and pedagogical introduction to all the concepts that are needed to understand modern cosmology in general and Higgs inflation in particular. These include, on the one side, the Friedmann models, the inflationary scenario, and cosmological perturbation theory, and, on the other side, the Standard Model of particle physics, effective actions, and renormalization group equations. This part of the book is self-contained and can be read as an accessible introduction into the field itself, even if one is not interested in Higgs inflation.

In the second half, the mechanism of Higgs inflation is discussed at length, and its merits and shortcomings are critically evaluated. This part also addresses the more general question of the frame equivalence in cosmology, in particular that of the JORDAN and the EINSTEIN frame. It is shown that an equivalence is not present at the one-loop level of quantum field theory, and consequences are drawn from this result. I have so far not seen a clearer discussion of this point in the literature.

The monograph concludes with an important section on quantum cosmology. After a concise introduction into this field, it is shown that a certain boundary condition leads to a wave function that is peaked around a sufficiently high initial value of the inflaton field, which is necessary for inflation to occur in the first place.

The book is recognizable for its interdisciplinary character and should be of interest for cosmologists and particle physicists. I am sure it will find many readers.

University of Cologne, Cologne, June 2013 Prof. Dr. Claus Kiefer

Abstract

In this thesis, we investigate the cosmological model of non-minimal Higgs inflation. The basic idea is to unify the "low-energy" sector of the Standard Model with the high-energy phase of inflation during the early universe. Both sectors are described by different theories and rely on the existence of a fundamental scalar field. The scalar Higgs boson is an integral part of the Standard Model in order to maintain gauge invariance while explaining the origin of the masses of elementary particles. Likewise, in cosmology the scalar inflaton is necessary to describe inflation, the early phase of accelerated expansion of the universe. The basic assumption of the Higgs inflation scenario is that the two scalar particles are manifestations of one and the same particle, the *Higgs-inflaton*. Consistency with observational data of the cosmic microwave background radiation requires the assumption of a strong non-minimal coupling of the Higgs-inflaton to gravity. It turns out that quantum corrections of the heavy Standard Model particles mainly determine the cosmological parameters of the model and are crucial for the predictions. Moreover, we resort to the renormalization group flow in order to connect the energy scale of the electroweak vacuum with the inflationary energy scale, separated by many orders of magnitude. We derive a range of possible values for the mass of the Higgs-inflaton. These predictions can be tested by future experiments at the Large Hadron Collider (LHC) and by the satellite PLANCK.

We further investigate an application of the non-minimal Higgs inflation model in the context of quantum cosmology. Via the path integral approach, we calculate the cosmological probability distribution for the universe to tunnel "from nothing to existence". A sharp peak in this distribution—selecting a specific value of the inflaton field—can be interpreted as the initial condition for inflation. The predicted value can be tested by the detection of primordial gravitational waves.

In order to technically realize these models, we calculate the divergent part of the one-loop off-shell effective action of a *O(N)*-symmetric multiplet of scalar fields non-minimally coupled to gravity in a closed form. Due to the functional nature of the couplings, the result is very general and can be used as a reference for many different cosmological applications.

In cosmology it is convenient to transform between two specific field parametrizations, denoted *Jordan frame* and *Einstein frame*. There is an ongoing discussion whether these two frames are physically equivalent. Mathematically, both frames are equivalent at the classical level. We explicitly show that this does not hold anymore at the quantum level. Moreover, we identify this failure of equivalence to be part of a more general problem: the parametrization dependence of the conventional off-shell effective action. This reduces the cosmological debate between Jordan frame and Einstein frame to one problem: the non-covariance of the conventional formalism.

Acknowledgments

I would like to thank all those who supported me in completing this thesis. In particular, I would like to express my gratitude to those who guided and accompanied me during the last years:

First of all, I am indebted to the Supervisor of my thesis, Prof. Dr. Claus Kiefer. I did not only benefit from his wide range of knowledge, but I also had the privilege to learn from numerous discussions with him about fundamental conceptual questions of the cosmos and quantum theory. He introduced me to his longterm collaborators and offered me the possibility to actively participate in joint projects, while also providing me with the freedom to pursue my own scientific ideas. Moreover, he encouraged me to present the results of our common work at various international conferences and generously supported my participation.

I owe gratitude to my collaborators, who kindly shared their knowledge with me. Most notably, I am deeply obliged to Prof. Dr. Alexander Yu. Kamenshchik, who always patiently answered all my questions. I strongly benefited from his extensive skills and his helpful advice. I am also grateful for the opportunity of working with Prof. Dr. Andrei O. Barvinsky, an expert on the mathematical techniques used in this thesis. Moreover, I am indebted to Prof. Dr. Alexei A. Starobinsky for illuminating conversations. Being a leading expert in cosmology, his rich ideas and helpful comments strongly influenced the direction of our common work.

I would also like to express my gratitude to Prof. Dr. Friedrich W. Hehl, who provided me with literature about computer algebra systems and gave me valuable advice.

It is a pleasure to thank Priv.-Doz. Dr. Rochus Klesse for interesting and stimulating discussions, especially on the process of decoherence.

I am thankful to Prof. Dr. Dominik J. Schwarz and Prof. Dr. Martin Reuter for kindly accepting to be the co-referees of this thesis.

I am grateful to my friend and long-term office mate Dr. Friedemann Queißer for the numerous interesting discussions during our coffee breaks. Furthermore, my thanks go to Manuel Krämer for his very accurate orthographic proofreading of large parts of this thesis.

I am thankful to Prof. Dr. Steven Christensen for sending me the latest version of the MathTensor package and to Dr. Andreas Sindermann for taking care of its proper installation.

Finally, I owe special thanks to the "Evangelisches Studienwerk Villigst e.V." for the financial support of my doctoral project.

Aside from the scientific, technical and financial support I have experienced during the years, I would like to thank Simin Askari and my parents Gabriele and Burkhart Steinwachs, who constantly encouraged me to follow my interests and always provided me with their help and advice.

Contents

Chapter 1
Introduction

Nowadays, particle physics as well as cosmology are on the edge of new experimental progress. During the next years, the Large Hadron Collider at CERN as well as satellites like PLANCK will advance to energy scales and to a level of accuracy that have been unreached so far. From a theoretical perspective, the hitherto covered experimental scales are mainly described by the Standard Model of Particle Physics and the cosmological ΛCDM model that is ultimately based on Einstein's theory of General Relativity.

However, theoretical considerations as well as the so far unexplained microscopic origin of observed phenomena, like e.g. dark energy, suggest that the Standard Model and General Relativity do not have a fundamental character. Rather, they have to be considered as effective theories, valid only within a limited range. In this thesis, we will adopt this pragmatic viewpoint of an effective field theory. We expect that there is an underlying, more fundamental theory, but we are not interested in its details. Instead, we will focus on the well-established theories of the Standard Model and cosmology and look for a minimal extension in order to unify them in accordance with experimental data.

When trying to connect the Standard Model with inflationary cosmology, it is natural to look for similarities. Both theories share a common feature. An integral part of both theories is a fundamental scalar field. In the Standard Model, it is called Higgs boson and it is responsible for the gauge-invariant generation of masses, whereas in cosmology it is called inflaton and it is a necessary means to describe the phase of accelerated expansion in the early universe. None of these scalar fields has been detected directly so far. It is thus natural to ask whether we can go beyond this formal similarity and identify both particles just as two manifestations of one and the same particle, the *Higgs-inflaton*. This would not only reduce the number of hypothetical particles, but it would offer the possibility to extend the Standard Model to a quantum field theory valid up to inflationary energy scales. In order to make the model compatible with current observational constraints, it is indispensable to assume in addition a *strong non-minimal coupling* of the Higgs-inflaton to gravity.

C. F. Steinwachs, *Non-minimal Higgs Inflation and Frame Dependence in Cosmology*, Springer Theses, DOI: 10.1007/978-3-319-01842-3_1,
© Springer International Publishing Switzerland 2014

Since the Standard Model is a quantum field theory, we have to include quantum corrections. The appropriate tool for such an analysis is the effective action. Moreover, since we intend to connect two theories at energy scales that are separated by many orders of magnitude, we have to account for the dependence of the physical parameters on the energy scale. Therefore, we have to consider the renormalization group improvement. The renormalization group flow allows to establish the connection between the low energies at the symmetry breaking scale of the Standard Model and the high energies during inflation.

This approach is obviously limited by a cut-off scale, where the effective description breaks down and the effects of the more fundamental theory become dominant. Thus, in order to make reliable predictions, it is of crucial importance to find this limiting scale. In terms of the effective action, this means that the influence of higher-dimensional operators is not negligible anymore, as they are supposed to be reliably suppressed by powers of this cut-off.

In this thesis, we will try to cover all—physical as well as mathematical—aspects of the non-minimal Higgs inflation scenario and its further applications.

Since we intend to bring together the formalism of the Standard Model and that of inflationary cosmology, we will present in detail an elaborate review of their basic concepts. We will derive and explain all formulas required for the further analysis.

More precisely, in Chap. 2, we will summarize the principles of cosmology and the mechanism of inflation. We will present a derivation of all the inflationary parameters whose quantum modification will become the basis for numerical predictions of the non-minimal Higgs inflation scenario discussed in Chap. 6.

In Chap. 3, we will explain how the different fundamental interactions and particles are organized in the Standard Model. In addition, we will explain the phenomenon of spontaneous symmetry breaking and show how the Higgs mechanism leads to the generation of masses. The derived formulas for the connection between the masses of the elementary particles and the Higgs field will provide the basis for the matter sector of the non-minimal Higgs inflation model.

In Chap. 4, we will introduce the concept of regularization, renormalization and the renormalization group. Furthermore, we will supply the mathematical knowledge and the technical tools to calculate the divergent part of the one-loop effective action. These techniques are essential for the calculations in Chap. 5.

After this extensive introductory part, we will analyse the mathematical aspects, the physical predictions and a quantum cosmological application of the non-minimal Higgs inflation model in the subsequent chapters.

In Chap. 5, we will use the Schwinger–DeWitt formalism in order to calculate the divergent part of the one-loop effective action for a $O(N)$-symmetric multiplet of scalar fields non-minimally coupled to gravity. We will first consider the simpler case of a single scalar field minimally coupled to gravity, which will serve as a guideline and an important cross-check for the more general model.

In cosmology it is convenient to work with two particular field parametrizations, denoted by *Jordan frame* and *Einstein frame*. We will investigate how the effective action depends on the choice of this parametrization. Moreover, the universal results of Chap. 5 will serve as a basis for the quantum investigation of the non-minimal Higgs inflation model in Chap. 6.

The core of this thesis, Chap. 6, deals with the derivation of physical predictions from the non-minimal Higgs inflation model. The numerical predictions can be tested by future experiments.

Finally, in the Chap. 7, we will apply the results of Chap. 6 within a path integral approach to the quantum cosmological tunnelling scenario. The more fundamental character of this application and its adequate integration into the underlying theoretical framework also requires the preliminary discussion of some conceptual questions.

We will mostly set $\hbar$, c and G equal to one. However, at some places, where we feel that it is necessary to emphasize certain aspects connected to the powers of one or more of these quantities, we will temporarily re-introduce these symbols.

Chapter 2
Cosmology

From the four types of interactions present in our world, only electromagnetism and gravity are long-range interactions. The weak and the strong interactions are so short-ranged that they cannot be responsible for the large-scale behaviour of our universe. However, despite its long range character, electromagnetism cannot be the dominant force in an electrically neutral universe. The only long range interaction that remains is gravity, whose source is energy density. Thus, a theory of the cosmos must be based on a theory of gravity. The best gravitational theory we have so far is Einstein's theory of General Relativity [9, 10]. In contrast to the Standard Model, which is a *quantum* field theory, General Relativity is a *classical* field theory. Aside from this, the distinctive feature of gravity compared to the other fundamental interactions is that it is not a theory defined *on* space-time, but a theory *of* space-time itself. The physical foundation of General Relativity is the equivalence principle, which states that gravity *uniformly* couples to *all* kind of energy density. Mathematically, space-time is described by a pseudo-Riemannian manifold $\mathcal{M}$ and the gravitational field by the metric field $g_{\mu\nu}(x)$ on $\mathcal{M}$. The gravitational interaction manifests itself geometrically as curvature of space-time.

2.1 General Relativity

General Relativity has replaced Newton's theory of gravity, but we can recover it in the limit of small velocities and vanishing curvature. General Relativity is a well-tested theory [31]. However, Einstein's theory is not only more accurate than Newton's, but also compels us to modify the fundamental concepts of space and time. The geometrical interpretation of gravity as curvature of space-time suggests that the gravitational action must contain information about the curvature. The action is a scalar functional and the simplest scalar that can be formed of the Riemannian curvature tensor $R_{\sigma\rho\mu\nu}$ is the Ricci scalar R, see Appendix A for a definition of curvature quantities. The Ricci scalar R is already of second order in the derivatives

C. F. Steinwachs, *Non-minimal Higgs Inflation and Frame Dependence in Cosmology*, Springer Theses, DOI: 10.1007/978-3-319-01842-3_2,
© Springer International Publishing Switzerland 2014

of the metric field $g_{\mu\nu}$. In order to obtain the correct Newtonian limit, the action must in addition be proportional to Newton's constant G. From dimensional considerations in four space-time dimensions, it follows that this prefactor must be proportional to $\kappa := 8\pi G/c^4$ or $\kappa = 1/M_{\rm P}^2$ in units $c = \hbar = 1$. Thus, scalar contractions of higher powers in the curvature would already correspond to higher derivative theories. For a general invariant volume element $\sqrt{-g}\,\mathrm{d}^4\mathrm{x}$, the simplest action is therefore given by

$$S_{\rm g} := \frac{c^4}{16\pi G}\int \mathrm{d}^4x\,\sqrt{-g}\,\left(R - 2\,\Lambda\right) + \text{surface terms}, \tag{2.1}$$

with the cosmological constant Λ and surface terms which we neglect in the following. This action is denoted Einstein–Hilbert action and describes the geometrical degrees of freedom. The matter content is encoded in

$$S_{\rm m} := \int \sqrt{-g}\,\mathrm{d}^4x\;\mathcal{L}_{\rm m}. \tag{2.2}$$

We define the total gravitational and matter action as

$$S := S_{\rm g} + S_{\rm m}. \tag{2.3}$$

The first functional derivative of (2.1) with respect to the metric field $g_{\mu\nu}(x)$ yields

$$\frac{\delta S_{\rm g}[g]}{\delta g_{\alpha\beta}} = \int \frac{\mathrm{d}^4x}{2\,\kappa}\left[\frac{\delta\sqrt{-g}}{\delta g_{\alpha\beta}}\,\left(R - 2\,\Lambda\right) + \sqrt{-g}\,\frac{\delta\,R}{\delta g_{\alpha\beta}}\right]. \tag{2.4}$$

Using the formulas derived in Appendix B.4, we find[1]

$$\frac{\delta R(x')}{\delta g_{\alpha\beta}(x)} = \Big[-\,R^{\mu\nu} + \nabla^\mu\nabla^\nu - g^{\mu\nu}\Box\Big]\delta^{\alpha\beta}_{\mu\nu}\,\delta(x,x'). \tag{2.5}$$

Inserting this into (2.4) and using again the formulas of Appendix B.4 yields

$$\frac{\delta S_{\rm g}[g(x')]}{\delta g_{\alpha\beta}(x)} = \int \mathrm{d}^4x\,\frac{\sqrt{-g}}{2\,\kappa}\,\left(\frac{1}{2}\,R\,g^{\mu\nu} - R^{\mu\nu} - \Lambda\,g^{\mu\nu} + \nabla^\mu\nabla^\nu - g^{\mu\nu}\Box\right)\delta^{\alpha\beta}_{\mu\nu}\,\delta(x,x'). \tag{2.6}$$

The last two terms in (2.6) are total derivatives and can be converted into surface terms. Integrating out the delta function in (2.6), the variation of (2.3) yields

$$\frac{\delta\,S}{\delta\,g_{\alpha\beta}} = \frac{\sqrt{-g}}{2\,\kappa}\,\left(\frac{1}{2}\,R\,g^{\alpha\beta} - R^{\alpha\beta} - \Lambda\,g^{\alpha\beta}\right) + \frac{\delta\,S_{\rm m}}{\delta\,g_{\alpha\beta}}. \tag{2.7}$$

[1] We use the sign convention: signature $(-1,1,1,1)$, $+R^\alpha{}_{\beta\gamma\delta} = \Gamma^\alpha{}_{\beta\delta,\gamma} - \ldots$ and $R^\sigma{}_{\mu\sigma\nu} = +R_{\mu\nu}$.

We define the energy-momentum tensor by

$$T^{\alpha\beta} := \frac{2}{\sqrt{-g}} \frac{\delta S_{\mathrm{m}}}{\delta g_{\alpha\beta}}. \tag{2.8}$$

The extremal condition $\delta S / \delta g_{\alpha\beta} = 0$ leads to the Einstein field equations

$$R_{\mu\nu} - \frac{1}{2} g_{\mu\nu} R + g_{\mu\nu} \Lambda = \kappa T_{\mu\nu}. \tag{2.9}$$

They describe the interplay between the matter content and the geometry of space-time and form a set of ten coupled non-linear partial differential equations. With the definition of the Einstein tensor

$$G_{\mu\nu} := R_{\mu\nu} - \frac{1}{2} g_{\mu\nu} R, \tag{2.10}$$

we can write the field equations (2.9) in a more compact form

$$G_{\mu\nu} + \Lambda g_{\mu\nu} = \kappa T_{\mu\nu}. \tag{2.11}$$

Using the contracted Bianchi identities and the symmetries of the Riemann tensor derived in Appendix A, we find the covariant identity

$$\nabla^{\mu} G_{\mu\nu} = 0. \tag{2.12}$$

In combination with the metric compatibility relation $\nabla_{\alpha} g_{\mu\nu} = 0$, this implies the *covariant* conservation law for the energy-momentum tensor

$$\nabla^{\mu} T_{\mu\nu} = 0. \tag{2.13}$$

2.1.1 Symmetry Reduction

We are interested in solutions of (2.11) which can describe, at least approximately, our observable universe. Empirical observations suggest that the spatial part of our universe is homogeneous and isotropic averaged over large scales $\gtrsim$100 Mpc. Since there is no evidence why we should have a distinguished position in space, we should not only demand isotropy with respect to *our* position, but with respect to *all* locations in the universe. This additional assumption is called "cosmological principle". It implies that on large scales the universe looks the same for *all* observers, no matter at which *spatial* point they are placed.

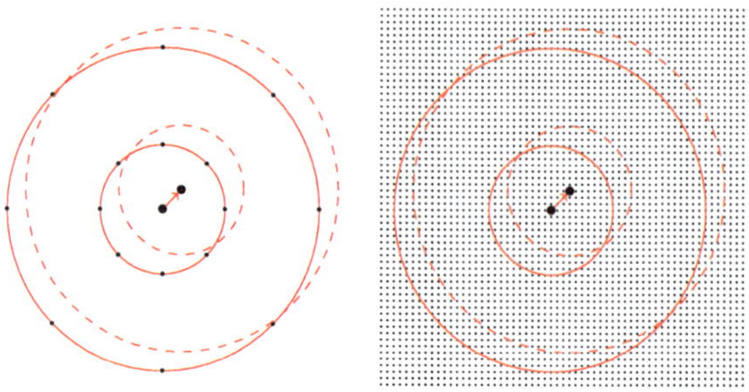

Fig. 2.1 The cosmological principle. The *black dots* represent galaxies. *Left* Isotropy only around the centre. *Right* Isotropy with respect to all points

2.2 Friedmann–Robertson–Walker Universe

By assuming a globally hyperbolic manifold with topology[2] $\mathcal{M} = \mathbb{R} \times \Sigma$, we can follow the Arnowitt–Deser–Misner-method [2] and decompose a four-dimensional manifold $\mathcal{M}$ into slices of three-dimensional hypersurfaces Σ_t at constant time t

$$ds^2 = \left[-N^2(\mathbf{x},\, t) + N_i(\mathbf{x},\, t) N^i(\mathbf{x},\, t)\right] dt^2 + 2\, N_i(\mathbf{x},\, t)\, dt\, dx^i + q_{ij}(\mathbf{x},\, t) dx^i\, dx^j \tag{2.14}$$

with spatial indices $i, j = 1, \ldots, 3$, the lapse function $N(\mathbf{x},\, t)$, the shift vector $N_i(\mathbf{x},\, t)$ and the three-dimensional spatial metric $q_{ij}(\mathbf{x},\, t)$ on Σ_t.

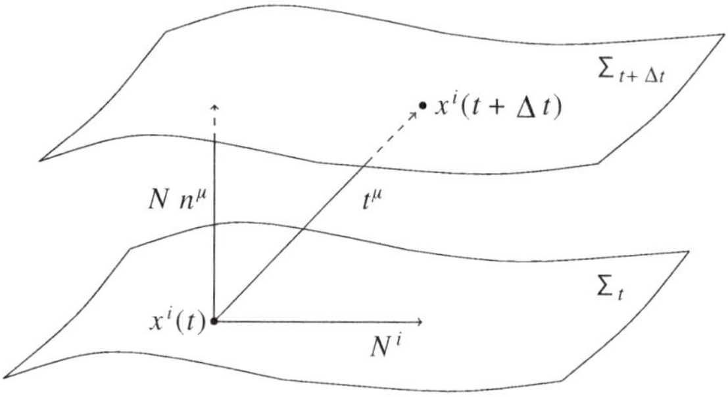

Fig. 2.2 We can decompose the full metric $g_{\mu\nu} = q_{\mu\nu} - n_\mu n_\mu$ into a part $q_{\mu\nu}$ that projects on Σ_t and a part orthogonal to Σ_t. The projector $q_{\mu\nu}$ satisfies $q_{\mu\nu} n^\nu = 0$ and $q_{\mu\nu} q^{\nu\rho} = q_\mu^{\ \rho}$, so that we can identify $q_{\mu\nu}$ with the induced three-dimensional metric q_{ij} on Σ_t. A *comoving* observer flows from $x^i \in \Sigma_t$ to $x^i \in \Sigma_{t+\Delta t}$ by means of the displacement vector t^μ. We have decomposed $t^\mu = N\, n^\mu + N^\mu$ into a component n^μ orthogonal and a component N^μ tangential to Σ_t. It is clear that N^μ is also a three-dimensional object, so that we can identify $N^\mu \equiv N^i$. The normal n^μ is a time-like unit vector $n_\mu n^\mu = -1$ so that the relation $N = -t^\mu n_\mu$ follows

[2] A globally hyperbolic manifold is equivalent to the existence of a Cauchy surface, i.e. given some initial data on a 3-hypersurface Σ_t, the evolution is uniquely determined by the equations of motion. There are also theoretical attempts to investigate whether our universe has a different topology, e.g. a torus, on the basis of experimental data, see e.g. [3, 6, 7, 18].

The symmetries of spatial homogeneity and isotropy are connected to three-dimensional translation and rotation invariance. Translation invariance means that $N(\mathbf{x}, t)$, $N_i(\mathbf{x}, t)$, $q_{ij}(\mathbf{x})$ in the decomposition (2.14) have to be independent of $\mathbf{x}$. Rotational invariance requires $N_i(t) \equiv 0$. Therefore, we can write (2.14) as

$$\mathrm{d}s^2 = -N^2(t)\,\mathrm{d}t^2 + q_{ij}(t)\mathrm{d}x^i\,\mathrm{d}x^j \tag{2.15}$$

Due to these symmetries, the ten independent degrees of freedom contained in the symmetric metric tensor $g_{\mu\nu}(\mathbf{x}, t)$ reduce to one single degree of freedom, the scale factor $a(t)$ which is a function of time t only. In contrast to $a(t)$, the choice of the lapse function $N(t)$ has no physical meaning. A different choice of time can be parametrized by a different choice of the function $N(t)$. It is a pure gauge freedom, reflecting the fact that General Relativity is invariant under a reparametrization of time. Therefore, the lapse function $N(t)$ is no dynamical degree of freedom. Mathematically, it corresponds to a Lagrange multiplier in the action. Variation of the action with respect to N does not yield an equation of motion for N but a *constraint* equation instead. In fact, the homogeneous and isotropic line element (2.15) describes a maximally symmetric space-time which has $d\,(d+1)/2$ linearly independent Killing vectors, see (B.13). It is therefore a space of constant curvature parametrized by the real number $\mathfrak{K}$. The spaces of constant curvature can be divided into three different classes $\mathfrak{K} < 0$, $\mathfrak{K} = 0$, $\mathfrak{K} > 0$. These classes can be represented by the three values $\mathfrak{K} = -1, 0, 1$ for which the three-dimensional spatial slices Σ_t take the form of a three-dimensional hyperboloid, cube or sphere. According to these considerations, we can write (2.15) as

$$\mathrm{d}s^2 = -N^2(t)\,\mathrm{d}t^2 + a^2(t)\left(\frac{\mathrm{d}r^2}{1-\mathfrak{K}r^2} + r^2\,\mathrm{d}\Omega^2\right). \tag{2.16}$$

This line element describes a Friedmann–Robertson–Walker (FRW) space-time in spherical coordinates (t, r, θ, ϕ), i.e. $\mathrm{d}\Omega^2 := \mathrm{d}\theta^2 + \sin^2\theta\,\mathrm{d}\phi^2$.

2.2.1 Physical Versus Comoving Distance

The use of comoving quantities will simplify many considerations. We imagine for a moment a one dimensional spatially flat universe as a rubber band. The galaxies are attached to this rubber band at fixed positions, labeled by *fixed* coordinates $x_1, x_2, \ldots, x_n$. Changing the scale factor $a(t)$ means to stretch or shrink the rubber band, i.e., to separate or squeeze the galaxies. Since each galaxy has a fixed coordinate label x^i, homogeneity requires that the comoving *coordinate difference* $\Delta x = |x_i - x_j|$ is the same for all galaxies and remains the same for all times. The coordinates are "streaming" with the galaxies. However, the actual *physical distance* between the galaxies grows or shrinks according to the expansion or shrinking of the scale factor $a(t)$

$$D = a(t)\,\Delta\,x. \tag{2.17}$$

Whenever we feel that it is important to distinguish between a physical or comoving quantity, we indicate them with a subscript "p" or "c" respectively.

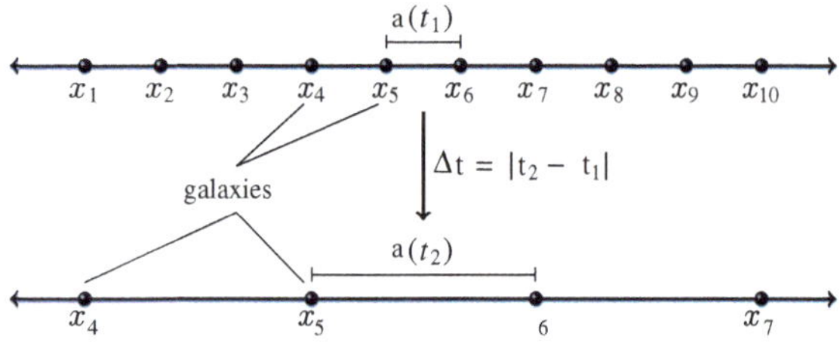

Fig. 2.3 The difference between the *comoving* coordinate difference Δx and the real *physical* distance between two neighbouring galaxies

2.2.2 Conformal Time

There is a very useful notion of time, called conformal time, which is defined by

$$\mathrm{d}\eta := \frac{N(t)\,\mathrm{d}t}{a(t)} \qquad \longleftrightarrow \qquad \eta_{\mathrm{f}} - \eta_{\mathrm{i}} = \int_{t_{\mathrm{i}}}^{t_{\mathrm{f}}} \frac{\mathrm{d}t\,N(t)}{a(t)}. \tag{2.18}$$

In terms of η, the FRW line-element (2.16) can be written as

$$\mathrm{d}s^2 = a^2(\eta)\left\{-\mathrm{d}\eta^2 + \left[\mathrm{d}\chi^2 + f(\chi)(\mathrm{d}\theta^2 + \sin^2\theta\,\mathrm{d}\phi^2)\right]\right\}, \tag{2.19}$$

$$\text{with} \quad f(\chi) := \begin{cases} \sinh^2\chi & \text{for } \mathfrak{K} = -1 \\ \chi^2 & \text{for } \mathfrak{K} = 0 \\ \sin^2\chi & \text{for } \mathfrak{K} = +1. \end{cases} \tag{2.20}$$

In the flat case $\mathfrak{K} = 0$, we can identify χ with the radial coordinate r and (2.19) is related to the Minkowski line element by the conformal factor $a^2(\eta)$. The causal structure is fixed by null geodesics which satisfy $\mathrm{d}s^2 = 0$. Thus, for radial null geodesics we can draw the usual light cone structure as for Minkowski space.

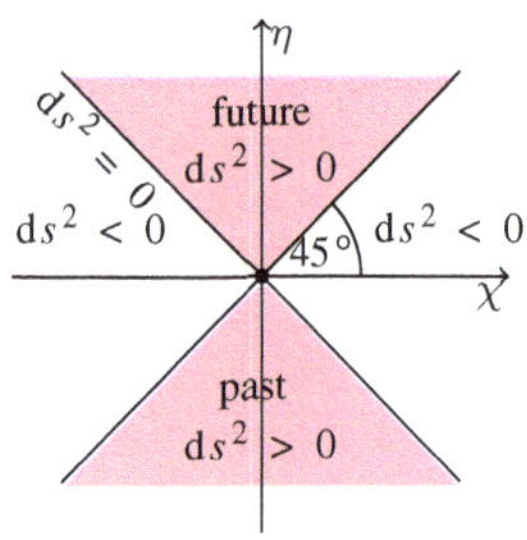

Fig. 2.4 Causal structure for radial null geodesics in conformal time η

2.2.3 Friedmann Equations

So far, we have only considered the kinematical aspects of the FRW universe. In order to describe the dynamics, we have to perform the symmetry reduction either at the level of the action (2.3) or directly for the Einstein equations (2.11). The geometrical part, including $R^{\rho}{}_{\mu\nu\sigma}$, $R_{\mu\nu}$, R and $\sqrt{-g}$, can be calculated for the line element (2.16). The square root of the determinant is given by

$$\sqrt{-g} = N\,a^3\,\frac{r^2 \sin\theta}{\sqrt{1-\mathfrak{K}\,r^2}}. \tag{2.21}$$

In the following, no sum is performed over identical Latin indices unless it is explicitly stated. The non-vanishing components of the Christoffel symbols are

$$\Gamma^0_{00} = \frac{\dot{N}}{N}, \quad \Gamma^i_{i0} = \frac{\dot{a}}{a}, \quad \Gamma^0_{ii} = \frac{\dot{a}}{a\,N^2}\,q_{ii}, \tag{2.22}$$

where we have used the spatial part of the metric from the line element (2.16)

$$q_{ij} = \mathrm{diag}\left(\frac{a^2(t)}{1-\mathfrak{K}\,r^2},\; a^2(t)\,r^2,\; a^2(t)\,r^2\sin^2\theta\right). \tag{2.23}$$

It is useful to express the curvature quantities with equal number of co- and contravariant indices to eliminate the explicit dependence on the coordinates. The only non-vanishing components of the Riemann tensor with two indices raised are

$$R^{0i}{}_{0i} = \frac{1}{N^2}\frac{\ddot{a}}{a} - \frac{\dot{N}}{N^2}\frac{\dot{a}}{a}, \quad R^{ij}{}_{ij} = \frac{\mathfrak{K}}{a^2} + \frac{\dot{a}^2}{N^2\,a^2}, \quad \text{for } i \neq j\,. \tag{2.24}$$

The non-vanishing components of the Ricci tensor with one index raised are

$$R^0{}_0 = \frac{3}{N^2}\frac{\ddot{a}}{a} - 3\,\frac{\dot{N}}{N^3}\frac{\dot{a}}{a}, \quad R^i{}_i = \frac{1}{N^2}\frac{\ddot{a}}{a} + \frac{2}{N^2}\frac{\dot{a}^2}{a^2} - \frac{\dot{N}}{N^3}\frac{\dot{a}}{a} + 2\,\frac{\mathfrak{K}}{a^2}. \tag{2.25}$$

By performing the sum over all components (2.25), we obtain the Ricci scalar

$$R = R^0{}_0 + \sum_{i=1,..,3}\left[R^i{}_i\right] = 6\left(\frac{1}{N^2}\frac{\ddot{a}}{a} + \frac{1}{N^2}\frac{\dot{a}^2}{a^2} - \frac{\dot{N}}{N^3}\frac{\dot{a}}{a} + \frac{\mathfrak{K}}{a^2}\right). \tag{2.26}$$

The symmetry assumptions require that the energy-momentum tensor has to acquire a special form. If we imagine the galaxies as freely streaming through the universe, this suggests that $T_{\mu\nu}$ should be the energy-momentum tensor of a perfect fluid. In the FRW universe, we can identify n^{μ} in Fig. 2.2, with the four-velocity u^{μ}, tangent to the time-like world line of a family of comoving observers. The four-velocity is defined as the four-vector $u^\mu := (\gamma, \mathbf{v})^T$, with the three-velocity $\mathbf{v}$ and the Lorentz factor $\gamma = 1/\sqrt{1-\mathbf{v}^2}$. A comoving observer is in his rest frame for each event so that $\gamma = 1$ and $\mathbf{v} = 0$ and $u^\mu = (1, \mathbf{0})^T = \delta^0_\mu$ for all times. With respect to the decomposition depicted in Fig. 2.2, $T_{\mu\nu}$ has to acquire the form

$$T_{\mu\nu} = \rho(t)\, u_\mu u_\nu + p(t)\, q_{\mu\nu}. \tag{2.27}$$

Using $u^\mu q_{\mu\nu} = 0$ and $u^\mu u_\mu = -1$, the coefficients $\rho(t)$ and $p(t)$ are determined by

$$\rho(t) := u^\mu\, u^\nu\, T_{\mu\nu} \quad \text{and} \quad p(t) := \frac{1}{3} q^{\mu\nu}\, T_{\mu\nu} \tag{2.28}$$

and can be identified as the energy density $\rho(t)$ and the pressure $p(t)$. They can only be functions of time due to the FRW symmetry. In addition, the two functions $\rho(t)$ and $p(t)$ are related by an equation of state, to be chosen independently:

$$\omega(t) := \frac{p(t)}{\rho(t)}. \tag{2.29}$$

The specific form of ω depends on the nature of the matter content described in $T_{\mu\nu}$. The energy-momentum tensor is further constrained by certain energy conditions. We will only consider the weak and strong energy conditions here. For a perfect fluid, these conditions can be parametrized by the following inequalities between ρ and p

$$\text{weak:}\quad \rho \geq 0 \quad \text{and} \quad \rho + p \geq 0 \;\Leftrightarrow\; \rho \geq 0 \quad \text{and} \quad \omega \geq -1, \tag{2.30}$$

$$\text{strong:}\quad \rho + p \geq 0 \quad \text{and} \quad \rho + 3\,p \geq 0 \;\Leftrightarrow\; \rho + p \geq 0 \quad \text{and} \quad \omega \geq -\frac{1}{3}. \tag{2.31}$$

By thinking of the galaxies as test particles flowing with the fluid, we can imagine a *comoving* observer as being attached to one of these galaxies. Inserting the symmetry-reduced expressions for the Ricci scalar and the energy-momentum tensor into the field equations (2.9), we obtain two equations, corresponding to the (00) and the (ii) components of (2.11). These are the *Friedmann equations*

$$H^2 = \frac{\dot{a}^2}{a^2} = \frac{\kappa}{3}\rho - \frac{\mathfrak{K}}{a^2} + \frac{\Lambda}{3}, \tag{2.32}$$

$$\dot{H} + H^2 = \frac{\ddot{a}}{a} = -\frac{\kappa}{6}(\rho + 3\,p) + \frac{\Lambda}{3} = -\frac{\kappa}{6}\rho\,(1 + 3\omega) + \frac{\Lambda}{3}. \tag{2.33}$$

We have introduced the Hubble parameter $H(t)$, defined by

$$H(t) := \frac{\dot{a}(t)}{a(t)}. \tag{2.34}$$

If we had derived the Eq. (2.32), (2.33) directly from the symmetry-reduced action, the constraint equation (2.32) would have been the result of the variation with respect to the Lagrange multiplier $N(t)$, whereas the equation of motion for $a(t)$ (2.33) would have been the result of varying with respect to the dynamical degree of freedom $a(t)$.

2.2.4 Epochs of the Universe

As a consequence of (2.12), we find by means of (2.32), (2.33)

$$\dot{\rho} + 3\,H\,(\rho + p) = \dot{\rho} + 3\,H\,\rho\,(1+\omega) = 0. \tag{2.35}$$

This has the form of a continuity equation. For $\omega = $ const., we obtain a differential equation in ρ. Using the ansatz $\rho(a) \propto a^x$, the solution in terms of a is

$$\rho(a) \propto a^{-3\,(\omega+1)}. \tag{2.36}$$

If multiple types of matter are present simultaneously, the individual densities ρ_i simply add up $\rho = \sum_i \rho_i$ and (2.36) holds individually for each type of matter. In order to obtain an explicit solution for $a(t)$ in terms of t, we substitute the solutions $\rho(a)$ into the Friedmann equations (2.32). Using again a power law ansatz $a(t) \propto t^y$, leads to

$$a(t) \propto \begin{cases} t^{\frac{2}{3\,(\omega+1)}} & \text{for } \omega \neq -1 \\ e^{H\,t} & \text{for } \omega = -1. \end{cases} \tag{2.37}$$

In order to obtain $a(\eta)$ with respect to conformal time, we use definition (2.18) and substitute (2.37). Neglecting constants of integration and prefactors yields

$$\eta \propto \begin{cases} t^{\frac{(\omega+\frac{1}{3})}{(\omega+1)}} & \text{for } \omega \neq -1 \\ -\frac{e^{-H\,t}}{H} & \text{for } \omega = -1. \end{cases} \tag{2.38}$$

Inverting this relation and re-substituting it into (2.37), we obtain the scale factor

$$a(\eta) \propto \eta^{\frac{2}{3\,(\omega+\frac{1}{3})}}. \tag{2.39}$$

For the most important choices of ω, we list the results in the following table:

In the last column we have calculated the *comoving* Hubble radius in terms of η

$$r_{\mathrm{H_c}} := (a\,H)^{-1} = \left(a\,\frac{\dot{a}}{a}\right)^{-1} = (\dot{a})^{-1} = \left(\frac{a'}{a}\right)^{-1} = \frac{a}{a'} =: \mathcal{H}^{-1} \tag{2.40}$$

Table 2.1 Overview of important quantities for different types of matter with $\omega =$ const.

	ω	$\rho(a)$	$a(t)$	$a(\eta)$	$(a\,H)^{-1}$
vacuum	-1	const.	e^t	η^{-1}	$-\eta$
radiation	$1/3$	a^{-4}	$t^{1/2}$	η	η
matter	0	a^{-3}	$t^{2/3}$	η^2	$\eta/2$

The behaviour of the energy densities collected in Table 2.1 in an expanding universe and their dilution with growing scale factor are illustrated in the following two figures.

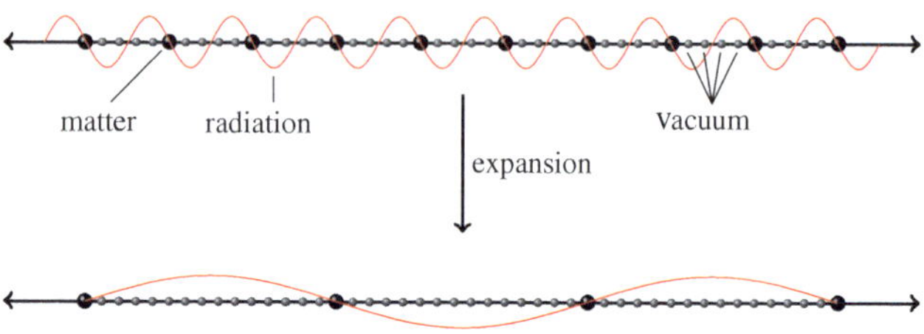

Fig. 2.5 Different types of energy densities in an expanding universe. Matter density is symbolized by the *fat black dots* and represent galaxies. Radiation density is represented by the *red wave* and vacuum density is illustrated by the constantly "re-filling" small red dots

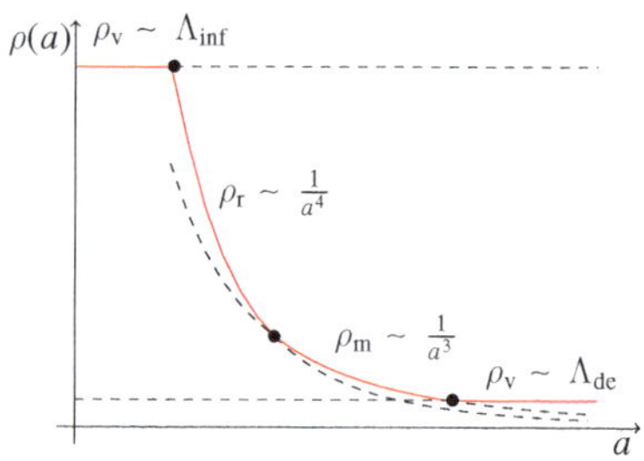

Fig. 2.6 Epochs of cosmic evolution. During the expansion of the universe, $\rho_m \sim 1/a^3$ decreases with increasing volume $V \sim a^3$, $\rho_r \sim 1/a^4$ with an additional factor of $a(t)$ due to the redshift of $\lambda_p = a(t)\,\lambda_c$ and ρ_v remains constant

The history of our universe can be divided into different epochs which are named according to the dominating energy density contribution at the respective time. In the early universe, the energy density $\rho_{\Lambda_{\rm infl}}$ was very large and nearly constant. This is the era of inflation to be discussed later. When inflaton ends, the energy is transferred into particle creation and thermic energy. This post-inflationary epoch is dominated by radiation $\rho_r \propto 1/a^4$. The universe expands and cools down so that finally stable bound atoms can form. At some point $\rho_r = \rho_m$ the matter energy density ρ_m becomes the dominant contribution. In the future, ρ_m will be diluted completely. Observations of the accelerated expansion of the universe [25] suggest that we are currently at the turning point $\rho_m = \rho_v$, entering again into a phase dominated by a constant vacuum energy. This type of vacuum energy is called dark energy and can be described by a cosmological constant Λ_{de}. However, Λ_{de} is much smaller than the vacuum energy during inflation $\rho_{\Lambda_{\text{infl}}}$.

2.2.5 Observable Quantities

We divide (2.32) by H^2 and rewrite the first Friedmann equation in a dimensionless form by normalizing the different types of energy density with respect to a critical density $\rho_{\text{crit}} = 3H^2/\kappa$

$$1 = \Omega_\rho + \Omega_\Lambda + \Omega_{\mathfrak{K}}, \tag{2.41}$$

with

$$\Omega_\rho := \sum_i \frac{\kappa\,\rho_i}{3\,H^2} = \sum_i \frac{\rho_i}{\rho_{\text{crit}}}, \quad \Omega_\Lambda := \frac{\Lambda}{3\,H^2}, \quad \text{and} \quad \Omega_{\mathfrak{K}} := -\frac{\mathfrak{K}}{H^2\,a^2}. \tag{2.42}$$

Here, the index i labels different kinds of matter and radiation energy densities with individual equations of state. We can even subdivide the ρ_i's further by analysing different particle species separately, e.g. radiation composed of photons and neutrinos $\rho_{\text{r}} = \rho_\gamma + \rho_\nu$ and matter of baryons and cold dark matter $\rho_M = \rho_{\text{B}} + \rho_{\text{CDM}}$. On the one hand, we can consider (2.41) as a consistency relation which must be satisfied by any cosmological model. One the other hand we can make predictions for new types of energy such as dark energy. The form (2.41) is especially useful to constrain the parameters of models by experimental data. From (2.36), we can easily find the solution for a constant ω

$$\Omega_{\rho_i} = \Omega_{\rho_{i,0}} \left(\frac{H_0}{H}\right)^2 \left(\frac{a}{a_0}\right)^{-3(1+\omega)}, \tag{2.43}$$

where we have expressed Ω_{ρ_i} in terms of quantities that we measure today, indicated by a subscript "0". Using the definition of the cosmological redshift z

$$(1+z) := \frac{a(t_{\text{observation}})}{a(t_{\text{emission}})} = \frac{a_0}{a}, \tag{2.44}$$

and (2.43), we can formulate the Friedmann equation (2.41) in terms of measurable quantities

$$\left(\frac{H}{H_0}\right)^2 = \sum_i \Omega_{\rho_i,0}\,(1+z)^{3(\omega_i+1)} + \Omega_{\mathfrak{K},0}\,(1+z)^2 + \Omega_{\Lambda,0}. \tag{2.45}$$

By expanding the scale factor a in powers of $(t - t_0)$ with the value a_0 it has today

$$a(t) = a_0 \Big(1 + H_0 \, (t - t_0) - \frac{1}{2} q_0 \, H_0^2 \, (t - t_0)^2 + \ldots \Big), \qquad (2.46)$$

we recover Hubble's law as the linear correction term with the Hubble constant H_0 observed today. Hubble's law states that the velocities of the galaxies moving away from us are proportional to their distance. The parameter of proportionality is $H(t)$. The quadratic correction describes the acceleration of the universe. By convention the sign is chosen negative, so that $q_0 > 0$ corresponds to deceleration. The deceleration parameter q_0, as measured today, is defined by

$$q_0 := -\frac{\ddot{a}}{a \, H^2} \Big|_{t=t_0}. \qquad (2.47)$$

Using (2.33) and (2.42), we can re-express the deceleration parameter as

$$q_0 = \frac{1}{2} \sum_i \Omega_{\rho_i,0} \, (1 + 3\omega_i) - \Omega_{\Lambda,0}. \qquad (2.48)$$

Regardless of the curvature parameter $\mathfrak{K}$, we can distinguish by a measurement of $\Omega_{\rho,0}$ and $\Omega_{\Lambda,0}$, whether our universe is accelerating or decelerating today. Of course, we could reformulate this equation in such away that it effectively depends on $\Omega_{\mathfrak{K}}$ by using the constraint (2.41). We could also include the contribution of Ω_Λ in the sum over the ordinary energy contributions in (2.48), showing again that the equation of state for the vacuum energy is $\omega = -1$. We need at least one such negative contribution in order to have a negative q_0 corresponding to the acceleration of the universe observed today.

2.3 Inflation

The general idea of inflation is that the universe underwent a phase of accelerated expansion. This picture is adequate to solve several problems of cosmology ("flatness", "horizon" and "monopole" problems) [12, 27]. However, nowadays its main success and its major experimental evidence consist of its capability to explain the emergence of the large-scale structure of our present universe from tiny quantum fluctuations in the early universe. We will first give a qualitative overview and afterwards describe how inflation can be modelled by a scalar field. Then, we analyse the space-time structure of de Sitter space and consider the scenario of slow-roll inflation [1, 19]. Finally, we investigate small quantum perturbations on the inflationary background and calculate the cosmological parameters for these fluctuations observed today [13, 14, 21, 28].

2.3.1 Qualitative Preliminary Considerations

Let us assume that the universe undergoes an exponential expansion $a(t) \sim \exp\left(H_\mathrm{p} t\right)$ with a constant physical Hubble parameter H_p. By investigating the Friedmann equation (2.32), the term associated with the spatial curvature $\propto \mathfrak{K}/a^2$ becomes negligible very quickly and the universe flattens out. This makes the distinction between a closed, an open and a flat universe hard to detect and can explain why we observe a nearly flat universe today. The following illustrations consider the horizon problem which lies at the very heart of the inflationary paradigm. All other problems mentioned above can be traced back to this elementary problem. The horizon problem can roughly be summarized as follows: Today, we observe a nearly isotropic cosmic microwave background radiation (CMB). If we look into the sky, we look in the past. The identical temperature of the radiation from all directions in the sky suggests that these regions must have been in causal contact at some earlier time in order to establish thermodynamical equilibrium. In order to figure out the moment of time when the radiation was sent out, we must investigate the cosmic history. As already mentioned in Sect. 2.2.4, at some point in the past the hot baryon-photon plasma has cooled down and the free electrons were caught by the ionized nucleons to form stable atoms. The mean free path of the photons, which have been scattered with the fast electrons in the hot plasma, increases after the free electrons have disappeared so that the photons decouple and are freely streaming. In this period of conversion, the surface of last scattering has been formed. The free photons emanating from this surface are the photons we detect today in the CMB. Due to the cosmic expansion, they are red-shifted so that their associated temperature is small today.

Fig. 2.7 Recombination. Via Thomson scattering the photons collide with the electrons in the hot ionized baryon-photon plasma. During the expansion of the universe the plasma cools down and it becomes energetically favourable for the electrons and the ionized nucleons to form stable atoms. The photons do not scatter anymore, their mean free path becomes nearly infinite and the universe becomes transparent

We observe an isotropic CMB with an accuracy of the temperature contrast of about $\Delta T/T \sim 10^{-5}$. In order to illustrate the problem, we can analyse the following causal diagram. Figure 2.8 shows the past light cone in terms of conformal time η. The points A and B each correspond to a sphere S^2, depicted in Fig. 2.9. Each such sphere, in turn, corresponds to a homogeneous patch of the universe. At the time of recombination η_r, the surface of last scattering (grey line in the two-dimensional illustration of Fig. 2.8) has formed and has sent out the radiation from different homogeneous patches that we measure in the CMB. The horizon problem appears when we try to explain the observed isotropic radiation of the CMB radiation. In the projection onto our sky, the two patches associated with A and B correspond to two antipodal points. Without inflation, the initial singularity would have started with the "Hot Big Bang" at $\eta = 0$. Thus, the two past light cones of A and B could have never been in causal contact and therefore would have never established thermal equilibrium to explain the observed isotropy of the CMB. This puzzle can be solved by inflation. Inflation extends the conformal time to negative infinity so that the past light cones could have intersected (large red triangle in Fig. 2.8). Thus, it is required that inflation lasts long enough, $\eta_{\text{inf}} = \eta_{\text{min}}$, in order to ensure that the past light cones of the antipodal points A and B intersect at some moment of time. It is expected that quantum gravitational effects become dominant, somewhere between the minimal η_{min}, necessary to solve the horizon problem and the initial singularity $\eta = -\infty$. The left-hand side of Fig. 2.9 shows the surface of last scattering, if there would have been no inflation. We would observe many different patches covering the sky, all with different radiation indicated by the different colours. The grey sphere corresponds to the surface of last scattering projected on our observable sky with the Earth in its centre. Each S^2 sphere lies on the grey sphere and corresponds to a local patch that would be out of causal contact without inflation. Thus, without inflation, there should be a huge number of different patches, which is, however, not observed. It is the mechanism of inflation that allows all these patches to be in causal contact at some early time $\eta < 0$. While being in causal contact, these patches could have established a thermodynamical equilibrium. This situation is depicted on the right-hand side of Fig. 2.9. The big red sphere indicates thermodynamical equilibrium and explains why we observe the same CMB from every direction in the sky.

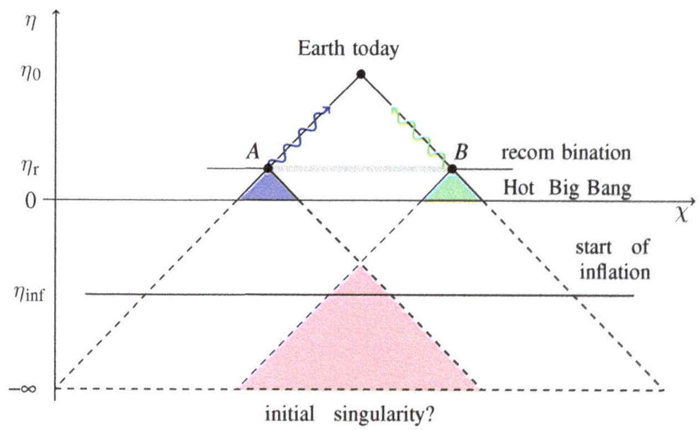

Fig. 2.8 Horizon problem. The past light cone of the Earth today in conformal time η. The isotropic CMB suggests that the patches associated with the size of the past light cones of A and B must have been in causal contact. However, this becomes possible only during a phase of inflation $\eta \to \eta_{\text{inf}}$, when the two light cones intersect

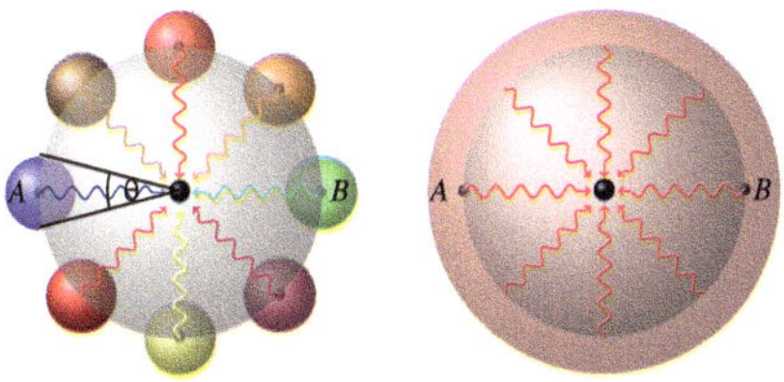

Fig. 2.9 Surface of last scattering. The *black dot* representing the Earth is surrounded by the surface of last scattering symbolized by the *grey sphere*. *Left* Without inflation we would observe radiation with different temperatures (indicated by the different colours) from different directions in the sky. Each sphere corresponds to a small homogeneous patch of the universe. *Right* All regions have been in causal contact during inflation and established thermodynamical equilibrium explaining the observed isotropy of the CMB

2.3.2 Scalar Field Model of Inflation

The qualitative overview of the last section clearly demonstrated the advantages of an inflationary phase. However, if this process is realized in nature we should be able to describe inflation at least phenomenologically by a concrete model. By looking at the definition of the scale factor in terms of conformal time η in (2.1), it is obvious that for a phase of inflation we need a scale factor $a(\eta) \propto -1/\eta$ that pushes the initial singularity from $\eta = 0$ to $\eta = -\infty$. From (2.1) we see that this corresponds to a constant equation of state $\omega = -1$ or to a vacuum energy density ρ_v that can be described in terms of a large cosmological constant Λ_inf. However, from the fact of our existence, we can conclude that inflation have had come to an end in order to allow for the evolution of structures. Thus, Λ_inf cannot be exactly constant. We consider a scalar field φ with the following action

$$S_\varphi = \int \mathrm{d}^4x\,\sqrt{-g}\left(-\frac{1}{2}\,g^{\mu\nu}\partial_\mu\varphi\partial_\nu\varphi - V(\varphi)\right). \tag{2.49}$$

From the CMB, we know that our observable universe is approximately isotropic and homogeneous which is why we assume a homogeneous scalar field $\varphi(t, \mathbf{x}) = \varphi(t)$. Assuming that the scalar field $\varphi(t) = \varphi_0$ is almost constant $\dot{\varphi}_0 \approx 0$ during the time interval of inflation implies that $\mathcal{L}_{\varphi_0} \approx V(\varphi_0) \approx$ const. Therefore, the vacuum energy density can be written as

$$\rho_\mathrm{v} = M_\mathrm{P}^2\,\Lambda_\mathrm{inf} := V(\varphi_0). \tag{2.50}$$

During inflation, we can effectively express the cosmological constant in terms of the scalar field potential $V(\varphi_0)$. The key idea is now to associate Λ_{inf} also for later times and in this way to allow a *dynamical*

$$\Lambda_\varphi(t) := \frac{V(\varphi(t))}{M_{\text{P}}^2}. \tag{2.51}$$

We know from observations that at present time scales the variation of Λ_φ is negligible and that Λ_φ is very small. In order to allow for a phase of inflation, we have to require that the potential has the appropriate shape, i.e. that it is very flat during inflation, and becomes more and more steep later on, in order to exit the phase of inflationary dynamics. This scenario is realized in the "slow-roll" approximation, which we will discuss in Sect. 2.3.4. We will now derive the Friedmann equations (2.32) and (2.33) in terms of the scalar field quantities. The energy-momentum tensor of the scalar field action can be calculated by identifying the scalar field action (2.49) with the matter action (2.2) and by using the definition of the energy-momentum tensor (2.8)

$$T^\varphi_{\mu\nu} = \partial_\mu\varphi\,\partial_\nu\varphi - \left(\frac{1}{2}\,\partial_\sigma\varphi\,\partial^\sigma\varphi + V(\varphi)\right)\,g_{\mu\nu}. \tag{2.52}$$

For a homogeneous scalar field $\varphi(t)$, we obtain by means of the FRW decomposition

$$T^\varphi_{\mu\nu} = \left(\frac{1}{2}\dot{\varphi}^2 + V(\varphi)\right)\,\delta^0_\mu\delta^0_\nu + \left(\frac{1}{2}\dot{\varphi}^2 - V(\varphi)\right)\,\delta^i_\mu\delta^j_\nu q_{ij}. \tag{2.53}$$

By comparison with the energy-momentum tensor of a perfect fluid (2.27), we find

$$\rho_\varphi = \frac{1}{2}\dot{\varphi}^2 + V(\varphi),\quad p_\varphi = \frac{1}{2}\dot{\varphi}^2 - V(\varphi)\quad\text{and}\quad \omega_\varphi = \frac{\frac{1}{2}\dot{\varphi}^2 + V(\varphi)}{\frac{1}{2}\dot{\varphi}^2 - V(\varphi)}. \tag{2.54}$$

Assuming in addition that the scalar-field induced Λ_φ is the only vacuum energy density, i.e. $\Lambda \equiv 0$, and is dominant, i.e $\Omega_{\Lambda_\varphi} \gg \Omega_M$, we find the Friedmann equations

$$H^2 = \left(\frac{\dot{a}}{a}\right)^2 = \frac{\kappa}{3}\,\rho_\varphi - \frac{\mathfrak{K}}{a^2} = \frac{\kappa}{3}\left(\frac{1}{2}\dot{\varphi}^2 + V(\varphi)\right) - \frac{\mathfrak{K}}{a^2}, \tag{2.55}$$

$$\dot{H} + H^2 = \frac{\ddot{a}}{a} = -\frac{\kappa}{6}\,(\rho_\varphi + 3\,p_\varphi) = \frac{\kappa}{3}\left(V(\varphi) - \dot{\varphi}^2\right). \tag{2.56}$$

We obtain the Klein–Gordon equation by varying (2.49) with respect to $\varphi(t)$

$$\ddot{\varphi} + 3\frac{\dot{a}}{a}\,\dot{\varphi} + V' = 0 \quad\text{with}\quad V' := \frac{\partial V(\varphi)}{\partial\varphi}. \tag{2.57}$$

Equation (2.56) implies that for an *accelerated* expansion a positive $\ddot{a}$ is required

$$(\rho_\varphi + 3p_\varphi) < 0 \quad \leftrightarrow \quad V(\varphi) > \dot{\varphi}^2. \tag{2.58}$$

This is a necessary and sufficient condition for inflation. If we assume in addition that the weak energy condition $\rho > 0$ (2.30) is satisfied, this requires a *negative* pressure $p < 0$ during inflation. It is exactly this negative pressure which is the reason for the phase of accelerated expansion. Equation (2.58) implies that during inflation the potential should dominate over the kinetic term. If the even stronger condition $V(\varphi) \gg \dot{\varphi}^2$ holds, then we have a nearly constant φ and (2.54) becomes

$$\rho_\varphi = -p_\varphi \quad \text{and} \quad \omega_\varphi \approx -1 \quad . \tag{2.59}$$

According to Table 2.1, a constant $\omega = -1$ leads to an *exponential* phase of expansion. Based on the structure of space-time, this is denoted "exact de Sitter inflation".

2.3.3 De Sitter Space

De Sitter space $dS_{(4)}$ is a four-dimensional space with coordinates $\{x^0,\ x^1,\ x^2,\ x^3\}$ and can be described as sections of an embedded hyperboloid in a five-dimensional Minkowski space $\mathcal{M}_{(5)}$ with coordinates $\{X^0,\ X^1,\ X^2,\ X^3,\ X^4\}$. The five-dimensional line element

$$\mathrm{d}s^2 := \eta_{IJ}\, \mathrm{d}X^I\, \mathrm{d}X^J \tag{2.60}$$

is defined by the five-dimensional Minkowski metric $\eta_{IJ} := \mathrm{diag}(-1,\ 1,\ 1,\ 1,\ 1)$. De Sitter space $dS_{(4)}$ can be described by the hyperboloid equation in $\mathcal{M}_{(5)}$

$$-(X_0)^2 + (X_1)^2 + (X_2)^2 + (X_3)^2 + (X_4)^2 = C^2. \tag{2.61}$$

Topologically, $dS_{(4)}$ is equivalent to $\mathbb{R} \times \mathcal{S}^3$, corresponding to one real time $t \in \mathbb{R}$ dimension and a compact three-dimensional spatial sphere $\mathcal{S}^3$. The constant C must have the dimension of a length. In a cosmological context it therefore seems natural to identify it with the only characteristic length scale, the Hubble radius H^{-1}. De Sitter space is a maximal symmetric space with constant curvature (as four-dimensional flat Minkowski space $\mathcal{M}_{(4)}$). The generators of the isometries in $d = 4$ dimensions are the $d(d+1)/2 = 10$ linear independent Killing vectors. The Ricci scalar of $dS_{(4)}$ is (cf. (2.32) with $\rho = \mathfrak{K} = 0$)

$$R = \frac{d\,(d-1)}{C^2} = 12\, H^2. \tag{2.62}$$

$dS_{(4)}$ is a solution of the vacuum Einstein equations (2.11) with a cosmological constant

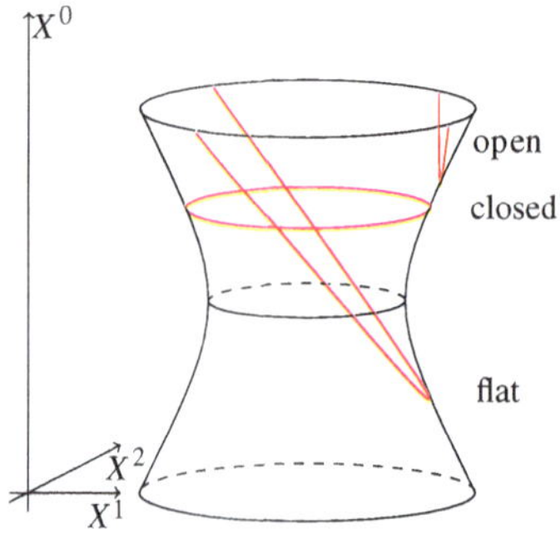

Fig. 2.10 Different slices of the de Sitter hyperboloid embedded in $\mathcal{M}_{(5)}$. The red lines correspond to hypersurfaces of constant time t

$$\Lambda = \frac{(d-2)(d-1)}{2\,C^2} = 3\,H^2. \tag{2.63}$$

The slices shown in Fig. 2.10 correspond to different choices of coordinates $X^I \in \mathcal{M}_{(5)}$ and lead to different induced metrics $\gamma_{\mu\nu}$ on $dS_{(4)}$ defined by (2.61)

$$\gamma_{\mu\nu} := \frac{\partial X^I}{\partial x^\mu}\frac{\partial X^J}{\partial x^\nu}\eta_{IJ}, \quad I,\ J, \ldots = 0, \ldots, 4. \tag{2.64}$$

This does not mean that de Sitter space is different for different coordinate choices X^I, it just means that de Sitter space is covered by different sets of coordinates x^μ. Not all choices for $X^I \in \mathcal{M}_{(5)}$ lead to a set of coordinates $x^\mu \in dS_{(4)}$ that covers the whole de Sitter space. In most of the physical applications for inflation, only a tiny patch of the complete $dS_{(4)}$ is needed and it is rather a matter of taste which different slicing in Fig. 2.10 we choose. In the flat slicing, $\gamma_{\mu\nu}$ has the simplest form. We choose the coordinates [26]

$$\begin{aligned} X^0 &:= \frac{1}{H}\sinh(H\,t) + \frac{H}{2}\,e^{H\,t}\,\mathbf{x}^2, \quad X^4 := \frac{1}{H}\cosh(H\,t) - \frac{H}{2}\,e^{H\,t}\,\mathbf{x}^2, \\ X^i &= e^{H\,t}\,x_i, \quad (i = 1, \ldots, 3,\ \mathbf{x}^2 = \sum_{i=1}^{3}(x_i)^2) \end{aligned} \tag{2.65}$$

and use (2.64) to calculate the induced metric

$$\gamma_{00} = -1, \quad \gamma_{ii} = e^{2\,H\,t}, \quad \gamma_{0i} = \gamma_{ij} = 0, \quad \text{for} \quad i \neq j\,. \tag{2.66}$$

Therefore, the de Sitter line element for the flat slicing is given by

$$\mathrm{d}s^2_{\text{flat}} = \gamma_{\mu\nu}\,\mathrm{d}x^\mu \mathrm{d}x^\nu = -\mathrm{d}t^2 + e^{2\,H\,t}\,\delta_{ij}\,\mathrm{d}x^i \mathrm{d}x^j\,. \tag{2.67}$$

The closed slicing leads to the induced line element [26]

$$\mathrm{d}s^2_{\text{closed}} = -\mathrm{d}t^2 + \frac{\cosh^2(H\,t)}{H^2}\left(\mathrm{d}\chi^2 + \sin^2 \mathrm{d}\Omega^2\right), \tag{2.68}$$

which has the advantage that it covers the complete de Sitter space. The open slicing can be obtained in a similar way.

2.3.4 Slow-Roll Inflation

In Sect. 2.3.2, we have seen that $V \gg \dot{\varphi}^2$ was required to obtain a quasi de Sitter phase $\omega \approx -1$ of inflation. We will now make this more quantitative. From now on, we will assume a spatially flat $\mathfrak{K} = 0$ universe, which is favoured by observation. However, for $a \sim \exp(H\,t)$, the term $\propto \mathfrak{K}/a^2(t)$ in (2.55) quickly becomes negligible in any case. The *quality* of the slow-roll approximation can be quantified by the two dimensionless slow-roll parameters

$$\varepsilon := -\frac{\dot{H}}{H^2} \quad \text{and} \quad \delta := -\frac{1}{H}\frac{\ddot{\varphi}}{\dot{\varphi}}. \tag{2.69}$$

The physical meaning of these parameters becomes transparent, when expressing the Friedmann equations (2.55) and (2.56) in terms of the slow-roll parameters (2.69). Equation (2.33) takes the form

$$\frac{\ddot{a}}{a} = \dot{H} + H^2 = H^2\,(1 - \varepsilon). \tag{2.70}$$

In order to have acceleration $\ddot{a} > 0$, $\varepsilon < 1$ is required. If even $\varepsilon \ll 1$, (2.70) becomes $\ddot{a} = H^2\,a$ with the pure de Sitter solution $a = \exp(H\,t)$ in the limit $\varepsilon \to 0$. The advantage of this parametrization is that we have a *small* dimensionless parameter ε which is suited for an expansion around the pure de Sitter solution. We define the moment of time t_{end} when inflation comes to an end by the end of the phase of accelerated expansion

$$\varepsilon(t_{\text{end}}) := 1. \tag{2.71}$$

If we compare (2.70) with (2.56), we can express ε in terms of ω_φ

$$\varepsilon = 1 + \frac{\kappa\,\rho_\varphi}{6\,H^2}\,(1 + 3\,\omega_\varphi) = \frac{3}{2}\,(\omega_\varphi + 1), \tag{2.72}$$

where we have used (2.55) in the last equality. Subtracting (2.56) from (2.55) yields

$$\varepsilon = \frac{\kappa}{2}\frac{\dot{\varphi}^2}{H^2}. \tag{2.73}$$

So far, we just have reparametrized quantities in terms of ε and all expressions are still *exact*. At this point, we have to make use of the slow-roll approximation in order to write ε in terms of the potential $V(\varphi)$. Neglecting $\ddot{\varphi}$ and $\dot{\varphi}^2$ terms in the Klein–Gordon equation (2.57) and the Friedmann equation (2.55) and using $\kappa = 1/M_{\rm P}^2$ leads to

$$3\,H\,\dot{\varphi} \simeq -V' \quad \text{and} \quad H^2 \simeq \frac{1}{3}\frac{V(\varphi)}{M_{\rm P}^2}. \tag{2.74}$$

Differentiating the second relation in (2.74) with respect to time t and using the first relation in (2.74) leads to the relation

$$\dot{H} = -\frac{1}{2}\frac{\dot{\varphi}^2}{M_{\rm P}^2}. \tag{2.75}$$

Using the first relation in (2.74) in order to eliminate $\dot{\varphi}$ in (2.75) and inserting this in the definition for ε in (2.69), we can express $\varepsilon(\varphi)$ in terms of V and V'

$$\varepsilon \simeq \frac{M_{\rm P}^2}{2}\left(\frac{V'}{V}\right)^2 =: \varepsilon_{\rm V}. \tag{2.76}$$

The physical meaning of the second slow-roll parameter δ is connected with the acceleration of φ. Provided that the potential is flat enough, i.e. $\varepsilon \ll 1$, it is a measure of how long inflation lasts. It is convenient to express δ also in terms of V. Differentiating the first relation in (2.74) with respect to time t and using (2.75) to eliminate $\dot{H}$, we obtain

$$\frac{\ddot{\varphi}}{\dot{\varphi}} \simeq -\frac{V''}{3\,H} - \frac{1}{2\,M_{\rm P}^2}\frac{\dot{\varphi}^2}{H}. \tag{2.77}$$

Inserting this into the definition (2.69) for δ and using again both approximated expressions in (2.74), we obtain the second slow-roll parameter δ in terms of $V(\varphi)$

$$\delta \simeq M_{\rm P}^2\,\frac{V''}{V} - \frac{1}{2}\,M_{\rm P}^2\,\frac{V'^2}{V^2} =: \delta_{\rm V} - \varepsilon_{\rm V} \tag{2.78}$$

with

$$\delta_{\rm V} := M_{\rm P}^2\,\frac{V''}{V}. \tag{2.79}$$

We could define even more slow-roll parameters including higher derivatives of the potential, but for our purpose the two slow-roll parameters (2.76) and (2.79) are sufficient. We note that *only during slow-roll* the following relations between the parameters (2.69) and the parameters in terms of the potential (2.76) and (2.79) hold

$$\varepsilon \simeq \varepsilon_{\rm V} \quad \text{and} \quad \delta \simeq \delta_{\rm V} - \varepsilon_{\rm V}. \tag{2.80}$$

The slow-roll inflation ($\varepsilon \ll 1$ and $|\delta| \ll 1$) constraints the otherwise arbitrary potential $V(\varphi)$. It must be sufficiently flat for a sufficient long time in order to guarantee that φ rolls down the hill slowly as long as inflation is needed.

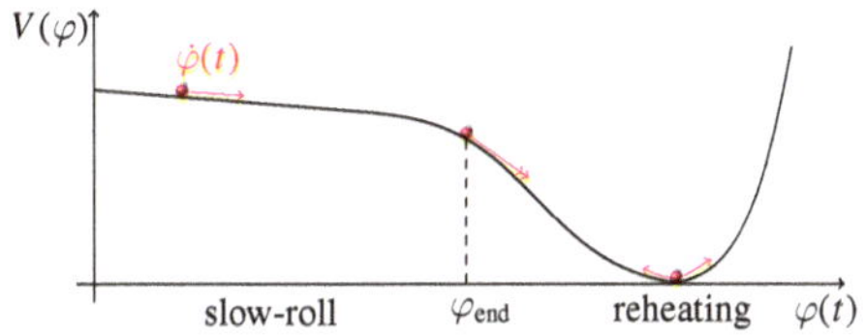

Fig. 2.11 Typical slow-roll potential

After the phase of slow-roll, $V(\varphi)$ becomes steeper, φ "speeds up", the slow-roll approximation breaks down and φ starts to oscillate around the minimum φ_0 in the harmonic potential

$$V''(\varphi)\Big|_{\varphi=\varphi_0} =: m_\varphi^2. \tag{2.81}$$

Since the universe cools down when expanding, at some point the coupling of the inflaton φ to other kinds of matter cannot be neglected any longer. Via this interaction, the inflaton transforms its energy into particle creation and thermic energy, starting a phase of reheating. Finally, the inflaton comes to rest and settles in the minimum of $V(\varphi)$.

2.4 Cosmological Perturbations

The mechanism of inflation successfully explains the observed isotropy of the CMB. However, there are two obstacles yet to overcome. First of all, more precise measurements of the CMB revealed that there are small anisotropies in the temperature contrast $\Delta T/T \sim 10^{-5}$ and we have used inflation to explain just the opposite: the smoothing-out of space-time. Thus, we have to explain the origin of these anisotropies. Secondly, we observe that matter is not homogeneously and isotropically distributed locally. We need at least some small density fluctuations to explain the gravitational clumping that ultimately leads to structure formation. We can solve both problems by taking into account quantum theory. So far, we have only considered the classical FRW evolution of the inflationary background. In quantum theory, we have to take into account quantum fluctuations of the fields. In the inflationary scenario we have considered the symmetry reduced versions $a(t)$ and $\varphi(t)$ of the metric field $g_{\mu\nu}(\mathbf{x}, t)$ and the inflaton field $\varphi(\mathbf{x}, t)$. On top of this classical (averaged) FRW background, we consider the fluctuations

$$\delta g_{\mu\nu}(\mathbf{x}, t) := g_{\mu\nu}(\mathbf{x}, t) - \bar{g}_{\mu\nu}(t) \quad \delta\varphi(\mathbf{x}, t) := \varphi(\mathbf{x}, t) - \bar{\varphi}(t). \tag{2.82}$$

The isotropic FRW metric and the homogeneous inflaton field are the background quantities, marked by a bar. The **x**-dependent fluctuations represent small inhomogeneities.

2.4.1 Generation of Density Fluctuations: Basic Mechanism

First, we will show the basic mechanism of how small quantum fluctuations can lead to density fluctuations by an illustrative example. Later we will present a more rigorous derivation. For a moment, we focus only on the fluctuations $\delta\varphi(\mathbf{x},\, t)$ of the inflaton field φ. We consider again our typical slow-roll potential, but this time with a three-dimensional extension including the three spatial directions **x**.

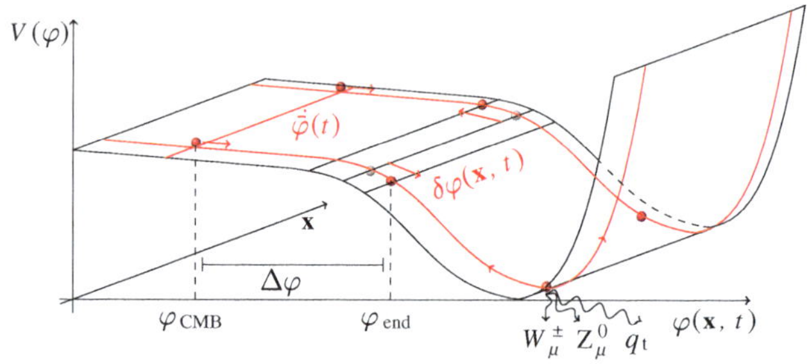

Fig. 2.12 Slow-roll potential with small perturbations of the inflaton field. The accelerated and delayed decay of φ due to small quantum corrections translates perturbations of the inflaton field $\delta\varphi$ into density perturbations $\delta\rho$

The fluctuations $\delta\varphi(\mathbf{x},\, t)$ are space-dependent and should be approximately random. Therefore, it is assumed that they obey (nearly) Gaussian statistics. This in turn means that the fluctuations are at each point independent objects and thus each Fourier mode $\delta\varphi_k(t)$ represents an independent random variable. The motion of the inflaton field $\varphi(\mathbf{x}, t) = \bar{\varphi}(t) + \delta\varphi(\mathbf{x},\, t)$ can be described as a superposition of the dominant classical slow-roll evolution of the background $\dot{\bar{\varphi}}(t)$ and the random perturbations characterized by small quantum oscillations $\delta\varphi(\mathbf{x},\, t)$ around the mean value $\bar{\varphi}$. The small "quantum kicks" can either drive the field upwards or downwards the hill of the potential. As Fig. 2.12 suggests, the inflaton rolls down earlier in some regions **x** than in other regions **x**′. If inflation ends, the inflaton rolls down into the minimum of $V(\varphi)$ and decays. The free energy is transformed into particle creation and thermal energy, starting the phase of reheating.[3] As a consequence, the universe becomes dominated by radiation. As we know from Table 2.1 and Fig. 2.6, the energy *density* $\rho_{\text{v}} \simeq V(\varphi)$ remains constant during inflation. In contrast, the matter and radiation densities dilute with an expanding scale factor. Thus, in space-time regions

[3] If $V(\varphi)$ was the Higgs potential of the Standard Model, the inflaton would settle in the electroweak vacuum, generating the mass of gauge bosons and fermions.

where inflation ends earlier, the inflaton field decays earlier and the energy density in these regions dilutes earlier, leading to an underdense region. In regions where inflation lasts longer, the field decays later, leading to slightly overdense regions. The mechanism of inflation thereby translates small quantum fluctuations of the scalar field into density perturbations. Finally, the universal attraction of gravity will amplify these small density perturbations and can explain the formation of structure.

2.4.2 Decomposition of Different Types of Fluctuations

In the last section we have roughly outlined the basic consequence of fluctuations in the inflaton field. The picture is, however, not complete. We must also include perturbations of the metric field and discuss the issue of gauge invariance. In the following sections, we will closely follow the treatment of [22] and [29].

In order to treat a general fluctuation $\delta\Phi(\mathbf{x},\,\eta)$ (with Φ denoting a collection of different fields, including the metric field), we have to eliminate the spatial dependence by switching to Fourier space. Thus, we trade the infinitely many degrees of freedom associated with the space-time points $\mathbf{x}$ of $\delta\Phi(\mathbf{x},\,\eta)$ for a set of infinitely many modes $\delta\Phi_{\mathbf{k}}(\eta)$, depending solely on (conformal) time and the mode $\mathbf{k}$. Further simplifications are due to the following observations: First of all, the measured CMB temperature contrast $\Delta T/T \simeq 10^{-5}$ suggests that only *small perturbations* $|\delta\Phi| \ll 1$ can be responsible for the observed anisotropy and this, in turn, justifies the approximation linear in the perturbations $\delta\Phi_{\mathbf{k}}(\eta)$. Secondly, translation invariance of the background combined with the linear evolution implies that different modes involve *individually*, i.e. couplings between different modes $\mathbf{k}$ and $\mathbf{k}'$ do not occur. Thirdly, we can classify the perturbations according to their behaviour under three-dimensional rotations in scalar-, vector- and tensor-type (S, V, T) perturbations. Linear evolution and rotational invariance of the background ensures that there will be no mixing between different types of perturbations during their evolution, i.e. the linearised Einstein equations only relate modes of equal type [4]. Ultimately, this allows us to consider each type of perturbation (S, V, T) and each mode $\mathbf{k}$ individually.

We decompose the metric perturbations into the three types

$$\delta\, g_{\mu}\nu := \delta_{\mathrm{S}}\, g_{\mu}\nu + \delta_{\mathrm{V}}\, g_{\mu}\nu + \delta_{\mathrm{T}}\, g_{\mu}\nu. \tag{2.83}$$

However, even if we can investigate the three types of perturbations individually, there still remains the issue of gauge transformations. An infinitesimal coordinate transformation

$$x^{\mu} \to \tilde{x}^{\mu} = x^{\mu} + \xi^{\mu} \tag{2.84}$$

induces a change in the perturbations. In order to illustrate the problem, we consider a perfectly homogeneous FRW universe with density $\rho = \rho(\eta)$. Performing a coordinate transformation of the time variable $\eta \to \tilde{\eta} = \eta + \xi^{0}(\mathbf{x},\,\eta)$ will induce a change in the density perturbation which is given by the Lie dragging (see Appendix B.12) along the time direction $\xi^0(\mathbf{x}, \eta)$

$$\delta\rho \rightarrow \widetilde{\delta\rho} = \delta\rho + \mathcal{L}_{\xi^0}\,\rho = \delta\rho + \rho'\,\xi^0(\mathbf{x}, \eta). \tag{2.85}$$

Thus, although initially $\delta\rho = 0$, by performing a coordinate transformation of time, we induce an inhomogeneous FRW universe $\widetilde{\delta\rho} = \rho'\,\xi^0(\mathbf{x}, \eta) \neq 0$. These perturbations are pure gauge artefacts. In order to get rid of these non-physical extra modes, we can either fix a specific gauge or we can construct and work with gauge-invariant variables exclusively. We will now investigate the three types of metric perturbations (2.83) individually. The tensor perturbations $\delta_{\mathrm{T}}\, g_{\mu\nu}$ are by construction gauge-invariant, which is why we can assign a direct physical meaning to them. They can be identified with gravitational waves

$$\delta_{\mathrm{T}}\, g_{\mu\nu} := a^2(\eta)\, h_{ij}. \tag{2.86}$$

Here, h_{ij} is a symmetric, traceless and divergence-free tensor and $i, j, \ldots = 1, 2, 3$. Otherwise, the divergence and trace part would contribute to vector and scalar perturbations, respectively. The tensor perturbations have two degrees of freedom according to the two different polarizations of a gravitational wave (six independent components of $h_{ij} = h_{ji}$, minus three components $h_{ij}{}^{;j} = 0$ minus one component $h_i{}^{;i} = 0$). Similarly, the vector modes can be parametrized by two divergence-free three-vectors C_i and D_i. If they were not divergence-free, the divergence would contribute to the scalar perturbations. Counting degrees of freedom, we find six components of the vectors C_i and D_i minus two constraint equations ($C_i{}^{;i} = D_i{}^{;i} = 0$), yielding four degrees of freedom.

$$\delta_{\mathrm{V}}\, g_{\mu\nu} := a^2(\eta) \begin{pmatrix} 0 & -C_i \\ -C_j & D_{i,j} + D_{j,i} \end{pmatrix}. \tag{2.87}$$

However, it turns out that the $\delta_{\mathrm{V}}\, g_{\mu\nu}$ have only decaying solutions in an expanding universe and thus are not important for our further analysis, which is why we will neglect them in what follows. The scalar perturbations are most important, since they are the only ones that couple to matter fluctuations[4] and therefore are responsible for the formation of structure in our observable universe. The mixing between metric and matter degrees of freedom is determined by $\delta G_{\mu\nu} = \kappa \delta T_{\mu\nu}$. In order to get rid of the gauge dependence, we parametrize the scalar perturbations of $g_{\mu\nu}$ by four scalar functions ϕ, ψ, B and E

$$\delta_{\mathrm{S}}\, g_{\mu\nu} := a^2(\eta) \begin{pmatrix} -2\,\phi & -B_{,i} \\ -B_{,i} & 2\,(\psi\,\delta_{ij} + E_{,ij}) \end{pmatrix}. \tag{2.88}$$

We can therefore write the flat, scalar and tensor perturbed FRW line element as

[4] The density and pressure fluctuations contained in $\delta T_{\mu\nu}$ are scalar perturbations.

$$\mathrm{d}s^2 = a^2(\eta)\Big\{-(1+2\,\phi)\,\mathrm{d}\eta^2 - 2\,(\partial_i\,B)\,\mathrm{d}x^i\mathrm{d}\eta + \Big[(1+2\psi)\,\delta_{ij} + 2\,\partial_i\partial_j E + h_{ij}\Big]\mathrm{d}x^i\mathrm{d}x^j\Big\}. \tag{2.89}$$

Counting the total number of degrees of freedom for all types of perturbations, we end up with ten as expected. However, we have not yet taken into account the gauge transformations. Since coordinates have no physical meaning, we are free to perform a coordinate transformation eliminating four degrees of freedom. This shows that the real *physical* degrees of freedom should be six and not ten. In the following, we will focus on scalar perturbations only. Then, the result for the tensor perturbations can easily be obtained in a similar way.

First, we have to figure out how the scalar perturbations behave under gauge transformations and how we can get rid of these non-physical degrees of freedom. We decompose the generator of the infinitesimal coordinate transformation (2.84)

$$\xi^\mu(\eta,\mathbf{x}) = \delta_0^\mu \xi^0 + \delta_i^\mu \xi^i \tag{2.90}$$

into its temporal and spatial components. The three-vector ξ^i can be further decomposed into a transversal and a longitudinal part (a gradient of a scalar)

$$\xi^i = \xi_{\mathrm{T}}^i + \delta^{ij}\xi_{,j} \tag{2.91}$$

Decomposing the gauge degrees of freedom (2.90) according to their behaviour under spatial rotations, we notice that from the four components of ξ^μ only the time component ξ^0 and the longitudinal component ξ influence the scalar modes. The remaining two degrees of freedom are contained in ξ_{T}^i (three spatial components minus one condition of transversality $\partial_j \xi_{\mathrm{T}}^j = 0$) which only influence the vector modes. Therefore, we have to investigate how the scalar perturbations ϕ, ψ, B and E behave under the transformations

$$\eta \to \tilde{\eta} = \eta + \xi^0 \quad \text{and} \quad x^i \to \tilde{x}^i = x^i + \delta^{ij}\partial_j\xi. \tag{2.92}$$

The scalar degrees of freedom ϕ, ψ, B and E transform under (2.92) as

$$\begin{aligned} &\phi \to \tilde{\phi} = \phi + \mathcal{H}\,\xi^0 + (\xi^0)', \quad &&\psi \to \tilde{\psi} = \psi + \mathcal{H}\,\xi^0, \\ &B \to \tilde{B} = B + \xi^0 - \xi', &&E \to \tilde{E} = E + \xi. \end{aligned} \tag{2.93}$$

We can construct gauge-invariant quantities by linear combinations of (2.93) and characterize the *physical* scalar metric perturbations by the Bardeen potentials

$$\Phi := \phi - \frac{1}{a}\,[a\,(B - E')]' \quad \text{and} \quad \Psi := \psi - \mathcal{H}\,(B + E'). \tag{2.94}$$

In general, the scalar perturbations of the energy momentum tensor $\delta T^S_{\mu\nu}$ can also be characterized by four scalar functions. However, the linearised Einstein equations and the Bianchi identities yield additional constraint equations restricting the number of independent degrees of freedom. Confining ourselves to a single scalar field minimally coupled to gravity in a FRW background, only one physical scalar degree of freedom remains which fully characterizes the scalar perturbations of metric and matter (see [22, 29, 20] for a detailed treatment). An especially useful choice is the Mukhanov–Sasaki variable

$$v := a\left(\delta\varphi_{\rm gi} + \frac{\bar{\varphi}'}{\mathcal{H}}\,\Phi\right), \tag{2.95}$$

with the gauge-invariant scalar field perturbation constructed by using (2.93)

$$\delta\varphi_{\rm gi} := \delta\varphi - \bar{\varphi}'\,(B + E') \tag{2.96}$$

and the behaviour of $\delta\varphi$ under the gauge transformation (2.92)

$$\delta\varphi \rightarrow \widetilde{\delta\varphi} = \delta\varphi + \xi^0\bar{\varphi}'. \tag{2.97}$$

In [22] it has been shown that starting with the action (2.3), identifying (2.2) with (2.49) and expanding S around a FRW background up to second order in the perturbations, S_2 can be written (up to total divergences) as an effective action for the perturbation v

$$S_{\rm eff} := \int {\rm d}\eta\,{\rm d}^3x\,\mathcal{L}_{\rm eff} := \frac{1}{2}\int {\rm d}\eta\,{\rm d}^3x\left(v'^2 - \partial_i\,v\,\partial_j\,v\,\delta^{ij} + \frac{z''}{z}\,v^2\right). \tag{2.98}$$

This is the action of a free scalar field v in Minkowski space. The only relict of curved space-time is the effective *time-dependent* mass term

$$m^2_{\rm eff} := \frac{z''}{z} \quad \text{with} \quad z := \frac{a}{\mathcal{H}}\,\bar{\varphi}'. \tag{2.99}$$

The scalar v is related to a geometrically meaningful gauge-invariant quantity $\mathcal{R}$

$$v = z\,\mathcal{R}. \tag{2.100}$$

The quantity $\mathcal{R} = \psi + \mathcal{H}\,\delta\varphi/\bar{\varphi}'$ is called comoving curvature perturbation and describes the gravitational potential on spatial hypersurfaces (surfaces of constant φ). As an observable, it is especially suited, since it has a geometrical meaning. It is related to the intuitive picture for the delayed scalar field perturbations $\delta\varphi$ of Sect. 2.4.1 for the spatially flat case $\psi = 0$. The identification of $\mathcal{R}$ with the spatial curvature follows from the equation $\delta R^{(3)} = 4/a^2\,\nabla^2\,\psi$ for the perturbation of the spatial three-dimensional Ricci scalar $\delta R^{(3)}$. For super-horizon scales $\mathcal{R}$ coincides with the curvature perturbation $-\zeta = \Psi + \mathcal{H}\,\delta\,\rho/\bar{\rho}'$.

2.4.3 *Quantization of Fluctuations*

In order to derive the primordial power spectrum, we have to quantize the scalar field v. We follow the usual canonical quantization scheme and first calculate the conjugated momentum π_v

$$\pi_v := \frac{\partial\,\mathcal{L}_{\text{eff}}}{\partial\,v'} = v'. \tag{2.101}$$

When promoting v and π_v to operators $\hat{v}$ and $\hat{\pi}_v$, they must satisfy

$$\begin{aligned}[\hat{v}(\mathbf{x},\,\eta),\,\hat{v}(\mathbf{x}',\,\eta')]_{\eta=\eta'} &= [\hat{\pi}_v(\mathbf{x},\,\eta),\,\hat{\pi}_v(\mathbf{x}',\,\eta')]_{\eta=\eta'} = 0,\\ [\hat{v}(\mathbf{x},\,\eta),\,\hat{\pi}_v(\mathbf{x}',\,\eta')]_{\eta=\eta'} &= i\,\delta^{(3)}(\mathbf{x}-\mathbf{x}').\end{aligned} \tag{2.102}$$

Now we expand the field operator $\hat{v}$ in eigenmodes of the Klein–Gordon equation for the field v. This equation can be obtained by variation of (2.98) and reads in Fourier space

$$v_k'' + \left(k^2 - \frac{z''}{z}\right)\,v_k = 0, \tag{2.103}$$

with the time- and mode-dependent frequency

$$\omega_k(\eta) := k^2 - \frac{z''(\eta)}{z(\eta)}. \tag{2.104}$$

The field operator can be decomposed in creation and annihilation operators

$$\hat{v} = \frac{1}{2\,(2\pi)^{3/2}}\int \mathrm{d}^3k\,\left(e^{i\,\mathbf{k}\,\mathbf{x}}\,v_k^*(\eta)\,a_{\mathbf{k}} + e^{-i\,\mathbf{k}\,\mathbf{x}}\,v_k(\eta)\,a_{\mathbf{k}}^\dagger\right), \tag{2.105}$$

with the two independent solutions v_k and v_k^* which have to satisfy the reality conditions $v_k = v_{-k}^*$ (since v is a real field). If we neglect for a moment the time-dependence of (2.99), Eq. (2.103) reduces to that of a simple harmonic oscillator

$$v_k'' + k^2\,v_k = 0. \tag{2.106}$$

Since (2.106) is a second-order differential equation, we must specify two boundary conditions in order to obtain a *unique* solution. One of these conditions can be expressed by the proper normalization condition of the modes v_k, so that the commutator relations between the creation and annihilation operators $a_{\mathbf{k}}^\dagger$ and $a_{\mathbf{k}}$ acquire the standard form

$$[\hat{a}_{\mathbf{k}}, \hat{a}_{\mathbf{k}'}] = [\hat{a}^{\dagger}_{\mathbf{k}}, \hat{a}^{\dagger}_{\mathbf{k}'}] = 0, \quad [\hat{a}_{\mathbf{k}}, \hat{a}^{\dagger}_{\mathbf{k}'}] = \delta^{(3)}(k - k'). \tag{2.107}$$

The correct (time-independent) choice of the Wronskian W is therefore

$$W[v_k, v_k^*] := i\left(v_k^* v_k' - v_k (v_k^*)'\right) = 1 = \langle v|v\rangle. \tag{2.108}$$

It can be verified directly by substitution that this choice indeed connects the commutator algebra (2.102) with (2.107). However, in order to obtain a unique solution of (2.106), we need a second condition. In terms of the annihilation and creation operators it is natural to implement this condition by fixing the state of lowest energy $|0\rangle$ by

$$\hat{a}|0\rangle := 0. \tag{2.109}$$

To see how (2.109) translates into a condition for the mode function v_k, we can express the Hamiltonian for the harmonic oscillator in terms of $\hat{a}$ and $\hat{a}^{\dagger}$ and perform $\hat{H}|0\rangle$

$$\hat{H}|0\rangle \propto \left[(v_k')^2 + \omega_k^2 v_k^2\right] \hat{a}^{\dagger}_{\mathbf{k}}\hat{a}^{\dagger}_{\mathbf{k}}|0\rangle + E_0|0\rangle, \tag{2.110}$$

with the ground state energy E_0. In order for $|0\rangle$ to be an eigenstate of $\hat{H}$, the combination in the bracket of the first term must vanish. This leads to the condition

$$v_k' = \pm i\,\omega_k v. \tag{2.111}$$

Substituting this back into (2.108), selects the negative solution of (2.111) for the condition of positive norm $\langle v|v\rangle \geq 0$. Integrating (2.111) with the time-independent $\omega_k = k^2$, we obtain a unique properly normalized positive frequency solution of (2.106)

$$v_k(\eta) = \frac{1}{\sqrt{2k}} e^{-i\,k\,\eta}. \tag{2.112}$$

So far, we have completely neglected the time-dependence of (2.104). The solution (2.112) just corresponds to a plane wave in Minkowski space. However, all interesting information about the curvature of space-time is encoded in $\omega_k(\eta)$. In the presence of the time-dependent effective mass (2.99), there is simply no unique prescription of the vacuum state as defined in (2.109) anymore [23]. This is a general feature of curved space-time and in the context of a FRW universe, the spatial expansion leads to particle creation [24]. Moreover, it can be shown that even an uniformly *accelerated* observer in a flat and empty Minkowski space-time will detect particles [8, 11, 30]. This just reflects the fact that the definition of the vacuum is an observer-dependent statement in general. Nevertheless, due to the symmetries of de Sitter space, we can at least in the inflationary scenario hope for a physically well-founded assumption that finally leads to a meaningful unique solution. We consider a comoving observer in the limit $\eta \to -\infty$ far before inflation started. All relevant comoving scales $\lambda_c \sim 1/k$ were well inside the Hubble radius H^{-1}. This is equivalent to the condition $k \gg a\,H$.

Physically, this corresponds to the fact that the modes which are deep within the Hubble radius do not "feel" the curvature (or in the FRW case, the expansion of space). Using this for exact de Sitter space $\varepsilon \to 0$ in (2.104) leads again to the time-independent frequency

$$\lim_{\substack{\varepsilon\to 0\\ \eta\to -\infty}} \omega_k(\eta) = \lim_{\substack{\varepsilon\to 0\\ \eta\to -\infty}} \left(k^2 - \frac{z''}{z}\right) = \lim_{\eta\to -\infty} \left(k^2 - \frac{2}{\eta}\right) = k^2. \tag{2.113}$$

Thus, we find again (2.112) as asymptotic solution in de Sitter space. In order to obtain a unique physically meaningful solution for the *general* equation (2.103), we *demand* that a solution must satisfy the *asymptotic boundary condition*

$$\lim_{\eta\to -\infty} v_k = \frac{1}{\sqrt{2k}} e^{-ik\eta} \tag{2.114}$$

instead of (2.111). This is called the Bunch-Davies boundary condition [5]. In general, it is difficult to solve e.q. (2.103) explicitly with the boundary conditions (2.108) and (2.114). However, we wish to describe inflation and are mainly interested in the solution for a *quasi* de Sitter case $\varepsilon \ll 1$. First, we can try to find an exact solution for exact de Sitter space $\varepsilon = 0$. Using $\lim\limits_{\varepsilon\to 0} z''/z = 2/\eta$, Eq. (2.103) becomes

$$v_k'' + \left(k^2 - \frac{2}{\eta}\right) v_k = 0. \tag{2.115}$$

Equation (2.115) can be solved exactly by the solution

$$v_k = A \frac{1}{\sqrt{2k}} e^{-ik\eta} \left(1 - \frac{i}{k\eta}\right) + B \frac{1}{\sqrt{2k}} e^{ik\eta} \left(1 + \frac{i}{k\eta}\right). \tag{2.116}$$

The boundary conditions (2.108) and (2.114) uniquely select $A = 1$ and $B = 0$

$$v_k = \frac{1}{\sqrt{2k}} e^{-ik\eta} \left(1 - \frac{i}{k\eta}\right). \tag{2.117}$$

2.5 Horizon Crossing and Observable Quantities

There are two complementary but equivalent ways of thinking about inflation by either considering only *comoving* quantities or by considering only *physical* quantities.

When formulating everything in comoving quantities, it is convenient to think of an underlying comoving *coordinate* grid and a constant comoving perturbation wavelength λ_c. Formulated in physical quantities, we should think of a rigid *coor-*

dinate grid and a stretched physical wavelength $\lambda_p = a(t)\,\lambda_c$. Similarly, but in a sense opposite to the wavelength or coordinate grid, we can consider the comoving and physical Hubble radius. We have required that the physical Hubble parameter H_p should be constant during the rapid expansion. This leads to a constant physical Hubble radius $r_p := H_p{}^{-1}$ (cf. Fig. 2.13). In contrast, the comoving Hubble radius $r_c := H_c^{-1} = [a(t)\,H_p]^{-1}$ decreases with growing $a(t)$. In the comoving picture inflation means to shrink the comoving horizon to a small homogeneous piece of the original patch. In the physical picture it is the underlying space-time itself that expands under the constant physical Hubble radius. Both pictures are equivalent and their use is a matter of taste, but they are often the origin of confusion.

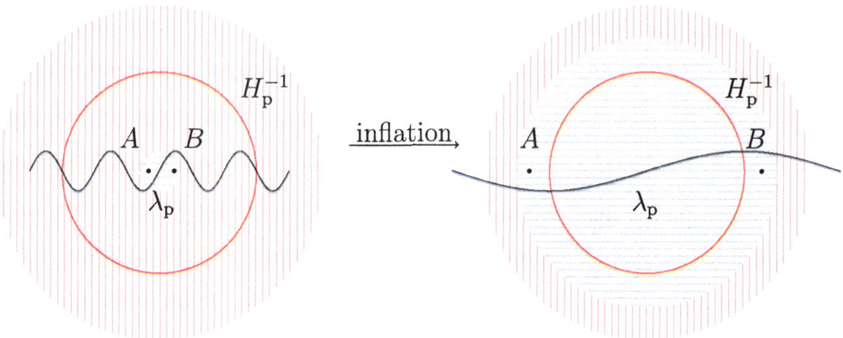

Fig. 2.13 Inflation in the physical picture. The physical Hubble radius remains constant, while the physical wavelength gets stretched. The *black horizontal pattern* represents a homogeneous patch, while the *red vertical pattern* represent different patches. The points *A* and *B* were in causal contact before inflation, have crossed the horizon during inflation and might re-enter in a post-inflationary era, when the Hubble radius grows faster than the scale factor

For modes well inside the horizon ($k/a\,H_p = r_{H_p}/\lambda_p \gg 1$), the time-dependent mass term in (2.103) can be dropped and we recover the oscillating equation of flat space-time with the plane wave solutions

$$v_k(\eta) \propto \frac{1}{\sqrt{2k}}\,e^{-i\,k\,\eta} \quad \text{for} \quad \lambda_p \ll r_{H_p}. \tag{2.118}$$

This is consistent because we would not expect curvature (expansion of space-time) to play a role for scales λ_p much smaller than r_{H_p}. For the opposite case, when the modes have left the horizon and are much bigger than the physical Hubble radius ($k \ll a\,H_p$), we can neglect the k^2 term in (2.103). Then the solution for the growing modes is

$$v_k \propto z \quad \text{for} \quad \lambda_p \gg r_{H_p}. \tag{2.119}$$

Relations (2.119) and (2.100) imply that the comoving curvature scalar remains constant

$$\mathcal{R} = \frac{v}{z} = \text{const.} \quad \text{for} \quad \lambda_{\text{p}} \gg r_{\text{H}_{\text{p}}}. \tag{2.120}$$

This aspect is crucial for the estimation of the power spectrum of the fluctuations. It guarantees that the modes outside the Hubble radius will "freeze" and are in particular independent of the physics taking place well inside the horizon (e.g. they are unaffected by the details of the reheating phase). We can therefore connect the perturbations at re-entry with the perturbations at first horizon crossing. The power spectrum of the $\mathcal{R}_k$ can be calculated by the Fourier transform

$$\hat{\mathcal{R}}(\eta, \mathbf{x}) = \frac{1}{(2\,\pi)^{3/2}} \int \mathrm{d}^3 k\, \hat{\mathcal{R}}_{\mathbf{k}}(\eta)\, e^{i\,\mathbf{k}\mathbf{x}} \quad \text{with} \quad \hat{\mathcal{R}}_{\mathbf{k}}(\eta) = \frac{v_k(\eta)}{z}\, a_{\mathbf{k}} + \frac{v_k^*(\eta)}{z}\, a^{\dagger}_{-\mathbf{k}}. \tag{2.121}$$

The power spectrum is defined by the two-point correlation function

$$< 0|\hat{\mathcal{R}}_{\mathbf{k}}(\eta)\, \hat{\mathcal{R}}^{\dagger}_{\mathbf{k}'}(\eta)|0 > =: \frac{2\,\pi^2}{k^3} \mathcal{P}_{\mathcal{R}}\, \delta^{(3)}(\mathbf{k} - \mathbf{k}') \quad \text{with} \quad \mathcal{P}_{\mathcal{R}} = \frac{k^3}{2\pi^2} \frac{|v_k(\eta)|^2}{z^2}. \tag{2.122}$$

To illustrate the behaviour of the modes during the different epochs of cosmological evolution, we will switch to the *comoving picture*. Figure 2.14 shows the *comoving* Hubble radius $r_{\text{H}_\text{c}} = (a\,H)^{-1}$ in a plot of *comoving* length scale L_c against conformal time η.

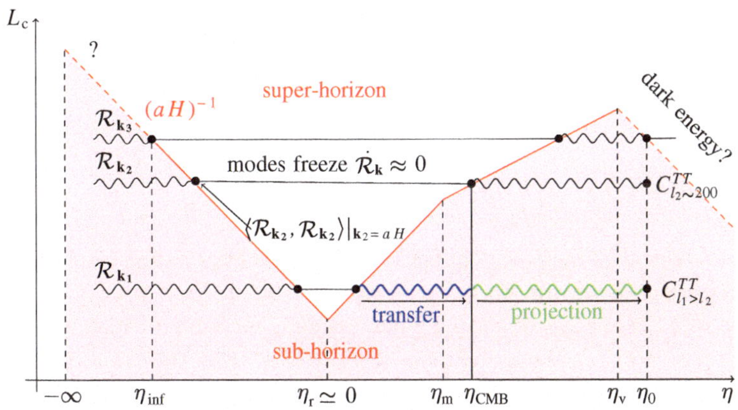

Fig. 2.14 Mode crossing. Different comoving scalar curvature modes cross the comoving Hubble radius twice at different epochs in our observable universe. The idea of this figure has been borrowed from [4]

Different curvature modes $\mathcal{R}_{\mathbf{k}^\mathbf{i}}$, $i = 1, \ldots, 3$ with fixed comoving wavelengths $\lambda_c^i \sim 1/k^i$ cross the horizon at different times η_{cross} determined by the relation

$$k_{\text{cross}} := a(\eta_{\text{cross}})H(\eta_{\text{cross}}). \tag{2.123}$$

All scalar curvature modes $\mathcal{R}_{\mathbf{k}_i}$ start deep inside the sub-horizon region as free oscillating waves when the space looks like flat Minkowski space (they do not feel the expansion yet). This is the physical motivation for the Bunch-Davies vacuum condition (2.114). As soon as the modes cross the horizon at their individual time-scale k^i_{cross} and enter the super-horizon region, the modes freeze until re-entry. Decoherence can explain the quantum-to-classical transition [15–17] and once the modes re-enter the horizon, they become quasi-classical perturbations subjected to all possible physical processes within the horizon. They set the initial conditions for the density fluctuations which evolve and amplify under the influence of gravity. We can describe the post-inflationary evolution by complicated "transfer functions". The modes re-enter according to the principle "first out – last in" .

Since these small irregularities are projected onto our sky, it is useful to expand the power spectrum of perturbations in spherical harmonics in order to obtain an angular power spectrum depending on the multipole order of the spherical harmonics. The angle θ can roughly be related to the multipole l by $l \sim \pi/\theta$. Thus, small angular separations correspond to large multipoles l in the angular power spectrum. Today, we observe with increasing resolution correlations with an angular separation θ.

The mode $\mathcal{R}_{\mathbf{k}_2}$ in Fig. 2.14 re-enters the horizon at the time of recombination η_{CMB}, where the photons decoupled from the oscillating photon-baryon plasma. It corresponds to the first acoustic peak in the observed angular power spectrum and can be related to the size of the horizon at the time of recombination. It therefore provides direct information about the curvature. Different locations of the first peak would correspond to different curvature. This allows us to conclude that our universe is indeed flat up to about one percent. Modes that have re-entered at *earlier* times correspond to *larger* angular moments l or smaller angular correlations $\theta \sim \pi/l$ on the sky.

2.5.1 Slow-Roll Power Spectrum

In order to calculate the power spectrum (2.122), we have to find a solution to the mode equation (2.103) in the slow-roll regime $\varepsilon \ll 1$. In contrast to the $\varepsilon = 0$ solution with $z''/z = 2/\eta$ of pure de Sitter space, we must approximate the effective mass term (2.99) z''/z to first order in the slow-roll parameters $\varepsilon \ll 1$, $|\delta| \ll 1$. We could exactly solve (2.103) in terms of Hankel functions of the first kind $H^{(1)}_{\nu_s}(-k\,\eta)$, by expressing (2.99) as

$$\frac{z''}{z} = \frac{\nu_s^2 - 1/4}{\eta^2}, \tag{2.124}$$

where ν_s is an expression linear in the *constant* ε and δ. Let us start by re-writing the definitions (2.69) in terms of conformal expressions $\mathcal{H} = a'/a = a\,H$

$$\varepsilon = -\frac{\dot{H}}{H^2} = 1 - \frac{\mathcal{H}'}{\mathcal{H}^2} \quad \text{and} \quad \delta = \varepsilon - \frac{1}{2\mathcal{H}}\frac{\varepsilon'}{\varepsilon} = 1 - \frac{\varphi''}{\mathcal{H}\varphi'}. \tag{2.125}$$

Equation (2.125) implies $\varepsilon' = 2\mathcal{H}(\varepsilon^2 - \varepsilon\,\delta)$ and we find that terms $\propto \varepsilon'$ and $\propto \delta'$ are of second order and can be neglected within our approximation. Using (2.99), we can calculate

$$\frac{z'}{z} = \frac{a'}{a} + \frac{\varphi''}{\varphi'} - \frac{\mathcal{H}'}{\mathcal{H}} = \mathcal{H}\left(1 + \frac{\varphi''}{\mathcal{H}}\varphi' - \frac{\mathcal{H}'}{\mathcal{H}^2}\right) = \mathcal{H}\,(1 + \varepsilon - \delta). \tag{2.126}$$

Differentiating once more and using (2.125), we find

$$\begin{aligned}\frac{z''}{z} &= \left(\frac{z'}{z}\right)' + \left(\frac{z'}{z}\right)^2 = \mathcal{H}'\,(1 + \varepsilon - \delta) + \mathcal{H}\,(\varepsilon' + \delta') + \mathcal{H}^2\,(1 + \varepsilon - \delta)^2 \\ &= \mathcal{H}^2\,(2 + 2\,\varepsilon - 3\,\delta) + O(2).\end{aligned} \tag{2.127}$$

We further observe that $\mathrm{d}\eta = \mathrm{d}a/a\,\mathcal{H}$, so that we can write

$$\frac{\varepsilon}{a\,\mathcal{H}}\,\mathrm{d}a = \varepsilon\,\mathrm{d}\eta = \mathrm{d}\eta - \frac{\mathcal{H}'}{\mathcal{H}^2}\,\mathrm{d}\eta = \mathrm{d}\eta + \mathrm{d}\left(\frac{1}{\mathcal{H}}\right) \quad \text{or} \quad \eta = -\frac{1}{\mathcal{H}} + \int \frac{\varepsilon}{a\,\mathcal{H}}\,\mathrm{d}a. \tag{2.128}$$

Since we have assumed that ε is approximately constant, we obtain

$$\eta = -\frac{1}{\mathcal{H}} + \varepsilon \int \mathrm{d}\eta \quad \Leftrightarrow \quad \mathcal{H} = -\frac{1}{\eta\,(1-\varepsilon)}. \tag{2.129}$$

Inserting this into the result (2.127), we can rewrite z''/z in the desired form

$$\frac{z''}{z} \approx \frac{2 + 2\varepsilon - 3\delta}{\eta^2\,(1-\varepsilon)^2} \approx \frac{2 + 2\varepsilon - 3\delta}{\eta^2\,(1-2\varepsilon)} \approx \frac{(2 + 2\varepsilon - 3\delta)(1 + 2\varepsilon)}{\eta^2} \approx \frac{2 + 6\varepsilon - 3\delta}{\eta^2}. \tag{2.130}$$

In each step we have successively approximated the expression, finally keeping only terms linear in the slow roll parameters. We substitute (2.130) in (2.124) and obtain ν_s to first order in ε and δ

$$\nu_\mathrm{s}^2 = \frac{9}{4} + 6\varepsilon - 3\delta \quad \Rightarrow \quad \nu_\mathrm{s} = \frac{3}{2} + 2\,\varepsilon - \delta + O(2). \tag{2.131}$$

The general solution of (2.103) for z''/z defined in (2.124) in terms of the Hankel function $H^{(1)}_{\nu_\mathrm{s}}(-k\,\eta)$ reads

$$v_k(\eta) = \frac{\sqrt{\pi}}{2}\,e^{i\,\pi/2\,(\nu_\mathrm{s}+1/2)}\,\sqrt{-\eta}\,H^{(1)}_{\nu_\mathrm{s}}(-k\eta). \tag{2.132}$$

Since we are interested in super-horizon scales $-k\,\eta = k/a\,H = r_{\mathrm{H_p}}/\lambda_{\mathrm{p}} \ll 1$ and we know that the scalar comoving curvature modes $\mathcal{R}_k$ freeze outside the Hubble radius $r_{\mathrm{H_p}}$, we can use $i\,H^{(1)}_{\nu_{\mathrm{s}}}(\alpha) \simeq \Gamma(\nu_{\mathrm{s}})/\pi\,(z/2)^{-\nu_{\mathrm{s}}}$ in the limit $\alpha \to 0$ [29], express $-\eta$ by (2.129) and finally find for the amplitude $|v_k(\eta)|$ the asymptotic expression

$$|v_k(\eta)| \simeq C(\nu_{\mathrm{s}})\,\frac{1}{\sqrt{2k}}\left(\frac{k}{a\,H}\right)^{-\nu_{\mathrm{s}}+1/2}, \qquad C(\nu_{\mathrm{s}}) := 2^{\nu_{\mathrm{s}}-3/2}\,\frac{\Gamma(\nu_{\mathrm{s}})}{\Gamma(3/2)}\,(1-\varepsilon)^{\nu_{\mathrm{s}}-1/2}. \tag{2.133}$$

We can insert (2.133) into the definition of the power spectrum (2.122) and obtain

$$\mathcal{P}_{\mathcal{R}}(k) = \frac{k^3}{2\pi^2}\,\frac{|v_k(\eta)|^2}{z^2} = \frac{k^2}{4\,\pi^2}\frac{C^2(\nu_{\mathrm{s}})}{z^2}\left(\frac{k}{a\,H}\right)^{1-2\,\nu_{\mathrm{s}}}. \tag{2.134}$$

Since $\nu_{\mathrm{s}} \approx 3/2$ during slow-roll, we can approximate $C(\nu_{\mathrm{s}}) \approx 1$. Substituting $z = a\,\varphi'/\mathcal{H} = a\,\dot{\varphi}/H$, we finally obtain the *nearly* scale invariant power spectrum

$$\begin{aligned} \mathcal{P}_{\mathcal{R}}(k) = \mathcal{P}_{\asymp}(k) &\simeq \left(\frac{H}{2\,\pi}\right)^2\left(\frac{H}{\dot{\varphi}}\right)^2\left(\frac{k}{a\,H}\right)^{3-2\,\nu_{\mathrm{s}}} \\ &\simeq \left(\frac{H}{2\,\pi}\right)^2\left(\frac{H}{\dot{\varphi}}\right)^2\Big|_{k=a\,H} = \frac{V(\varphi)}{24\,\pi^2\,M_{\mathrm{P}}^4\,\varepsilon_{\mathrm{V}}}\Big|_{k=a\,H}. \end{aligned} \tag{2.135}$$

In the second identity, we have used again the fact that the comoving curvature perturbations are constant in the super-horizon regime. This allows us to evaluate them at the point where they leave the Hubble radius $k = a\,H$ since they remain unchanged until they re-enter again. In the last identity we have used the slow-roll approximation to rewrite $\mathcal{P}_{\mathcal{R}}$ in terms of the inflaton potential.

2.5.2 Cosmological Parameters During Slow-Roll

As a measure for the scale dependence of (2.135), we define the spectral index

$$n_{\mathrm{s}} - 1 := \frac{\mathrm{d}\ln\mathcal{P}_{\mathcal{R}}(k)}{\mathrm{d}\ln k} = 3 - 2\,\nu_{\mathrm{s}} = 2\delta - 4\varepsilon = 2\delta_{\mathrm{V}} - 6\varepsilon_{\mathrm{V}}. \tag{2.136}$$

In the last line, we have used relation (2.80) to write (2.136) in terms of the slow-roll parameters expressed in terms of the potential. For tensor fluctuations we can follow exactly the same steps as for the scalar perturbations and start with a similar effective action as (2.98). Since tensor perturbations are already gauge-invariant by construction, we can directly use the h_{ij} of (2.89). The Fourier decomposition reads [29]

$$h_{ij}(\mathbf{x},\,\eta) = \frac{1}{(2\,\pi)^{3/2}}\int \mathrm{d}k\,\left(\hat{h}^{+}_{\mathbf{k}}(\eta)\,\varepsilon^{+}_{ij}(\mathbf{k}) + \hat{h}^{\times}_{\mathbf{k}}(\eta)\,\varepsilon^{\times}_{ij}(\mathbf{k})\right) \tag{2.137}$$

Tensor modes are gravitational waves and have two polarizations ε_{ij}^{+} and $\varepsilon_{ij}^{\times}$.

$$\hat{v}_{\mathbf{k}}^{+/\times}(\eta) = z\,\hat{h}_{\mathbf{k}}^{+/\times}(\eta) = \frac{M_{\mathrm{P}}\,a}{\sqrt{8\,\pi}}\,\hat{h}_{\mathbf{k}}^{+/\times}(\eta), \tag{2.138}$$

with $z = a\,M_{\mathrm{P}}/\sqrt{8\,\pi}$. Thus, the effective mass (2.99) term z''/z simplifies to a''/a

$$\frac{a''}{a} = \mathcal{H}^2\,(2-\varepsilon) = \frac{(2-\varepsilon)}{\eta^2\,(1-\varepsilon)^2} = \frac{2+3\varepsilon}{\eta^2} + O(2) \tag{2.139}$$

and the corresponding index ν_{t} in the solution (2.133) is given by

$$\nu_{\mathrm{t}}^2 = \eta^2\,\frac{a''}{a} + \frac{1}{4} = \frac{9}{4} + 3\,\varepsilon \quad\Rightarrow\quad \nu_{\mathrm{t}} = \frac{3}{2} + \varepsilon + O(2). \tag{2.140}$$

Substituting this with $z = a\,M_{\mathrm{P}}/\sqrt{8\,\pi}$ in the asymptotic solution (2.133) yields

$$\begin{aligned}\mathcal{P}_h(k) &:= 4\,\mathcal{P}_{h^{+/\times}}(k) := 4\,\frac{8\,\pi}{M_{\mathrm{P}}^2\,a^2}\,\frac{k^3}{2\,\pi^2}|v_k(\eta)|^2 \simeq \frac{16}{\pi}\left(\frac{H}{M_{\mathrm{P}}}\right)^2\left(\frac{k}{a\,H}\right)^{3-2\nu_{\mathrm{t}}}\\ &\simeq \frac{16\,G}{\pi}\,H^2\Big|_{k=a\,H}.\end{aligned} \tag{2.141}$$

The tensor spectral index is defined with a different convention compared to (2.136)

$$n_{\mathrm{t}} := \frac{\mathrm{d}\ln\mathcal{P}_h(k)}{\mathrm{d}\ln k} \simeq 3 - 2\,\nu_{\mathrm{t}} = -2\,\varepsilon. \tag{2.142}$$

If the spectral indices (2.136) and (2.142) are themselves independent of the mode k, they simply describe the tilt of the power spectra. If $n_{\mathrm{s}}(k)$ and $n_{\mathrm{t}}(k)$ are functions of the mode k, this effect can be quantified by the running of the spectral indices

$$\alpha_{\mathrm{s}} := \frac{\mathrm{d}\,n_{\mathrm{s}}}{\mathrm{d}\ln k} \quad\text{and}\quad \alpha_{\mathrm{t}} := \frac{\mathrm{d}\,n_{\mathrm{t}}}{\mathrm{d}\ln k}, \tag{2.143}$$

which are however already of second order in ε and δ. The tensor to scalar ratio

$$r := \frac{\mathcal{P}_{\mathcal{R}}}{\mathcal{P}_h} = 16\,\varepsilon \tag{2.144}$$

is a direct measure for the energy scale of inflation, since $\mathcal{P}_{\mathcal{R}} \simeq 10^{-9}$ is fixed by observation and $\mathcal{P}_h \sim H^2$. We note that the measurement of $\mathcal{P}_{\mathcal{R}}$ alone would not be sufficient, since it is a ratio of the two parameters H^4 and $\dot{\varphi}^2$ with $\dot{\varphi}$ being model dependent. Combination of (2.142) and (2.144) yields the consistency relation

$$r = -8\,n_{\mathrm{t}} \tag{2.145}$$

which is a *generic* prediction of all inflationary models. A detection of primordial gravitational waves would therefore provide direct information about the absolute energy scale of inflation. We will use these cosmological parameters when considering the non-minimal Higgs inflation scenario in Chap. 6.

References

1. Albrecht, A., Steinhardt, P.J.: Cosmology for grand unified theories with radiatively induced symmetry breaking. Phys. Rev. Lett. **48**, 1220 (1982)
2. Arnowitt, R., Deser, S., Misner, C.W.: The dynamics of general relativity. In: Witten, L. (ed.) Gravitation: An Introduction to Current Research, pp. 227–265. Wiley, New York (1962)
3. Aurich, R., Lustig, S., Steiner, F.: CMB anisotropy of the Poincaré dodecahedron. Class Quantum Gravity **22**, 2061 (2005)
4. Baumann, D.: TASI lectures on inflation. http://arxiv.org/abs/0907.5424 (2009) [160 pages] (cited on 19 Dec 2011)
5. Bunch, T., Davies, P.: Quantum field theory in de Sitter space: renormalization by point splitting. Proc. Roy. Soc. Lond. A **360**, 117 (1978)
6. Copi, C.J., Huterer, D., Schwarz, D.J., Starkman, G.D.: Large-angle anomalies in the CMB. Adv. Astron. **2010**, 847541 (2010)
7. Cornish, N.J., Spergel, D.N., Starkman, G.D., Komatsu, E.: Constraining the topology of the Universe. Phys. Rev. Lett. **92**, 201302 (2004)
8. Davies, P.C.W.: Scalar particle production in Schwarzschild and Rindler metrics. J. Phys. A **8**, 609 (1975)
9. Einstein, A.: Die Feldgleichungen der Gravitation. Sitzber. kgl.-preuß. Akad. Wiss. Berlin, Sitzung der phys.-math. Klasse, **XLVIII**, 844 (1915)
10. Einstein, A.: Die Grundlage der allgemeinen Relativitätstheorie. Ann. Phys. (Berlin), 4th series, **49**, 769 (1916)
11. Fulling, S.A.: Nonuniqueness of canonical field quantization in Riemannian space-time. Phys. Rev. D **7**, 2850 (1973)
12. Guth, A.: Inflationary universe: a possible solution to the horizon and flatness problems. Phys. Rev. D **23**, 347 (1981)
13. Guth, A.H., Pi, S.Y.: Fluctuations in the new inflationary universe. Phys. Rev. Lett. **49**, 1110 (1982)
14. Hawking, S.W.: The development of irregularities in a single bubble inflationary universe. Phys. Lett. B **115**, 295 (1982)
15. Kiefer, C., Lohmar, I., Polarski, D., Starobinsky, A.A.: Origin of classical structure in the universe. J. Phys. Conf. Ser. **67**, 012023 (2007)
16. Kiefer, C., Lohmar, I., Polarski, D., Starobinsky, A.A.: Pointer states for primordial fluctuations in inflationary cosmology. Class Quantum Gravity **24**, 1699 (2007)
17. Kiefer, C., Polarski, D., Starobinsky, A.A.: Quantum-to-classical transition for fluctuations in the early universe. Int. J. Mod. Phys. D **7**, 455 (1998)
18. Lachièze-Rey, M., Luminet, J.-P.: Cosmic topology. Phys. Rep. **254**, 135 (1995)
19. Linde, A.D.: A new inflationary universe scenario: a possible solution of the horizon, flatness, homogeneity, isotropy and primordial monopole problems. Phys. Lett. B **108**, 389 (1982)
20. Maldacena, J.M.: Non-gaussian features of primordial fluctuations in single field inflationary models. J. High Energy Phys. **05**, 013 (2003)
21. Mukhanov, V.F., Chibisov, G.V.: Quantum fluctuation and a nonsingular universe. JETP Lett. **33**, 532 (1981). (In Russian)
22. Mukhanov, V.F., Feldman, H.A., Brandenberger, R.H.: Theory of cosmological perturbations. Phys. Rep. **215**, 203 (1992)

23. Parker, L.: Particle creation in expanding universes. Phys. Rev. Lett. **21**, 562 (1968)
24. Parker, L.: Quantized fields and particle creation in expanding universes. I. Phys. Rev. **183**, 1057 (1969)
25. Perlmutter, S., et al. (Supernova Cosmology Project): Measurements of Ω and Λ from 42 high-redshift supernovae. Astrophys. J. **517**, 565 (1999)
26. Peter, P., Uzan, J.-P.: Primordial Cosmology. Oxford University Press, New York (2009)
27. Starobinsky, A.A.: Spectrum of relict gravitational radiation and early state of the universe. JETP Lett. **30**, 682 (1979)
28. Starobinsky, A.A.: Dynamics of phase transition in the new inflationary universe scenario and generation of perturbations. Phys. Lett. B **117**, 175 (1982)
29. Straumann, N.: From primordial quantum fluctuations to the anisotropies of the cosmic microwave background radiation. Ann. Phys. (Leipzig) **15**, 701 (2006)
30. Unruh, W.G.: Notes on black-hole evaporation. Phys. Rev. D **14**, 870 (1976)
31. Will, C.M.: Theory and Experiment in Gravitational Physics. Cambridge University Press, Cambridge (1993)

Chapter 3
Standard Model

The Standard Model of Particle Physics is a gauge theory of the group

$$U_{Y_w}(1) \times SU_L(2) \times SU_c(3). \tag{3.1}$$

The gauge group is a direct product of three groups. This implies that the different transformations associated with the subgroups do not mix with each other. Therefore, we will first briefly analyse them individually and assign to each subgroup a different type of interaction. All subgroups are Lie groups and thus also have a geometrical meaning as a manifold. In the Lagrangian language of field theory the fundamental particles (scalar, vector, spinor,...) are described as *local* field operators represented collectively by $\Psi(x)$. Interactions are then described as products of these field operators. The field operators transform in the *fundamental* representation of the corresponding symmetry groups.

A linear representation $\rho[G]$ is the realization of the abstract group action on some vector space $V(G)$. If we choose a basis in $V(G)$, we can express the field operators Ψ as vectors $\Psi^a \in V(G)$, $a = 1, ..., \dim G$. In other words, those particles which are subjected to a certain interaction arrange themselves as multiplets in the internal vector space $V(G)$. Interactions between the field operators Ψ are then described by linear operators in $V(G)$ that act on the internal indices $a, b, ...$ of the field operators Ψ^a. These linear operators are associated with the *adjoint* representation of the Lie group. It is convenient to label the different irreducible representations ρ of a Lie group G by the dimension of the representation $\dim_\rho[G]$. The product structure of (3.1) allows to describe the interaction properties of the particles in the Standard Model by analysing their transformation behaviour under the action of the individual subgroups. Each particle can then be characterized by three numbers that label the dimensionality of the irreducible representations in which the corresponding particle transforms

$$\left(\dim_\rho\left[SU_c(3)\right], \dim_\rho\left[SU_L(2)\right], \dim_\rho\left[U_{Y_w}(1)\right]\right). \tag{3.2}$$

Every Lie group has a trivial representation $\dim_\rho[G] = \mathbf{1}$, a singlet. Thus, a particle that is not subjected to the interaction described by the group G transforms as a

C. F. Steinwachs, *Non-minimal Higgs Inflation and Frame Dependence in Cosmology*, Springer Theses, DOI: 10.1007/978-3-319-01842-3_3,
© Springer International Publishing Switzerland 2014

singlet **1** under G. The fundamental representation which is associated with the field operators Ψ corresponds to a multiplet $\Psi^a \in V(G)$ that has the same dimension as the Lie group G, i.e. $a, \ldots = 1, \ldots \dim G$. The dimension of the adjoint representation corresponds to the number of independent generators of the Lie group. These generators are associated with the vector gauge bosons and we have to express them as $(\dim G \times \dim G)$ matrices in $V(G)$ in order to act on the multiplets $\Psi^a \in V(G)$. Gauge invariance requires that all terms contained in the Lagrangian density of the Standard Model transform as singlets under the three individual subgroups of (3.1). We formulate this requirement as in (3.2):

$$(\mathbf{1},\ \mathbf{1},\ \mathbf{1}). \tag{3.3}$$

Before discussing the structure of the Standard Model, we will collect the most important properties of the elementary particles in an overview table as a reference. The numerical data for the masses in this table was taken from [6].

Table 3.1 Table of properties of elementary standard model particles

Name	Symbol	s	Q	T_z	$\dim_\rho[G]$	M
Electron R	e^-	1/2	−1	0	(1, 1, −2)	0.511 MeV
Electron L	e^-	1/2	−1	−1/2	(1, 2, −1)	0.511 MeV
e-neutrino L	ν_e	1/2	0	+1/2	(1, 2, −1)	2.2 eV
Muon R	μ	1/2	−1	0	(1, 1, −2)	<105.7 MeV
Muon L	μ	1/2	−1	−1/2	(1, 2, −1)	<105.7 MeV
μ-neutrino L	ν_μ	1/2	0	+1/2	(1, 2, −1)	<0.17 MeV
Tau R	τ	1/2	−1	0	(1, 1, −2)	1.777 GeV
Tau L	τ	1/2	−1	−1/2	(1, 2, −1)	1.777 GeV
τ-neutrino L	ν_τ	1/2	0	+1/2	(1, 2, −1)	< 15.5 MeV
Down-quark R	q_d	1/2	−1/3	0	(3, 1, −2/3)	4.8 MeV
Down-quark L	q_d	1/2	−1/3	−1/2	(3, 2, 1/3)	4.8 MeV
Up quark R	q_u	1/2	2/3;	0	(3, 1, 4/3)	2.4 MeV
Up quark L	q_u	1/2	2/3	+1/2	(3, 2, 1/3)	2.4 MeV
Strange-quark R	q_s	1/2	2/3	0	(3, 1, −2/3)	1.27 GeV
Strange-quark L	q_s	1/2	2/3	−1/2	(3, 2, 1/3)	1.27 GeV
Charm-quark R	q_c	1/2	−1/3	0	(3, 1, 4/3)	104 MeV
Charm-quark L	q_c	1/2	−1/3	+1/2	(3, 2, 1/3)	104 MeV
Bottom-quark R	q_b	1/2	2/3	0	(3, 1, −2/3)	171.2 GeV
Bottom-quark L	q_b	1/2	2/3	−1/2	(3, 2, 1/3)	171.2 GeV
Top-quark R	q_t	1/2	−1/3	0	(3, 1, 4/3)	4.2 GeV
Top-quark L	q_t	1/2	−1/3	+1/2	(3, 2, 1/3)	4.2 GeV
Gluons	G_μ^α	1	0	0	(8, 1, 0)	<20 MeV
Photon	γ	1	0			$<10^{-18}$ eV
W^+-boson	W^+	1	+1			80.4 GeV
W^--boson	W^-	1	−1			80.4 GeV
Z-boson	Z^0	1	0			91.2 GeV

The meaning of the individual symbols[1] and the pattern of organization will become clear in the following sections. We will briefly recall the general structure of a gauge theory on the basis of the electromagnetic interaction. Then we will investigate the individual interactions of (3.1) separately and finally combine them.

3.1 Gauge Theories

A gauge theory is based on an underlying *continuous* symmetry of a system. A symmetry operation is characterized by an abstract mathematical group G. A system possesses a symmetry if it is invariant under the action of G. Noether's theorem guarantees that each continuous symmetry corresponds to a conserved quantity. We will distinguish between *rigid* symmetries[2] and *local* symmetries where the group parameters $g_i(x)$ are functions of space-time. Given a rigid symmetry of a system, the symmetry is *gauged* by promoting the group parameters to functions of space-time $g_i \rightarrow g_i(x)$. The gauge group associated with this local symmetry is a Lie group G. In general, the action functional will not be invariant under the *local* group transformations anymore because the derivatives in the kinetic part will induce extra terms associated with the space-time derivatives of the group parameters $\partial_\mu g_i(x)$. However, there is a way to maintain invariance under the gauge group by introducing a new field in the action functional. This field is called gauge field and its transformation behaviour has to be chosen in such a way that the unwanted extra terms cancel out exactly. In order to keep the notation compact, we can absorb the new gauge field in the definition of a *covariant* derivative. Since the underlying gauge group is a Lie group, the group can be associated with a manifold and the gauge field obtains a geometrical interpretation as a connection.

So far, we have introduced the gauge field at a purely kinematic level with the only objective to make the action functional invariant under local gauge transformations. However, to assign a physical meaning to this field, we have to describe its dynamics. We have to implement a kinetic term for the gauge field which is quadratic in its derivatives and is itself invariant under gauge transformations. It turns out that we can associate such a term with the (square of the) field strength of the gauge field. Geometrically, the field strength corresponds to the curvature of the Lie manifold. During the process of promoting a rigid symmetry to a local one, we have introduced a new type of *interaction* between the gauge field and the fields originally present in the action functional via the coupling contained in the covariant derivative. In the language of quantum field theory, we start with a dynamical description of *free* field quanta, obeying a rigid symmetry. By gauging this symmetry and demanding

---

[1] Antiparticles are denoted by a bar over the symbol, not to be confused with the Dirac conjugate. The irreducible representations of $U_{Y_w}(1)$ can be written as $\rho[U_{Y_w}(1)] = \exp[i\, Y_w\, \theta(x)]$ and are labelled by the eigenvalue Y_w (with the conjugated $-Y_w$). Thus, we just write the real number Y_w for $\dim_\rho[U_{y_w}(1)]$ instead of a bold symbol. In this notation the $U_{Y_w}(1)$ singlet corresponds to $Y_w = 0$.

[2] The group parameters g_i controlling the group action of a rigid symmetry are constant with respect to space-time.

invariance of the Lagrangian density under this local symmetry we are forced to introduce an interaction with a new field. This is the gauge field and all interactions among the field quanta can be traced back to the emission and absorption of a gauge field (later we will see that the gauge fields can also interact among themselves, depending on the properties of the gauge group).

The complete information of a Lie group is determined by its behaviour in the neighbourhood of the identity element. The infinitesimal transformations near the identity element form the algebra of the group. By successive application of these infinitesimal transformations we can build up finite group transformations. The algebra of a group, in turn, is determined by the infinitesimal generators of the group action. The number of independent group parameters corresponds to the number of generators. The individual subgroups of (3.1) are all *finite* Lie groups and thus have a finite number of generators. Physically, these generators correspond to the conserved charges of the system. It turns out that Nature is very successfully expressed in terms of gauge theories by identifying the gauge field as the force carrier mediating the interaction described by the gauge group. Everything we will discuss below will be restricted to the tree-level. Quantization of gauge theories can be achieved with similar methods as described in Chap. 4. We will limit ourselves to explain the structure of the different interactions and try to organize them into a clear pattern.

3.1.1 Electromagnetic Interaction $U_{EM}(1)$

In this section, we will explain the concept of a $U(1)$ gauge theory by Maxwell's theory of electrodynamics. In the covariant formulation of Maxwell's theory we can combine the scalar electric and vector magnetic potential in one four-potential

$$\vec{E} := -\vec{\nabla} A^0 - \frac{\partial}{\partial t}\vec{A}, \quad \vec{B} := \vec{\nabla} \times \vec{A} \quad \longrightarrow \quad A_\mu := \begin{pmatrix} A^0 \\ \vec{A} \end{pmatrix}. \tag{3.4}$$

Here, $\vec{E}$ and $\vec{B}$ are the electric and magnetic fields, respectively. The source of the electromagnetic interaction is the electric charge Q, expressed in units of the electric charge of an electron e. We can either have a positive $(+e)$ or a negative $(-e)$ charge. From Maxwell's theory we know that the photons can be described as transversal electromagnetic waves that carry energy. In addition, we know from Dirac's theory that for each electron there is a corresponding antiparticle, the positron. Electrons and positrons have spin $1/2$ and are therefore fermions. They possess a small mass of roughly 0.5 MeV. What do we need to describe the interaction between electrons? We can write down a Lagrangian for the field operator of the electron Ψ_e. Due to the spin $1/2$, Ψ_e transforms as a complex Dirac spinor. The complex conjugate $\Psi_e^\dagger$ can be interpreted as the field operator of the antiparticle with respect to Ψ_e. The dynamics is described by the Dirac equation

$$(i\gamma^\mu \partial_\mu - m_e)\Psi_e = 0, \tag{3.5}$$

with γ^μ being the (4×4)-Dirac matrices connecting the spinor representation of Ψ_e and the vector representation of ∂_μ and obeying the anti-commutator relations

$$\{\gamma^\mu, \gamma^\nu\} = 2\,\eta^{\mu\nu}. \tag{3.6}$$

By choosing a particular basis, these matrices can be represented as

$$\gamma^0 := \begin{pmatrix} \mathbb{1}_{2\times 2} & 0 \\ 0 & -\mathbb{1}_{2\times 2} \end{pmatrix}, \quad \gamma^i := \begin{pmatrix} \sigma^i & 0 \\ 0 & \sigma^i \end{pmatrix}, \quad i = 1, 2, 3 \tag{3.7}$$

with the (2×2)-Pauli spin matrices

$$\sigma^1 := \begin{pmatrix} 0 & 1 \\ 1 & 0 \end{pmatrix}, \quad \sigma^2 := \begin{pmatrix} 0 & -i \\ i & 0 \end{pmatrix}, \quad \sigma^3 := \begin{pmatrix} 1 & 0 \\ 0 & -1 \end{pmatrix}. \tag{3.8}$$

By multiplying (3.5) from the left with the adjoint field $\bar{\Psi}_e$, defined by

$$\bar{\Psi}_e := \gamma^0\, \Psi_e^\dagger, \tag{3.9}$$

we obtain the Lagrangian density for the free electron system

$$\mathcal{L}_e := i\,\bar{\Psi}_e\,\gamma^\mu \partial_\mu \Psi_e - m_e\,\bar{\Psi}_e\,\Psi_e. \tag{3.10}$$

The Lagrangian density (3.10) is invariant under a *rigid* $U(1)$ transformation. The field operator Ψ transforms in the fundamental representation of $U(1)$. This is just the multiplication with a constant phase factor

$$\Psi_e \to \Psi_e' = e^{i\,e\,\alpha}\,\Psi_e, \quad \Psi_e^\dagger \to \Psi_e^{\dagger\prime} = e^{-i\,e\,\alpha}\,\Psi_e^\dagger \quad \Rightarrow \quad \bar{\Psi}_e \to \bar{\Psi}_e' = e^{-i\,e\,\alpha}\,\bar{\Psi}_e. \tag{3.11}$$

The group parameter that regulates the transformation is the constant α. Now we promote this rigid symmetry to a local symmetry

$$\alpha \to \alpha(x). \tag{3.12}$$

We have to check if (3.10) is invariant under the *local* $U(1)$ transformations

$$\Psi_e \to \Psi_e' = e^{i\,e\,\alpha(x)}\,\Psi_e \quad \text{and} \quad \bar{\Psi}_e \to \bar{\Psi}_e' = e^{-i\,e\,\alpha(x)}\,\bar{\Psi}_e. \tag{3.13}$$

Substituting these new transformation laws into (3.10) gives

$$\mathcal{L}_e \to \mathcal{L}_e' := i\,\bar{\Psi}_e\,\gamma^\mu \partial_\mu \Psi_e - m_e\,\bar{\Psi}_e\,\Psi_e - e\,\bar{\Psi}_e \Psi_e\,\gamma^\mu \partial_\mu \alpha(x) \neq \mathcal{L}_e. \tag{3.14}$$

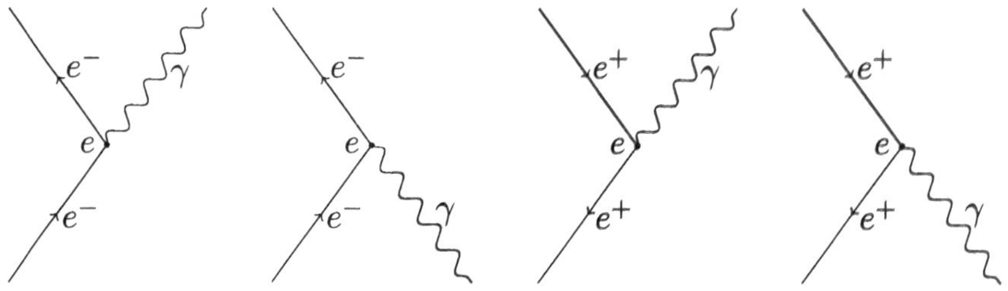

Fig. 3.1 Feynman diagrams for the basic interaction vertex in QED

The last term spoils the invariance. We have to add the gauge vector field A_μ to fix this. The required transformation behaviour of the gauge vector field is

$$A_\mu \to A'_\mu = A_\mu - \frac{1}{e}\,\partial_\mu \alpha(x). \tag{3.15}$$

Now we can define a new (covariant) derivative by

$$D_\mu := \partial_\mu + i\,e\,A_\mu, \tag{3.16}$$

and ask if the modified Lagrangian density

$$\mathcal{L}_e^{\text{gauge}} := i\,\bar{\Psi}_e\,\gamma^\mu\,D_\mu \Psi_e - m_e\,\bar{\Psi}_e\,\Psi_e \tag{3.17}$$

is invariant under the local gauge transformation. This is indeed the case, as can be checked easily by investigating how the kinetic term behaves under (3.13) and (3.15). Thus, we have succeeded to implement the local $U_{\text{EM}}(1)$ symmetry by adding another field. The physical meaning of the massless gauge field is clear. It corresponds to the vector four-potential (3.4) and hence can be identified with the photon γ. In fact, we have managed to obtain a complete description of electrodynamics containing all electromagnetic interactions by the demand of gauge invariance starting from a description of free electrons only. The photon mediates the electromagnetic interaction and all possible processes can be traced back to the single interaction vertex

$$\propto e\,\bar{\Psi}_e\,\gamma^\mu\,A_\mu\,\Psi_e \tag{3.18}$$

between the photon and the electron, induced via the covariant derivative in (3.16). In particular, we note that the interaction is proportional to the electromagnetic coupling constant e. Thus, the probability *amplitude* for the basic processes in quantum electrodynamics (QED) like the emission or absorption of a photon from an electron — described in the Feynman diagrams of Fig. 3.1 — is proportional to e. By convention, time runs from the bottom to the top of the diagram. Flipping a leg in its time direction corresponds to switching between particle and antiparticle,[3] or in the case of the gauge boson between emission and absorption. The *probability* for such a process to occur is proportional to the *square* of the amplitude e^2. In this way, we have recovered the fine structure constant (restoring units)

$$\alpha := \frac{e^2}{\hbar c} \sim \frac{1}{137}, \tag{3.19}$$

not to be confused with the group parameter $\alpha(x)$. However, there is still one piece missing. The Lagrangian density (3.17) does not contain any term that describes the dynamics of the gauge field A_μ itself and in order to prescribe physical reality to A_μ, we have to implement a kinetic term for A_μ. It has to be a term containing derivatives of A_μ. From Maxwell's theory, it is clear that such a term is connected to the antisymmetric field tensor

$$F_{\mu\nu} := \partial_\mu A_\nu - \partial_\nu A_\mu = -F_{\nu\mu}. \tag{3.20}$$

By comparison with (3.4), we see that $F_{0i} \sim E_i$ and $F_{ij} \sim \epsilon_{ijk} B_k$. However, this object has two space-time indices and is only linear in the derivatives of A_μ. In order to describe the dynamics of A_μ in (3.17), we need a term quadratic in $\partial_\mu A_\nu$. Furthermore, the kinetic term for A_μ must be a Lorentz scalar. Thus, we must get rid of the two remaining indices (μ, ν). The most natural attempt is to form the contracted quantity $F_{\mu\nu}F^{\mu\nu}$ which is a Lorentz scalar and is also quadratic in $\partial_\mu A_\nu$. Finally, we have to ensure that the new term $F_{\mu\nu}F^{\mu\nu}$ is also invariant under gauge transformations, otherwise it would spoil the whole construction. In order to convince ourselves that this is indeed the case, it is sufficient to investigate the behaviour of $F_{\mu\nu}$ under (3.15). Due to the commutation of partial derivatives, $F_{\mu\nu}$ and therefore also the product $F_{\mu\nu}F^{\mu\nu}$ are gauge-invariant. Finally, we can therefore write down the complete QED Lagrangian invariant under local $U_{\mathrm{EM}}(1)$ gauge transformations describing the interaction between electrons and photons

$$\mathcal{L}_{\mathrm{QED}} := i\,\bar{\Psi}_e\,\gamma^\mu\,D_\mu\Psi_e - m_e\,\bar{\Psi}_e\,\Psi_e - \frac{1}{4}\,F_{\mu\nu}F^{\mu\nu}. \tag{3.21}$$

Particles different from the electron also transform under $U_{\mathrm{EM}}(1)$, if they carry a multiple of the electric charge. However, despite its similarity, we want to emphasize that the naive identification of the $U_{Y_{\mathrm{w}}}(1)$ in the gauge group of the Standard Model with the electromagnetic $U_{\mathrm{EM}}(1)$ is not correct. The reason will become clear when describing the unification of the weak and the electromagnetic interaction by the electroweak interaction. In the Standard Model, it is more appropriate to consider the product $SU_L(2) \times U_{Y_{\mathrm{w}}}(1)$ rather than the single groups separately.

[3] This follows from the complex integration contour of the Feynman propagator in QED.

3.1.2 Strong Interaction $SU_c(3)$

The fundamental particles subjected to the strong interaction are the quarks q, which are fermions, and the gluons G_α, $\alpha = 1, ..., 8$, which are bosons. Since this interaction is formulated as a gauge theory of $SU_c(3)$, we can identify the gluons with the gauge bosons that mediate this interaction between the quarks. The new abstract quantum number associated with the strong interaction is called "colour" and has three possible states: red, green and blue — the origin of the name Quantum Chromodynamics (QCD). The necessity to introduce a new quantum number like colour follows from the Pauli exclusion principle for fermions. Since quarks are fermions, each q_i itself has four spinor components, whose indices we will suppress in the following. Moreover, there are two different types of quarks: up-type quarks u and down-type quarks d, which both carry non-integer fractions of the electric charge. In nature only objects with integer multiples of charge are observed. Two fermions cannot simultaneously be in the same quantum state without differing in at least one quantum number. Thus, for example a *baryon* made up of three identical quarks is not a possible state. However, we know from experiments that such compound objects do exist. Thus, there must be an additional quantum number, the colour, to resolve the contradiction with the Pauli principle. Since a quark can possess three different colours, we need to describe its behaviour under colour exchange by a three-dimensional colour multiplet. These three possible colour states of a quark are described as components of a $SU_c(3)$ triplet that transforms under the fundamental representation **3** of $SU_c(3)$. For every charged particle there is a corresponding antiparticle. In the case of quarks, the antiparticles are antiquarks $\bar{q}$ charged with an anticolour. Thus, it is possible to combine a quark and its antiquark to form a neutral colourless object called *meson*. Baryons and mesons are collectively denoted *hadrons*. Objects which transform as singlets under $SU_c(3)$ are invariant under the colour mixing strong interaction. Thus, these objects must be colourless objects. A baryon composed of three quarks with different colours is colourless, as indicated by Fig. 3.2.

Fig. 3.2 Three quarks with different colour charges form a colourless (*white*) object

Gauge invariance requires that only gauge singlets **1** can appear as terms in the action functional. Thus, all observed hadrons must be colourless.[4] Hadrons are objects composed of quarks. Due to confinement, a phenomenon discussed below, quarks have never been observed in any experiment as free particles (as asymptotic "in" and

[4] As an empirical fact, all observed objects in nature carry an integer multiple of the electric charge. This leads to the requirement that quarks have to carry a non-integer electric charge Q.

"out" states). Thus, all free particles must be colourless. An antiquark $\bar{q}$ transforms in the conjugated representation $\bar{\mathbf{3}}$ of $SU_c(3)$. If we want to see how a meson transforms under $SU_c(3)$, we have to find the decomposition of the product of representations into a direct sum of irreducible representations

$$\mathbf{3} \otimes \bar{\mathbf{3}} = \mathbf{1} \oplus \mathbf{8}. \tag{3.22}$$

The group-theoretical background can be found in many textbooks [3]. The calculations can be done with the help of Young tableaux, see Appendix C. We observe that the spectrum of the direct product of a quark and an antiquark contains a singlet or an octet. We can identify the singlet with the colourless meson. Baryons are composed of three quarks. In particular, the most important baryons are the building blocks of atomic nuclei, the neutron *udd* and the proton *uud*. In order to prove that within our group theoretical description there really does exist a baryon composed of three quarks which transforms as a singlet **1**, we have to find the irreducible representations of the direct product of three single quarks $\mathbf{3} \otimes \mathbf{3} \otimes \mathbf{3}$. Using again the results of Appendix C, we find

$$\mathbf{3} \otimes \mathbf{3} \otimes \mathbf{3} = \mathbf{1} \oplus \mathbf{8} \oplus \mathbf{8} \oplus \mathbf{10}. \tag{3.23}$$

The spectrum contains a singlet **1**, which we can identify with the baryon. However, as seen from the decompositions of the direct products (3.22) and (3.23), in addition to the singlet, there is always an octet **8** appearing in the spectrum. The **8** corresponds to the adjoint representation of $SU_c(3)$ which we have identified with the $3^2 - 1 = 8$ gauge bosons, the gluons. In contrast to electrodynamics, where the gauge boson of the abelian group $U_{\text{EM}}(1)$, the photon γ, does not carry any charge, the gluons themselves are not colourless objects. They carry colour charge which means that they can also interact among themselves by colour exchange. This is a consequence of the non-abelian character of $SU_c(3)$, making the strong interactions non-linear. Therefore, we must analyse if singlets can appear in the spectrum of two gluons

$$\mathbf{8} \otimes \mathbf{8} = \mathbf{1} \oplus \mathbf{8} \oplus \mathbf{8} \oplus \mathbf{10} \oplus \overline{\mathbf{10}} \oplus \mathbf{27}. \tag{3.24}$$

The important fact is that a colourless combination of two gluons, a singlet state really does exist. This object is called a "2-glueball" and goes beyond the hadron classification. Moreover, since the spectrum of two gluons contains among other representations again an octet **8**, the interaction of a 2-glueball with another gluon must again produce a singlet, a 3-glueball. This process can be iterated to produce higher N-glueballs. So far, no such glueballs have been detected directly. It is believed that glueballs are very heavy objects which makes them extremely hard to detect.

In analogy to (3.21), we can write down the QCD Lagrangian density

$$\mathcal{L}_{\text{QCD}} := i\, \bar{q}_a\, \gamma^\mu\, (D_\mu)_a^b q_b - m_q \delta^{ab}\, \bar{q}_a\, q_b - \frac{1}{4}\, G^\alpha_{\mu\nu} G^{\mu\nu}_\alpha, \tag{3.25}$$

where we have abbreviated the field operators for the quarks as q_a and explicitly displayed the indices $a, b, .. = 1, ..., 3$, labelling the components of the colour triplet and the indices $\alpha = 1, ..., 8$ labelling the generators of $SU_c(3)$. The basic interactions of QCD are illustrated by three- and four-point vertices in Fig. 3.3.

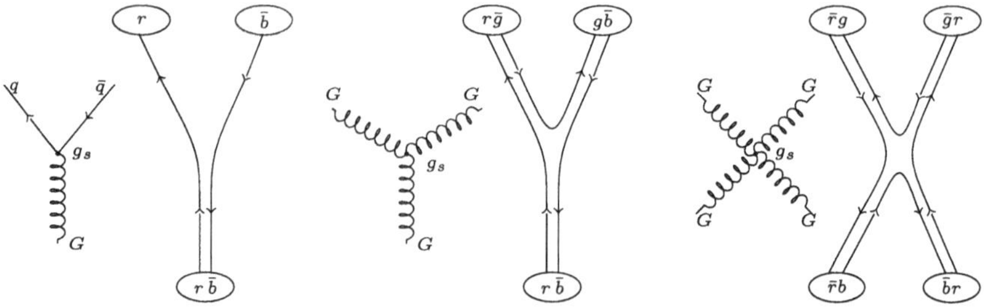

Fig. 3.3 The basic interaction vertices of QCD

As indicated in Fig. 3.3, it is convenient to think of the gluons as a pair of a quark and an antiquark with different colours. The covariant derivative $(D_\mu)_a^b\, q_b$ acting on a quark is defined by

$$(D_\mu)_a^b\, q_b = \left(\partial_\mu \delta_a^b - i\, g_s\, G_\mu^\alpha\, (T_\alpha)_a^b\right) g_b, \tag{3.26}$$

with the gluon vector gauge field G_μ^α and the generators T_{ab}^α of $SU_c(3)$, relating the fundamental representation (associated with a, b) and the adjoint representation (associated with α). The gluon field strength tensor follows from the commutator of the covariant derivative (3.26)

$$G_{\mu\nu}^\alpha := \partial_\mu G_\nu^\alpha - \partial_\nu G_\mu^\alpha + g_s\, f^{\alpha\beta\gamma}\, G_\mu^\beta\, G_\nu^\gamma, \tag{3.27}$$

with the structure constants $f^{\alpha\beta\gamma}$ defined by the commutator

$$[T^\alpha,\, T^\beta] = i\, f^{\alpha\beta\gamma}\, T^\gamma. \tag{3.28}$$

Although summation over $\alpha, \beta, ...$ is always assumed, the index position of $\alpha, \beta, ...$ is irrelevant. Equation (3.28) implies the antisymmetry of the $f_{\alpha\beta\gamma}$ and the Jacobi identity

$$f^{\alpha\beta\gamma} = -f^{\beta\alpha\gamma}, \quad f^{\alpha\beta\eta} f_{\eta\gamma\epsilon} + f^{\beta\gamma\eta} f_{\eta\alpha\epsilon} + f^{\gamma\alpha\eta} f_{\eta\beta\epsilon} = 0. \tag{3.29}$$

A particularly important representation of the generators T^α are the Gell-Mann matrices $T^\alpha = (\lambda^\alpha)_{ab}$. In this representation, the gluons G_μ are described as (3×3) matrices $G_\mu = G_\mu^\alpha T_\alpha = \frac{1}{2}\, G_\mu^\alpha (\lambda_\alpha)_{ab}$ in order to act on the three-dimensional vectors Ψ^a in the fundamental representation.

Finally, we will discuss the phenomenon of confinement. We will follow a rather heuristic argument, but it can be justified rigorously in the context of the renormalization group (RG). The strength of the strong interaction can be characterized by the coupling constant g_s. The beta function $\beta_{g_s}(E)$ describes the behaviour of g_s with respect to the energy scale E. Depending on the global sign of β_{g_s}, the coupling $g_s(E)$ grows or decreases with higher energies $E \to \infty$. In contrast to the electroweak couplings, $g_s(E)$ becomes negligible for very high energies

$$\lim_{E\to\infty} g_s(E) \to 0. \tag{3.30}$$

This phenomenon is called *asymptotic freedom* and explains that the bound states of quarks can only break up at very high energies. Only at these energies, the quarks can be considered as free particles. The phenomenon of *confinement* represents in some sense the complementary behaviour at low energies.[5] At low energies, the coupling g_s is so strong compared to the couplings of the other interactions, that only bound states of colourless objects can form.

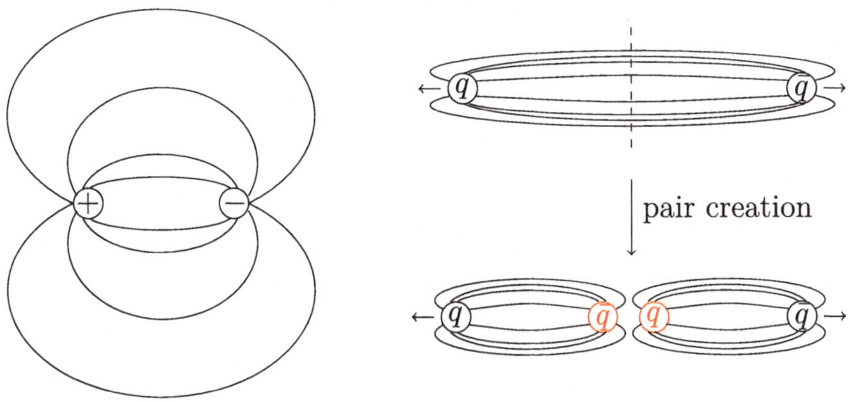

Fig. 3.4 Force lines. *Left* EM dipole. *Right* meson confinement and jet formation

This situation is analogous to charge-neutral compound objects like atoms in the context of the electromagnetic interaction. However, there is an essential difference. We can imagine the gluons as colour force lines spreading out of colour-charged quarks. As we have shown, the gluons act among themselves by colour interaction and thus the force lines also do. Force lines never intersect, but colour force lines can attract each other, leading to "force tubes". In contrast to the exterior field of a dipole, the colour field between two quarks does not fall off with larger distance. A measure for the field strength at a given point is the flux line density. Integrating the flux line density over a spatial volume V yields the field energy stored in V. Thus, for larger and larger distances r between the charges, the electromagnetic flux

[5] In fact, $g_s(E)$ diverges at very low energy scales E and perturbation theory breaks down.

line density falls off like $1/r^2$ according to the Coulomb law. On the contrary, the colour flux line density remains *constant* with growing distance between the quark and antiquark. Thus, the field energy stored in the flux tube grows linearly with r (in contrast to the $1/r$ behaviour of the electromagnetic potential). We can imagine an experiment where a high energetic particle collides with a meson. First of all, if the energy is too low, an "adiabatic kick" will just shift the whole meson as a compound object. But if the energy is high enough to affect the "inner structure", it might kick out a quark. This quark will be accelerated away from the antiquark. With increasing distance r between the quark-antiquark pair, it converts its kinetic energy into potential binding energy stored in the flux tube. At some point, it is energetically possible to spontaneously create another quark-antiquark pair, leading to two confined mesons as illustrated in Fig. 3.4. Depending on the energy of the initial scattering process, there can be a whole cascade of such events leading to the observed phenomenon called "jet". A similar process can also happen for baryons or any other colourless $SU_c(3)$ singlet. In electrodynamics, a separation of a compound charge neutral object would lead to free ionized (charged) particles instead. As we have seen, this is not possible for the colour interaction and as a consequence, no free colour-charged object has ever been observed directly. This explains why the strong interaction has such a short range.

3.1.3 Electroweak Interaction $SU_L(2) \times U_{Y_w}(1)$

In Sect. 3.1.2, we have already seen that the non-abelian character of $SU_c(3)$ compared to the abelian $U_{\text{EM}}(1)$ leads to drastic physical differences between QED and QCD and is responsible for phenomena like glueballs, asymptotic freedom and confinement. However, there are at least two additional empirical facts which go beyond the treatment so far. First of all, it has been observed that there is an additional weak force in nature which has a short range. In contrast to QED and QCD, where the gauge bosons are massless, the gauge bosons mediating the weak force are massive.[6] However, a mass term for the gauge bosons would violate the requirement of gauge invariance. Secondly, it has been noticed that this weak interaction is not symmetric under parity transformations (reflections in space).

3.1.3.1 Helicity, Handedness and Chirality

Before we will discuss these points in detail, we will clarify the notations of helicity, handedness and chirality. The spin magnitude $s := |\mathbf{s}|$ of an elementary particle is a constant. The helicity of an elementary particle can be defined as the (normalized) projection of the spin vector $\mathbf{s}$ on the momentum vector $\mathbf{k}$

---

[6] In QCD, the phenomenon of confinement was responsible for the short range interaction and not a mass of the force carrier — the gluons.

$$\frac{\mathbf{s}\cdot\mathbf{k}}{|\mathbf{s}|\,|\mathbf{k}|} = \cos\theta_{\mathrm{H}}(k). \tag{3.31}$$

We define the handedness of a particle according to its helicity

$$\cos\theta_{\mathrm{H}}(k) = \begin{cases} > 0 \; =: \text{right-handed} \\ < 0 \; =: \text{left-handed} \end{cases} . \tag{3.32}$$

The angle $\theta_{\mathrm{H}}(k)$ is a *frame-dependent* quantity since it depends on the magnitude of the momentum $k := |\mathbf{k}|$. Let us assume that a particle moving with velocity k in the $\mathbf{e}_z$ direction of the frame K has positive helicity $\cos\theta_{\mathrm{H}}(k) > 0$. If we boost to another reference frame $K \to K'$ moving in the opposite direction of $\mathbf{k} = k\,\mathbf{e}_z$ with a relative velocity $\mathbf{v} = -v\,\mathbf{e}_z$ between the two frames K and K', we have to use the relativistic formula to add velocities for k and $-v$ in order to obtain the new velocity k'.

$$k' = \frac{k - v}{1 - \frac{k\,v}{c^2}}. \tag{3.33}$$

If $|v| > |k|$ the velocity k' as seen from the K' frame is negative and therefore $\mathbf{k}'$ points in the opposite direction as $\mathbf{k}$ so that $\cos\theta_{\mathrm{H}}(k') < 0$ and the particle has negative helicity. However, if we consider a *massless* particle, it has to propagate with the speed of light $k = k' = c$ and according to (3.33) must have a *fixed* helicity. It has always *either* helicity $+1$ *or* helicity -1.

Chirality is a more abstract concept. According to the classification of Wigner [8], any elementary particle can be classified with respect to the irreducible representation of the Poincaré group[7] $ISO(1, 3)$, which is the semi-direct product of the Lorentz group $O(1, 3)$ and the group of translations $\mathbb{R}^4$

$$ISO(1, 3) \cong O(1, 3) \rtimes \mathbb{R}^4. \tag{3.34}$$

The different irreducible representations, in turn, can be labelled by two properties of the particle: its mass m and its spin s. Since we are interested in massive particles (the only massless particles in the Standard Model are the photon and the gluons), we must classify them according to their spin, i.e. their representations of the Lorentz group. The behaviour of particles under the Lorentz group $SO(1, 3)$ can be reduced to the classification of the finite-dimensional irreducible representations of the double-covering group $SU_L(2) \times SU_R(2)$. The irreducible representations of $SU_L(2)$ are characterized by a single number L with the eigenvalues $L(L+1)$ of the Casimir operator L_α. The same is true for $SU_R(2)$. Thus, the irreducible representations of $SU_L(2) \times SU_R(2)$ can be labelled by two numbers $(L,\, R)$. We are mainly interested in massive fermions $s = 1/2$ and therefore in the spinor representation of the Lorentz

---

[7] The notation $ISO(1, 3)$ stems from the fact that the Poincaré group is a ten parametric Lie group and the generators of the corresponding Lie algebra are the Killing vectors of Minkowski space-time. See the definition of a Killing vector (B. 13) and set $g_{\mu\nu} = \eta_{\mu\nu}$.

group. The notation with the L and R suggests that we can divide the representations into left-handed $(1/2,\ 0)$ and right-handed $(0,\ 1/2)$. This property is called chirality. In particular, fermions described by a Dirac spinor $\Psi = \Psi_R + \Psi_L$ have left- and right-handed components.

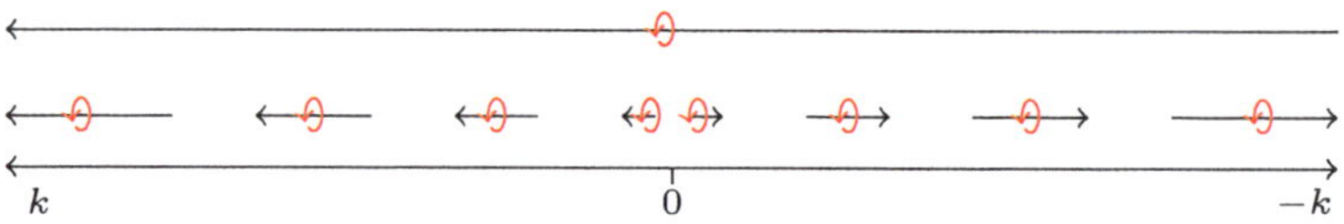

Fig. 3.5 Handedness. *Upper line* A right-handed massless particle propagating in the direction indicated by the arrow, with the velocity of light. *Lower line* A massive particle has a rest mass and therefore can be brought to rest. Moreover, it can be accelerated in the opposite direction. The direction of spin stays the same in both directions. Thus, when changing the direction of propagation, the right-handed particle turns into a left-handed particle. This shows that a mass term mixes right- and left-handed particles

The single components transform as right-handed spinors Ψ_R and as left-handed spinors Ψ_L, respectively. They can be obtained by projecting Ψ onto the chiral basis

$$\Psi_L := P_L\Psi = \frac{1}{2}\left(\mathbb{1}_{2\times 2} + \gamma_5\right)\Psi, \quad \Psi_R := P_R\Psi = \frac{1}{2}\left(\mathbb{1}_{2\times 2} - \gamma_5\right)\Psi. \tag{3.35}$$

The matrix γ^5 has eigenvalues ± 1 and can be written as a product of the Dirac matrices (3.7)

$$\gamma^5 := i\,\gamma^0\gamma^1\gamma^2\gamma^3. \tag{3.36}$$

3.1.3.2 The Beta-Decay

There are physical processes in nature that are not symmetric under the interchange of left- and right-handed spinors. The parity-violating property of the weak interaction was experimentally confirmed by [9]. The weak interaction does only involve left-handed particles. If we want to describe the weak interaction as a gauge theory, this means that right-handed particles have to transform as singlets under the corresponding gauge group. Since the weak interaction treats left- and right-handed spinors differently, we must really treat left- and right-handed massive fermions as *different individual particles*. The prototype of a weak interaction process is the beta-decay of a neutron into a proton, an electron and an anti-neutrino.

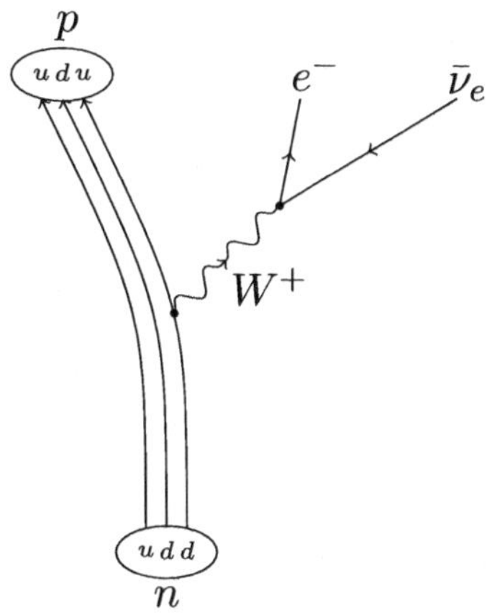

Fig. 3.6 The beta-decay. A neutron n decays into a proton p, an electron e^- and an anti-neutrino $\bar{\nu}_e$ by the emission of a W^+ boson

From Sect. 3.1.2, we know that the neutron and the proton can be described as a bound state of three quarks udd and udu. This shows that the weak interaction must somehow change a d-quark type into a u-quark type and an electron e^--lepton type into an electron-neutrino-lepton type ν_e.[8] In order to describe the weak interaction as a gauge theory, we follow the same procedure as for the strong interaction. We try to represent the left-handed particles that interact via the weak interaction as multiplets that transform in the fundamental representation of the gauge group. In view of the $u_{\text{L}} \leftrightarrow d_{\text{L}}$ and $e^-_{\text{L}} \leftrightarrow \nu_{\text{L, e}^-}$ exchange in the beta-decay, we need a gauge group with a two-dimensional fundamental representation. The simplest group is $SU_L(2)$. Thus, we group the left-handed up- and down-quark types and the left-handed electron- and electron-neutrino-lepton type as doublets with respect to $SU_L(2)$. Although the abstract mathematical description is similar to the ordinary spin $1/2$ treatment, the origin of the doublet structure is an *internal gauge symmetry*. We will therefore denote the quantum number associated with the doublet of left-handed particles *weak isospin* $T = 1/2$. Right-handed particles transform as singlets $\mathbf{1}$ under $SU_L(2)$ and have $T = 0$. Following the mathematical treatment of the spin algebra, we can distinguish the components of the left-handed doublet by the projection of the weak isospin T to the z-axis. T_{z} has $2T + 1 = 2$ different values $T_{\text{z}} = \pm 1/2$, labelling the upper and lower component of the doublet. However, we must also distinguish between the right-handed singlets u_{R} of the up- and d_{R} of the down-types. This can be done by the introduction of an additional quantum number, the *weak hypercharge* Y_{w}. It has to be the same for both components of the left-handed weak doublet, but has to be different for the right-handed particles. The weak hypercharge corresponds to the eigenvalue of the abelian $U_{Y_{\text{w}}}(1)$ phase transformation and has one generator B_μ. The group $SU_L(2)$ has three generators W^i_μ, $i = 1, ..., 3$ mediating the weak interaction.

[8] The anti-electron-neutrino can be thought of as the electron-neutrino moving backwards in time.

How can we relate electromagnetism U_{EM} with the electroweak interaction $SU_L(2) \times U_{Y_{\text{w}}}(1)$? We will discuss this question in Sect. 3.2.4, but we can already try to relate the electric charge Q with the weak isospin component T_{z} and the weak hypercharge Y_{w}. From the beta-decay, we know that the weak interaction does not only change the weak isospin component T_{z} by one, but also the electric charge Q. We can empirically relate Q, T_{z} and Y_{w} by a formula similar to the Gell-Mann–Nishijima formula [2, 7]

$$Q = T_{\text{z}} + \frac{Y_{\text{w}}}{2} \quad \leftrightarrow \quad Y_{\text{w}} = 2\,(Q - T_{\text{z}}). \tag{3.37}$$

The factor of $1/2$ in (3.37) is a convention and defines the normalization of Y_{w}.

3.2 Masses of Elementary Particles

Both Lagrangians for QED (3.21) and QCD (3.25) contain a mass term proportional to $m_\Psi\, \bar{\Psi}_e \Psi_e$ for electrons and quarks respectively. This is possible because in this case we deal with gauge interactions that do not distinguish between left- and right-handed fermions. Only the mass term for the gauge *bosons* A_μ and G^α_μ is forbidden in order not to spoil gauge invariance. However, we have to explain the masses of the fermions while ensuring invariance of the Standard Model action under the *full* gauge group (3.1).

3.2.1 Massive Fermions in the Standard Model?

Let us consider again the Lagrangian density for a *fermion* described by a Dirac spinor Ψ and its conjugate $\bar{\Psi}$ defined in (3.9)

$$\mathcal{L} = i\,\bar{\Psi}\gamma^\mu \partial_\mu \Psi - m_\Psi \bar{\Psi}\Psi. \tag{3.38}$$

According to (3.35), we can decompose a spinor by projecting out its left- and right-handed components. The kinetic part decomposes into a sum of two independent terms

$$i\bar{\Psi}\gamma^\mu\partial_\mu\Psi = i\bar{\Psi}_{\text{R}}\gamma^\mu\partial_\mu\Psi_{\text{R}} + i\bar{\Psi}_{\text{L}}\gamma^\mu\partial_\mu\Psi_{\text{L}}, \tag{3.39}$$

involving only left- *or* right-handed spinors and *no mixed* left- *and* right-handed spinors. This follows from the vanishing mixed components of the kinetic term

$$\begin{aligned} i\bar{\Psi}_{\text{R/L}}\gamma^\mu\partial_\mu\Psi_{\text{L/R}} &= \frac{i}{4}(1 \mp \gamma^5)\gamma^0\Psi^\dagger\gamma^\mu\partial_\mu(1 \pm \gamma^5)\Psi = \frac{i}{4}(1 \mp \gamma^5)(1 \pm \gamma^5)\,\gamma^0\Psi^\dagger\gamma^\mu\partial_\mu\Psi \\ &= [1 - (\gamma^5)^2]\,\bar{\Psi}\gamma^\mu\partial_\mu\Psi = 0, \end{aligned} \tag{3.40}$$

where we have used the following relations for the chirality matrix γ^5:

$$\{\gamma^\mu, \gamma^5\} = 0, \quad (\gamma^5)^2 = 1, \quad (\gamma^5)^\dagger = \gamma^5, \quad \bar{\Psi}_{L/R} = \gamma^0\, \Psi^\dagger_{L/R}. \tag{3.41}$$

However, the mass term in (3.38) mixes left- and right-handed components of Ψ

$$m_\Psi \bar{\Psi}\Psi = m_\Psi\,(\Psi_R\bar{\Psi}_R + \bar{\Psi}_R\Psi_L + \bar{\Psi}_L\Psi_R + \bar{\Psi}_L\Psi_L) = m_\Psi\,(\bar{\Psi}_R\Psi_L + \bar{\Psi}_L\Psi_R), \tag{3.42}$$

and the coupling between equally handed components vanishes:

$$m_\Psi\,\Psi_{R/L}\bar{\Psi}_{R/L} = \frac{m_\Psi}{4}\,(1 \mp \gamma^5)\gamma^0\Psi^\dagger(1 \mp \gamma^5)\Psi = \frac{m}{4}\,(1 \mp \gamma^5)(1 \pm \gamma^5)\Psi^\dagger\Psi = 0. \tag{3.43}$$

From (3.39) and (3.42), we can observe that left- and right-handed particles completely decouple in the absence of a mass term. In the case of electromagnetism, the left- and right-handed electrons e_L and e_R both carry the same electromagnetic charge Q. This means that the interaction $\propto m_e e_L \bar{e}_R$ between e_L and $\bar{e}_R$ via the mass term mixing (3.42) does respect charge conservation. In the language of gauge theory, this means that both, e_L and e_R, pick up the *same* phase factor under a local $U_{EM}(1)$ transformation

$$e_L \to e'_L = e^{i\,Q\,\alpha(x)}\,e_L, \quad e_R \to e'_R = e^{i\,Q\,\alpha(x)}\,e_R. \tag{3.44}$$

Thus, both transformation factors cancel in the mixed left- and right-handed combination

$$m_e\,e_L\,\bar{e}_R \to m_e\,e'_L\,\bar{e}'_R = m_e\,e^{i\,Q\,\alpha(x)}\,e_L\,e^{-i\,Q\,\alpha(x)}\bar{e}_R = m_e\,e_L\,\bar{e}_R. \tag{3.45}$$

This simply reflects the fact that $U_{EM}(1)$ does not distinguish between left- and right-handed so that a mass term does not violate gauge invariance. However, the electroweak interaction treats left- and right-handed components differently. For the sake of simplicity, we consider the different behaviour of left- and right-handed electrons with respect to the weak hypercharge phase transformation. The group parameter of $U_{Y_w}(1)$ is denoted by $\theta(x)$ and the weak hypercharge is determined by the eigenvalue Y_w.

$$e_L \to e'_L = e^{i\,Y_w\,\theta(x)}\,e_L = e^{-i\,\theta(x)}\,e_L, \quad e_R \to e'_R = e^{i\,Y_w\,\theta(x)}\,e_R = e^{-2i\,\theta(x)}\,e_R \tag{3.46}$$

According to (3.37), the left-handed electrons have eigenvalue $Y_w = -1$ and the right-handed electrons have $Y_w = -2$. Thus, the $U_{Y_w}(1)$ gauge contributions of a mass term

$$m_e\,\bar{e}_R\,e_L \to m_e\,\bar{e}'_R\,e'_L = m_e\,e^{-i\,\theta(x)}\,e_L\,e^{2i\,\theta(x)}\,\bar{e}_R = m_e\,\bar{e}_R\,e_L\,e^{i\,\theta x}, \tag{3.47}$$

do not cancel anymore (the conjugation $\bar{e}_{\mathrm{R}}$ induces a flip of sign $Y_{\mathrm{w}} = +2$). This shows that fermionic mass terms violate gauge invariance and have to be excluded in order to obtain a gauge-invariant Standard Model Lagrangian density. As a consequence, the most general Lagrangian density invariant under the full gauge group $SU_c(3) \times SU_L(2) \times U_{Y_{\mathrm{w}}}(1)$ can contain no mass term neither for the gauge bosons nor for the fermions.

3.2.2 The Massless Standard Model Lagrangian

We have arranged the whole particle content of the Standard Model according to the behaviour under the full gauge group (3.1). We briefly summarize the organization pattern of the fermions but suppress the fermion generation index $m = 1, ..., 3$ that labels the three identical copies of fermions with the same interaction pattern which only differ with respect to their mass. Fermions are defined by their spin $s = 1/2$. Leptons L can be characterized as fermions that transform as singlets under the colour-changing interaction $SU_{\mathrm{c}}(3)$, while quarks Q are grouped in colour triplets. Right-handed fermions L_{R} and Q_{R} can be characterized as singlets $T = 0$ under $SU_L(2)$, while left-handed fermions L_{L} and Q_{L} are grouped in flavour doublets $T = 1/2$. A further distinction can be made according to the different behaviour with respect to weak hypercharge $U_{Y_W}(1)$ phase transformations which can be read off from (3.37). For the left-handed fermions $T = 1/2$ and both components $T_z = \pm 1/2$ of the $SU_L(2)$ doublet share the same $U_{Y_{\mathrm{w}}}(1)$ eigenvalue Y_{w}. For the right-handed fermions which are organized as two different $SU_L(2)$ singlets, we can decide whether it is an "up-type" or a "down-type" (a "neutrino-type" or an "electron-type" respectively) by a different Y_{w} eigenvalue. Since we deal here with massless neutrinos,[9] we can distinguish between *five types of fermionic particles* characterized by their transformation behaviour under (3.1)

$$L_{\mathrm{L}} := \begin{pmatrix} \nu_{e\mathrm{L}} \\ e_{\mathrm{L}} \end{pmatrix}, \quad Q_{\mathrm{L}} := \begin{pmatrix} u_{\mathrm{L}} \\ d_{\mathrm{L}} \end{pmatrix}, \quad L_{\mathrm{R,d}} := e_{\mathrm{R}}, \quad Q_{\mathrm{R,u}} := u_{\mathrm{R}}, \quad Q_{\mathrm{R,d}} := d_{\mathrm{R}}. \tag{3.48}$$

Each of these five fermion types has two additional copies from the remaining two generations and to each particle there corresponds an antiparticle. In order to write down the most general Lagrangian density invariant under (3.1), we need to find suitable representations of the gauge vector fields G^{α}_{μ}, $\alpha = 1, ..., 8$, W^{i}_{μ}, $i = 1, ..., 3$ and B_{μ}. The dimension of these representations should coincide with the dimension of the fundamental representation of the corresponding gauge group, so that the gauge fields can act as matrices in the vector space spanned by the corresponding gauge

[9] Although, neutrino oscillations suggest that neutrinos could have a mass, we do not consider the possibility of right-handed neutrinos. A sterile (not participating in the gauge interaction) heavy (at the GUT scale) right-handed neutrino could generate a tiny mass for the left-handed neutrino by the so-called *seesaw mechanism*. Since neutrinos are electrically neutral, they are their own antiparticles and have to be described as Majorana particles. Whether there do exist Majorana particles or not can in principle be tested by neutrino-less double beta-decay experiments [5].

multiplets. The $SU_c(3)$ gluons $G_\mu = G^\alpha_\mu T_\alpha$ can be represented in terms of the three-dimensional representation by the Gell-Mann matrices. The gluons G_μ transform as $(\mathbf{8}, \mathbf{1}, 0)$ under (3.1). The $SU_L(2)$ gauge fields $W_\mu = W^j_\mu T_j$ can be represented in terms of the two-dimensional representation by the Pauli matrices (3.8). The gauge bosons W_μ transform as $(\mathbf{1}, \mathbf{3}, 0)$ under (3.1). The $U_{Y_\mathrm{w}}(1)$ gauge field B_μ can be characterized by a single eigenvalue Y_w and should act by multiplication of a phase factor $e^{i\,Y_\mathrm{w}\,\theta(x)}$. The B_μ transforms as $(\mathbf{1}, \mathbf{1}, 0)$ under (3.1). The invariance of the Standard Model Lagrangian density with respect to the gauge group (3.1) requires to substitute all partial space-time derivatives by the following covariant derivatives, where the prefactors can be read off from Table 3.1

$$
\begin{aligned}
D_\mu L_\mathrm{L} &:= \partial_\mu L_\mathrm{L} + \left(-\frac{i}{2}\, g\, W^j_\mu\, \tau_j + i\, g'\, B_\mu\right) L_\mathrm{L},\\
D_\mu Q_\mathrm{L} &:= \partial_\mu Q_\mathrm{L} + \left(-\frac{i}{2}\, g_\mathrm{s}\, G^\alpha_\mu\, \lambda_\alpha - i\, g\, W^j_\mu\, \tau_j - \frac{i}{3}\, g'\, B_\mu\right) Q_\mathrm{L},\\
D_\mu e_\mathrm{R} &:= \partial_\mu e_\mathrm{R} + 2\, i\, g'\, B_\mu\, e_\mathrm{R},\\
D_\mu u_\mathrm{R} &:= \partial_\mu u_\mathrm{R} + \left(-\frac{i}{2}\, g_\mathrm{s}\, G^\alpha_\mu\, \lambda_\alpha - i\, \frac{4}{3}\, g'\, B_\mu\right) u_\mathrm{R},\\
D_\mu d_\mathrm{R} &:= \partial_\mu d_\mathrm{R} + \left(-\frac{i}{2}\, g_\mathrm{s}\, G^\alpha_\mu\, \lambda_\alpha + i\, \frac{2}{3}\, g'\, B_\mu\right) d_\mathrm{R},
\end{aligned}
\tag{3.49}
$$

with the gauge couplings g_s of $SU_c(3)$, g of $SU_L(2)$ and g' of $U_{Y_\mathrm{w}}(1)$. The field strength tensors for each of the gauge fields G_μ, W_μ, B_μ read in terms of the bases λ_α, τ_j:

$$
\begin{aligned}
G^\alpha_{\mu\nu} &:= \partial_\mu G^\alpha_\nu - \partial_\nu G^\alpha_\mu + g_\mathrm{s}\, f^\alpha_{\beta\gamma}\, G^\beta_\mu\, G^\gamma_\nu,\\
W^j_{\mu\nu} &:= \partial_\mu W^j_\nu - \partial_\nu W^j_\mu + g\, \epsilon^j_{ik}\, W^i_\mu\, W^k_\nu,\\
B_{\mu\nu} &:= \partial_\mu B_\nu - \partial_\nu B_\mu.
\end{aligned}
\tag{3.50}
$$

The kinetic terms for G_μ, W_μ, B_μ can then be written as the quadratic invariants

$$
-\frac{1}{4}\, G^\alpha_{\mu\nu}\, G^{\mu\nu}_\alpha, \quad -\frac{1}{4}\, W^j_{\mu\nu}\, W^{\mu\nu}_j, \quad -\frac{1}{4}\, B_{\mu\nu}\, B^{\mu\nu}. \tag{3.51}
$$

Now, we can write the complete massless Standard Model Lagrangian density as

$$
\begin{aligned}
\mathcal{L}_\mathrm{SM} := -\sum_{\substack{\text{generations,}\\ \text{antiparticles}}} \Big\{ &\frac{1}{2}\bar{L}_\mathrm{L}\, \not{D}\, L_\mathrm{L} + \frac{1}{2}\bar{e}_\mathrm{R}\, \not{D}\, e_\mathrm{R} + \frac{1}{2}\bar{Q}_\mathrm{L}\, \not{D}\, Q_\mathrm{L} + \frac{1}{2}\bar{u}_\mathrm{R}\, \not{D}\, u_\mathrm{R} + \frac{1}{2}\bar{d}_\mathrm{R}\, \not{D}\, d_\mathrm{R}\\
&+ \frac{1}{4}\, G^\alpha_{\mu\nu}\, G^{\mu\nu}_\alpha + \frac{1}{4}\, W^j_{\mu\nu}\, W^{\mu\nu}_j + \frac{1}{4}\, B_{\mu\nu}\, B^{\mu\nu}\Big\},
\end{aligned}
\tag{3.52}
$$

with the Feynman slash notation $\not{D} := \gamma^\mu D_\mu$.

3.2.3 Spontaneous Symmetry Breaking

Although we have organized the whole particle content of the Standard Model in a clear pattern according to their behaviour under the gauge group (3.1), we have to explain the appearance of the observed masses and the emergence of electrodynamics. A basic ingredient in this explanation is the phenomenon of spontaneous symmetry breaking. Since this point is so important, we will analyse it in some detail and try to illustrate its basic features with the help of an analogy to a ferromagnetic system [1].

We can imagine a ferromagnet as built up by microscopic magnetic moments $\vec{\mu}_i$ arranged for simplicity and illustrative purposes on a two-dimensional grid. The magnetic moments $\vec{\mu}_i$ have a an arbitrary direction in the plane and the whole system possesses a *rigid* rotational symmetry. We further assume that there is an interaction between two neighbouring magnetic moments $\propto -\vec{\mu}_i \cdot \vec{\mu}_j$, which energetically favours the *relative* parallel alignment $\uparrow\uparrow$ of neighbouring magnetic moments. The opposite antiparallel alignment $\uparrow\downarrow$ is energetically the most "expensive" and all other intermediate angles between two neighbouring $\vec{\mu}$ are allowed in principle.

The whole system will tend to minimize its energy by aligning *all* magnetic moments parallel to each other. In this case, the microscopic magnetic moments add up to a macroscopic magnetization $\vec{M}$ pointing in a direction parametrized by length $|\vec{M}|$ and the angle Θ with respect to some fixed reference axis.

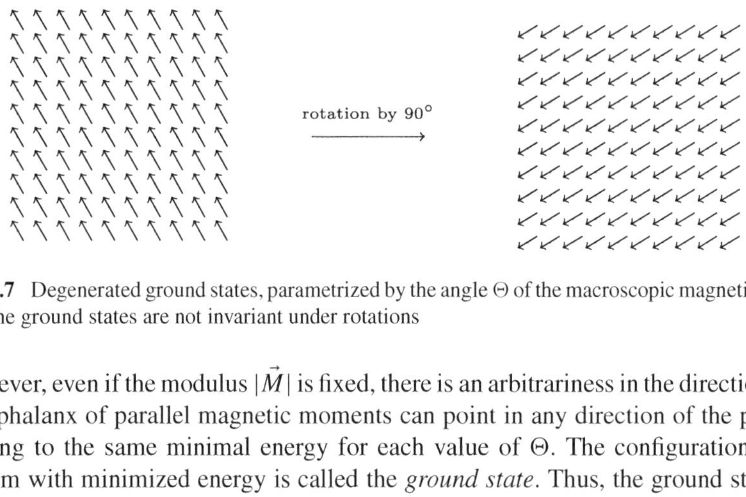

Fig. 3.7 Degenerated ground states, parametrized by the angle Θ of the macroscopic magnetization $\vec{M}$. The ground states are not invariant under rotations

However, even if the modulus $|\vec{M}|$ is fixed, there is an arbitrariness in the direction Θ. The phalanx of parallel magnetic moments can point in any direction of the plane, leading to the same minimal energy for each value of Θ. The configuration of a system with minimized energy is called the *ground state*. Thus, the ground state is degenerated and this degeneracy is parametrized by the arbitrary value of Θ. The degeneracy of the ground state is a reflection of the rotational symmetry. However, when approaching the state of minimal energy, the system has to decide for *one* particular ground state. The ground state itself is not invariant under rotations anymore, since a global rotation changes the direction of the macroscopic magnetization $\vec{M}$.

Now, we assume that the two-dimensional plane is extended to infinity with a complete random distribution of the $\vec{\mu}_i$, so that no direction is distinguished on average. This situation is realized in a ferromagnet by exceeding the Curie temperature

$T \geq T_c$. Then, the thermic fluctuations of the $\vec{\mu}_i$ completely dominate the system and inhibit the intentions of two spins to align parallelly to each other.

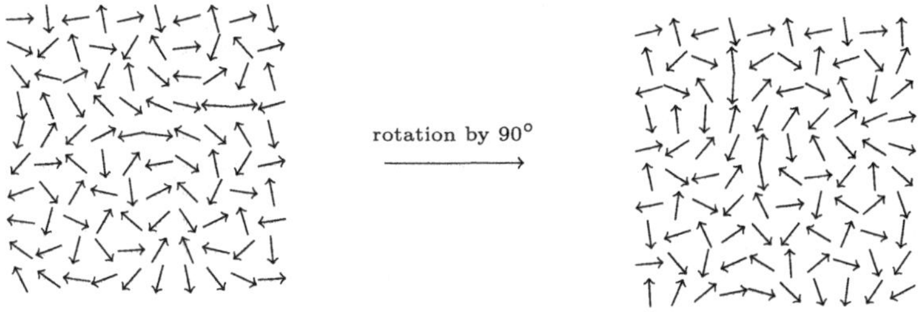

Fig. 3.8 Randomly distributed moments. Thermic fluctuations lead to an averaged $\vec{M} = 0$. The ground state of the system is invariant under rotations

In this case, the ground state of the system has an average magnetization $\vec{M} = 0$. The macroscopic parameter $\vec{M}$ which characterizes the state of the system is therefore invariant under rotations. Now we assume that we lower the temperature $T < T_c$ so that the interactions between neighbouring spins will dominate the system. The system will "freeze" in an unstable state and already a tiny perturbation will decide on the further fate of the system. Such a perturbation could be the placement of one extra $\vec{\mu}_i^*$ with a fixed direction at infinity.[10] This single magnetic moment $\vec{\mu}_i^*$ will induce an interaction on its neighbouring magnetic moments by trying to align them parallelly to its fixed direction. By induction the nearest-neighbour interaction spreads among all magnetic moments so that the complete system will finally try to minimize its energy by aligning *all* $\vec{\mu}_i$ in the direction of $\vec{\mu}_i^*$. Thus, the direction of the $\vec{M}$ is fixed by the boundary configuration $\vec{\mu}_i^*$, leading to a *spontaneous* breakdown of the rotational symmetry, illustrated in Fig. 3.9.

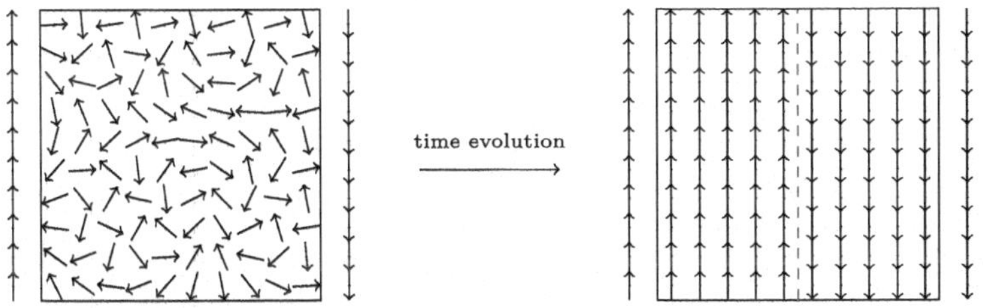

Fig. 3.9 *Left* Randomly distributed configuration with $T < T_c$ and fixed $\vec{\mu}_i$ at the boundary. *Right* The system will try to align all $\vec{\mu}_i$ parallelly to the boundary configuration. The opposite boundary configurations at both sides lead to the formation of a *domain wall*. For illustrative purposes, we have marked the domain wall as a sharp (*dashed*) border line between the two different ground states, but in general this will be an extended region

[10] This is nothing else but a boundary term reflecting the connection between unstable systems and their sensitivity to boundary conditions.

This behaviour is in strong contrast to *explicit* symmetry breaking. Explicit symmetry breaking occurs by *changing the interaction*. This might be realized by switching on an external magnetic field $\vec{B}$ pointing in a preferred direction. Thus, it is favourable for the $\vec{\mu}_i$ not only to align parallelly to each neighbouring $\vec{\mu}_j$, but also to align parallelly to the external field $\vec{B}$. The preferred direction of $\vec{B}$ explicitly breaks the invariance under rotations. It is also clear that once the external $\vec{B}$ field is switched on, the system is no longer sensitive to boundary conditions, since the energetic gain by the parallel alignment of two neighbouring magnetic moments $\vec{\mu}_i$ is negligible compared to the overwhelming gain in energy by alignment of *all* $\vec{\mu}_i$ in the direction of $\vec{B}$. The explicit breakdown of symmetry selects a *unique* ground state $\vec{M}||\vec{B}$.

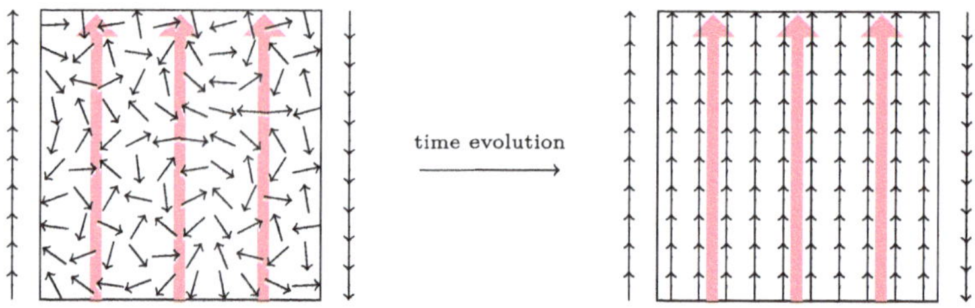

Fig. 3.10 Explicit symmetry breaking. The red arrows illustrate an external magnetic field $\vec{B}$. The initially randomly distributed $\vec{\mu}_i$ all align parallelly to $\vec{B}$

We can use the formation of *domain walls* as an indicator to decide whether a symmetry is spontaneously or explicitly broken,[11] cf. Fig. 3.9.

Transferring this analogy in the context of field theory, we have to replace the infinitely many discrete magnetic moments $\vec{\mu}_i$ by the continuous field operators $\Phi^a(x)$, $a = 1, \ldots, N$ and the local interaction $\vec{\mu}_i \cdot \vec{\mu}_j$ between nearest neighbours by products of local field operators like $\Phi^a(x)\,\Phi_a(x)$. The rotational symmetry of the system is implemented by the requirement that $\mathcal{L}(\Phi) = \mathcal{L}(\varphi)$ is invariant under a rigid $O(N)$ symmetry, so that it can only depend on the $O(N)$-symmetric quantity $\varphi^2 := \Phi^a\Phi^b\delta_{ab}$. More generally speaking, the interaction can then be described by the $O(N)$-invariant potential $V(\varphi)$. The ground state corresponds to the vacuum state and is also characterized as the state of minimum energy. In order to find this state, we have to look for the minima of $V(\varphi)$. The left-hand side of Fig. 3.11 shows a potential with a rotational symmetry

$$V(\varphi) := \frac{\lambda}{4}\,(\varphi^2 - \nu^2)^2. \tag{3.53}$$

The constant $\lambda \geq 0$ is assumed to be real and determines the interaction strength. In analogy with the magnetic moments, this determines the energy gain by the

[11] In the example of the ferromagnet, the formation of domain walls can be experimentally observed. They represent the boundaries between domains of different magnetizations $\vec{M}_i$, denoted Weiss domains.

alignment of two neighbouring spins. The constant ν^2 can be thought of as a parameter determining the shape of the potential. This corresponds to the temperature T of the magnetic system.

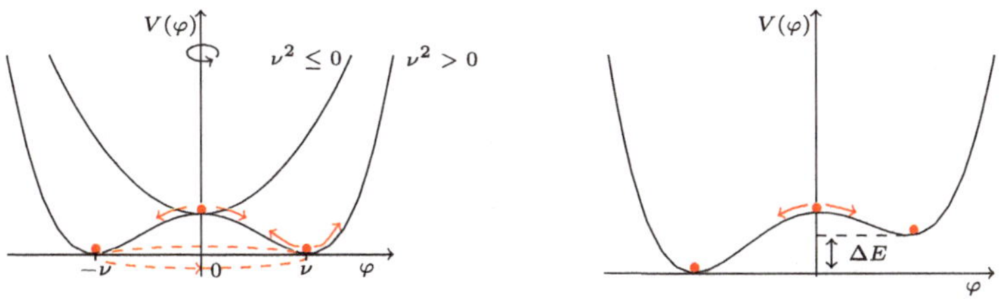

Fig. 3.11 *Left* Spontaneous symmetry breaking. *Right* Explicit symmetry breaking

For $\nu^2 \leq 0$, the potential is a parabola and no spontaneous symmetry occurs as it is for $T \geq T_c$. In contrast, for $\nu^2 > 0$, a circle of degenerated vacua forms, the system becomes unstable and has to choose a vacuum. This corresponds to the situation in which the magnetic moments have to choose a direction of alignment $\vec{M}$.

The right-hand side of Fig. 3.11 shows an antisymmetric potential

$$V^{\mathrm{expl}}(\varphi) = V(\varphi) + \alpha\,\varphi\,J. \tag{3.54}$$

for a single scalar field. The linear coupling with an external source J explicitly breaks the symmetry under reflections $\varphi \to -\varphi$ and a unique state of lowest energy is determined. In order to save the energy ΔE, the system will always dynamically choose the true ground state. Whether the field has to tunnel from the false vacuum to the true one or whether it can roll down classically depends on the dimension of the field space.

The rotational invariance of (3.53) suggests a parametrization of Φ^a in spherical coordinates with radial coordinate φ and $N-1$ angles θ_i, $i = 1, ..., N-1$. If we use the rotation invariance to rotate all angles to $\theta_i = 0$, the minimum of (3.53) is determined by the condition

$$\frac{\partial V(\varphi)}{\partial \varphi} = \lambda\,\varphi\,(\varphi^2 - \nu^2) = 0. \tag{3.55}$$

Thus, independently of the angular components, the minimum occurs at the radial coordinate $\varphi_0 = \nu$. The symmetry of the ground state is spontaneously broken from $O(N) \to O(N-1)$ because for all different values of θ_i, the energy is degenerated. In a quantum field theory, the quanta of the field operators are interpreted as particles. Therefore, we have to analyse what kind of particles correspond to φ and θ_i and what properties they have.

We have to expand the system around the ground state $\varphi = \nu$ and identify the fluctuations around this ground state with the field quanta. From the shape of the potential, it is clear that a fluctuation $\delta\theta_i$ in the angular directions θ_i does not cost any energy (except for a bit of kinetic energy). Thus, there is no oscillating behaviour around the ground state and therefore no mass associated with the particles $\delta\theta_i$. The scalar particles associated with the fluctuations $\delta\,\theta_i$ are the massless Goldstone bosons.

If we expand the radial component φ around the vacuum

$$\varphi := \nu + \delta\varphi := \nu + \sigma, \tag{3.56}$$

the fluctuation σ in the radial direction φ costs energy. In the vicinity of the minimum ν, the potential can be approximated by the harmonic potential

$$\frac{\partial^2 V}{\partial\,\varphi^2} = \lambda\,(3\,\varphi^2 - \nu^2). \tag{3.57}$$

The oscillation frequency around ν can be associated with the mass of σ

$$m_\sigma^2 = \left.\frac{\partial^2 V}{\partial\,\varphi^2}\right|_{\varphi=\nu} = 2\lambda\,\nu^2. \tag{3.58}$$

3.2.4 Higgs Mechanism and the Generation of Masses

The interaction in a gauge theory is described by the emission and absorption of a gauge boson (collectively called A_μ in this section). Gauge invariance of the complete Standard Model Lagrangian density (3.52) requires the vector gauge bosons to be massless. In addition, a fermionic mass term as in (3.38) would couple left- and right-handed particles. The electroweak interaction treats left- and right-handed fermions differently and as we have seen in Sect. 3.2.1, such terms would also spoil gauge invariance. In the previous section, we have explained how the spontaneous breakdown of a *rigid* symmetry leads to a shift of the vacuum expectation value $\langle\varphi\rangle := \varphi_0$ from $\langle\varphi\rangle = 0$ to $\langle\varphi\rangle = \nu \neq 0$ and the appearance of massless Goldstone bosons. However, such particles have not been found in any experiment. Now, we will discuss how the spontaneous breakdown of a *local* symmetry can solve all these problems at once. We just have to assume an additional particle, a scalar multiplet denoted as Higgs field [4]. In our description of the Standard Model, we have found a relation (3.37) that connects the projection of the weak isospin T_z and the weak hypercharge Y_w of $SU_L(2) \times U_Y(1)$ with the electric charge Q of U_{EM}. It seems necessary that the electromagnetic symmetry group $U_{EM}(1)$ must somehow emerge from the electroweak group $SU_L(2) \times U_{Y_w}(1)$. Moreover, from the discussion in (3.2.1), it is clear that a fermionic mass term can only be gauge-invariant, if it is effectively described as an interaction with a field Φ which compensates the ill behaviour under gauge transformations (3.47). Thus, we are supposed to look for a complex isospin doublet

$$\Phi := \begin{pmatrix} \phi^+ \\ \phi^0 \end{pmatrix}, \tag{3.59}$$

with $T = 1/2$ and $Y_w = 1$. Formula (3.37) then yields $Q = 1$ for the $T_z = +1/2$ component ϕ^+ and $Q = 0$ for the $T_z = -1/2$ component ϕ^0. The Higgs field Φ transforms as a singlet under $SU_c(3)$. Thus, its transformation behaviour under the full Standard Model group (3.1) can be summarized as $(\mathbf{1}, \mathbf{2}, 1)$. Gauge invariance of (3.52) demands that we couple the Higgs field Φ by means of the *covariant* derivative

$$D_\mu \Phi := \partial_\mu \Phi - i\, \frac{g}{2}\, W_\mu \Phi - i\, g'\, B_\mu \Phi. \tag{3.60}$$

We have to add to (3.52) the gauge-invariant Higgs Lagrangian density

$$\mathcal{L}_{\text{Higgs}} = \frac{1}{2}\, (D_\mu\, \Phi)^\dagger\, D^\mu\, \Phi - V(\varphi^2), \tag{3.61}$$

with the Higgs potential defined in (3.53) and $\varphi^2 = \Phi^\dagger \cdot \Phi = (\phi^+)^\dagger \phi^+ + (\phi^0)^\dagger \phi^0$. In order to further evaluate this expression, we represent the $SU_L(2)$ generators again in terms of the two-dimensional Pauli matrices (3.8)

$$D_\mu \Phi := \begin{pmatrix} \partial_\mu \phi^+ \\ \partial_\mu \phi^0 \end{pmatrix} - \frac{i}{2} \begin{pmatrix} g\, W_\mu^3 + g'\, B_\mu & g\, W_\mu^1 - i\, g\, W_\mu^2 \\ g\, W_\mu^1 + i\, g\, W_\mu^2 & -g\, W_\mu^3 + g'\, B_\mu \end{pmatrix} \begin{pmatrix} \phi^+ \\ \phi^0 \end{pmatrix}. \tag{3.62}$$

We are interested in the behaviour of the ground state of Φ. The rotational symmetry suggests that we can use a spherical representation of Φ and parametrize the *complex* doublet Φ as a $O(4)$-symmetric quartet of four *real* independent fields

$$\Phi = \begin{pmatrix} \phi^+ \\ \phi^0 \end{pmatrix} =: \begin{pmatrix} \phi^1 + i\phi^2 \\ \phi^3 + i\phi^4 \end{pmatrix} \to \begin{pmatrix} \phi^1 \\ \phi^2 \\ \phi^3 \\ \phi^4 \end{pmatrix} \to \begin{pmatrix} \theta_1 \\ \theta_2 \\ \theta_3 \\ \varphi \end{pmatrix}. \tag{3.63}$$

Spontaneous symmetry breaking yields one massive component, the Higgs field φ and three angular variables θ_1, θ_2 and θ_3, the Goldstone bosons. The degenerated minima of the potential $V(\varphi^2)$ at $\varphi = \nu$ lie on the hypersurface $\mathcal{S}^3$ parametrized by the angles $\theta_i, i = 1, ..., 3$. The invariance of (3.61) ensures that there is still a symmetry in internal space, which allows us to freely rotate the three angles θ_i. In contrast to the rigid rotational symmetry of Sect. 3.2.3, this is a *local* symmetry, i.e. we can perform rotations in every point x individually. In particular, there is a choice which rotates at every point all angles to zero $\theta_i = 0$. This type of gauge fixing is called unitary gauge and it simplifies the calculations and allows a clear physical interpretation of the results. In this gauge, we can write (3.62) as

$$D_\mu \Phi := \begin{pmatrix} 0 \\ \partial_\mu \sigma(x) \end{pmatrix} - \frac{i}{2} \begin{pmatrix} g\, W^3_\mu + g'\, B_\mu & g\, W^1_\mu - i\, g\, W^2_\mu \\ g\, W^1_\mu + i\, g\, W^2_\mu & -g\, W^3_\mu + g'\, B_\mu \end{pmatrix} \begin{pmatrix} 0 \\ \nu + \sigma(x) \end{pmatrix}. \tag{3.64}$$

Keeping only the quadratic terms in the kinetic term

$$\begin{aligned} -\frac{1}{2}\, (D_\mu \Phi)^\dagger\, D^\mu \Phi = -\frac{1}{2}\, \partial_\mu \sigma \partial^\mu \sigma - \frac{1}{8}\, (\nu + \sigma)^2\, g^2\, (W^1_\mu - i\, W^2_\mu)\, (W^{1\,\mu} + i\, W^{2\,\mu}) \\ -\frac{1}{8}\, (\nu + \sigma)^2\, (-g\, W^3_\mu + g'\, B_\mu)\, (-g\, W^{3\,\mu} + g'\, B^\mu) \end{aligned} \tag{3.65}$$

yields among higher terms two quadratic mass terms

$$-\frac{1}{2}\, (D_\mu \Phi)^\dagger\, D^\mu \Phi = -\frac{1}{8}\, \nu^2\, g^2\, |W^1_\mu - i\, W^2_\mu|^2 - \frac{1}{8}\, \nu^2\, (-g\, W^3_\mu + g'\, B_\mu)^2 + \ldots \tag{3.66}$$

First, we note that the terms W^1_μ and W^2_μ do not mix with the other components W^3_μ and B_μ. By a field redefinition, we can transform W^1_μ and W^2_μ to the diagonal basis

$$W^\pm_\mu := \frac{1}{\sqrt{2}} (W^1_\mu \mp i\, W^2_\mu). \tag{3.67}$$

The masses of the charged W^+ and W^- are equal and can be read off from (3.66)

$$M_{W^\pm} := \frac{1}{2}\, g\, \nu. \tag{3.68}$$

The two other components W^3_μ and B_μ also do mix only among each other. By changing to a new orthogonal basis, we can introduce the electrically neutral fields

$$Z^0_\mu := \frac{-g'\, B_\mu + g\, W^3_\mu}{\sqrt{g^2 + g'^2}} := W^3_\mu \cos\theta_\mathrm{W} - B_\mu \sin\theta_\mathrm{W} \tag{3.69}$$

$$A_\mu := \frac{g\, B_\mu + g'\, W^3_\mu}{\sqrt{g^2 + g'^2}} := W^3_\mu \sin\theta_\mathrm{W} - B_\mu \cos\theta_\mathrm{W} \tag{3.70}$$

with the definition of the Weinberg mixing angle

$$\cos\theta_\mathrm{W} = \frac{g}{\sqrt{g^2 + g'^2}}, \quad \sin\theta_\mathrm{W} = \frac{g'}{\sqrt{g^2 + g'^2}}. \tag{3.71}$$

We note that the second term of (3.66) can be expressed in terms of Z^0_μ only

$$-\frac{1}{8}\, \nu^2\, (-g\, W^3_\mu + g'\, B_\mu)^2 = -\frac{1}{8}\, \nu^2\, (g'^2 + g^2)\, Z^{0\,\mu} Z^0_\mu, \tag{3.72}$$

so that the mass of the Z^0_μ boson can be read off from (3.66) directly

$$M_Z^2 := \frac{1}{4}\nu^2\,(g^2 + g'^2). \tag{3.73}$$

In contrast, the field A_μ turns out to be massless. This is of course the photon we were looking for. Although the $SU_L(2) \times U_{Y_w}(1)$ symmetry of the ground state $\langle\Phi\rangle$ is broken, the ground state is still invariant under the electromagnetic subgroup $U_{EM}(1)$. The spontaneous symmetry breaking $SU_L(2) \times U_{Y_w}(1) \to U_{EM}(1)$ leads to the appearance of electrodynamics as an emergent low-energy phenomenon. The Goldstone bosons have disappeared in the unitary gauge. In order to check this procedure, we can count the degrees of freedom before and after the spontaneous breakdown of symmetry. Before, we have four scalar degrees of freedom plus four times two degrees of freedom for the massless vector gauge boson. After symmetry breaking, we have one scalar degree of freedom plus two degrees of freedom from the massless photon plus nine degrees of freedom, three for each of the three massive gauge vector bosons. The scalar degrees of freedom of the Goldstone bosons and the transversal degrees of freedom of the vector gauge bosons have merged and created a massive gauge boson. This was only possible because we considered a spontaneous breakdown of a *local* symmetry.

3.2.4.1 Masses of Fermions

So far, we have only generated the masses of the gauge bosons. We still have to analyse how the introduction of the Higgs field can yield mass terms for the fermions. The problem with the fermionic mass terms is that the mixing between left- and right-handed spinor components is not gauge-invariant. However, with the Higgs field Φ at our disposal, we can combine two mixed fermions without violating gauge invariance.[12] The new interaction terms are denoted as Yukawa interaction and acquire the form

Fig. 3.12 Yukawa interaction. Coupling of the Higgs field to left- and right-handed quarks

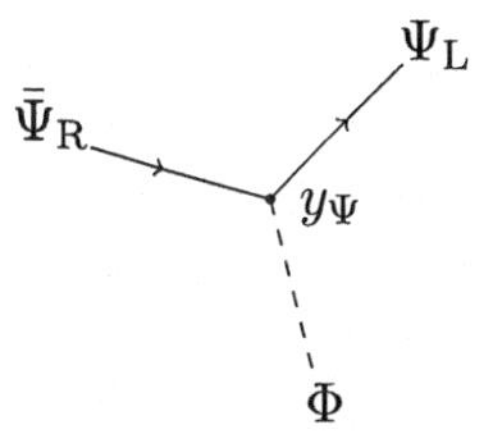

$$\propto y_\Psi\,\Phi\,\Psi_L\,\bar{\Psi}_R\,.$$

Let us consider the transformation behaviour of the Yukawa interaction term under the gauge transformation $U_{Y_w}(1)$ for the case of electrons two $\bar{e}_R$ and e_L

[12] Since the Yukawa coupling constant is dimensionless, it also respects the criterion of renormalizability, to be discussed in the next chapter.

$$y_e\,\Phi^\dagger\,\bar{e}_\mathrm{R}\,e_\mathrm{L} \rightarrow y_e\,\Phi^{\dagger\,\prime}\,\bar{e}^\prime_\mathrm{R}\,e^\prime_\mathrm{L} = y_e\,\Phi^\dagger\,e^{-i\,\theta(x)}\,\bar{e}_\mathrm{R}\,e^{2\,i\,\theta(x)}\,e_\mathrm{L}\,e^{-\,i\,\theta(x)} = y_e\,\Phi^\dagger\,\bar{e}_\mathrm{R}\,e_\mathrm{L}. \tag{3.74}$$

In contrast to (3.47), the mixed left- and right-handed fermionic mass term can be made gauge-invariant by the Yukawa type coupling to the Higgs field Φ. The Higgs field has the correct transformation behaviour to compensate the phase that has violated the gauge invariance in (3.47). After spontaneous symmetry breaking, the Higgs field Φ settles in the vacuum $\langle\Phi\rangle = \nu$ and expansion around the vacuum expectation value yields the mass term $\propto y_\Psi\,\nu\Psi_\mathrm{L}\bar{\Psi}_\mathrm{R}$. By comparison of this term with the structure of a conventional fermionic mass term $\propto m_\Psi\,\Psi_\mathrm{L}\bar{\Psi}_\mathrm{R}$, we can read off the masses of the different fermions $m_\Psi \propto y_\Psi\,\nu$. Each fermion has a *different* Yukawa coupling y_Ψ, but all masses are proportional to the vacuum expectation value $\langle\Phi\rangle = \nu \simeq 10^2$ GeV. It is an open question whether there are still undetected particles with very high masses. In fact, the RG analysis suggests that there is an intersection point for the RG trajectories of the running couplings $g_\mathrm{s}(t)$, $g'(t)$ and $g(t)$ at energies around a GUT scale 10^{16} GeV, which, in turn, is a hint that at this energy scale the three interactions could be unified to one fundamental interaction. In this sense, it seems very reasonable to assume that the natural mass scale for fundamental particles is very high. The fact that we empirically can observe only these "low energy" particles, which all need the Higgs mechanism to obtain a mass could have lead us to a very unnatural way of thinking about fundamental mass scales.

References

1. Coleman, S.: Aspects of symmetry. Cambridge University Press, Cambridge (1985)
2. Gell-Mann, M.: The interpretation of the new particles as displaced charge multiplets. Nuovo Cimento. **4**, 848 (1956)
3. Georgi, H.: Lie Algebras in Particle Physics: From Isospin to Unified Theories, 2nd edn. Westview Press, Colorado (1999)
4. Higgs, P.W.: Broken symmetries, massless particles and gauge fields. Phys. Lett. **12**, 132 (1964)
5. Klapdor-Kleingrothaus, H.V., Dietz, A., Harney, H.L., Krivosheina, I.V.: Evidence for neutrinoless double beta decay. Mod. Phys. Lett. A **16**, 2409 (2001)
6. Nakamura, K., et al.: (Particle data group): review of particle physics. J. Phys. G **37**, 075021 (2010)
7. Nakano, T., Nishijima, K.: Charge independence for V-particles. Prog. Theor. Phys. **10**, 581 (1953)
8. Wigner, E.P.: On unitary representations of the inhomogeneous Lorentz group. Ann. Math. **40**, 149 (1939)
9. Wu, C.S., Ambler, E., Hayward, R.W., Hoppes, D., Hudson, R.: Experimental test of parity conservation in beta decay. Phys. Rev. **105**, 1413 (1957)

Chapter 4
Effective Action and Renormalization Group

In this chapter, we will discuss the tools required for calculating the effective action and for performing the RG improvement. We will start with some general considerations about dimensionality, divergences and renormalization. Afterwards, we will show the connection between quantum corrections and the beta functions that determine the RG flow. Then we will present the derivation of the effective action in the path integral formalism and discuss the Faddeev–Popov method in the case of gravity. Finally, we will outline the basic steps in the general Schwinger–DeWitt algorithm, in order to calculate the divergent part of the one-loop effective action.

4.1 Classical Field Theory and Dimensional Analysis

For simplicity and clarity, we consider a simple model of a single scalar field ϕ in d dimensions in a generally curved manifold with metric $g_{\mu\nu}$ and classical action

$$S = \int \mathrm{d}^d x \sqrt{-g}\,\mathcal{L},\ \mathcal{L} = -\frac{1}{2}\,g^{\mu\nu}\,\partial_\mu\,\phi\,\partial_\nu\,\phi - \frac{1}{2}\,m^2\,\phi^2 - \sum_{n>1}^{\infty} g_n\,\phi^{2\,n}. \qquad (4.1)$$

The inverse metric $g^{\mu\nu}$ is defined by the relation $g_{\mu\alpha}\,g^{\alpha\nu} = \delta^\nu_\mu$, so that $g_{\alpha\beta}\,g^{\alpha\beta} = d$ and the volume element is defined as $|g| := |\det g_{\mu\nu}|$. We have assumed a discrete symmetry under reflection $\phi \to -\phi$, justifying the absence of odd powers in the infinite sum of polynomial interactions. The strength of the n interactions is determined by the n coupling constants g_n. We can divide (4.1) into a free part $\mathcal{L}_0$ and an interaction part $\mathcal{L}_{\mathrm{int}}$

$$\mathcal{L} = \mathcal{L}_0 + \mathcal{L}_{\mathrm{int}},\quad \mathcal{L}_0 := \frac{1}{2}\,\partial_\mu\,\phi\,\partial^\mu\,\phi - \frac{1}{2}\,m^2\,\phi^2,\quad \mathcal{L}_{\mathrm{int}} := -\sum_{n>1}^{\infty} g_n\,\phi^{2\,n}. \quad (4.2)$$

C. F. Steinwachs, *Non-minimal Higgs Inflation and Frame Dependence in Cosmology*, Springer Theses, DOI: 10.1007/978-3-319-01842-3_4,
© Springer International Publishing Switzerland 2014

The free part $\mathcal{L}_0$ is quadratic in the fields ϕ and $\mathcal{L}_{\text{int}}$ describes the remaining part. Now we will perform a dimensional analysis of the single terms. In units with $c = \hbar = 1$, a mass m with physical dimension $[m] = M$ can be written as inverse length $[m] = L^{-1}$. From now on, we will express the dimension of a quantity A unambiguously in terms of its mass dimension

$$[A]_{\text{M}} := \frac{\ln[A]}{\ln M}. \tag{4.3}$$

Since the action is dimensionless $[S]_{\text{M}} = 0$ in these units and the integration measure has the physical dimension $[\mathrm{d}^d x \sqrt{-g}\,]_{\text{M}} = -d$ in d space-time dimension, we can conclude $[\mathcal{L}]_{\text{M}} = d$. From $[m^2\,\phi^2]_{\text{M}} = 2 + 2\,[\phi]_{\text{M}}$ it follows that

$$[\phi]_{\text{M}} = \frac{d-2}{2}. \tag{4.4}$$

Before analysing the dimension for the kinetic term, we have to make one more choice. A line element $\mathrm{d}s^2 = g_{\mu\nu}(x)\,\mathrm{d}x^\mu\,\mathrm{d}x^\nu$ obviously has mass dimension $[\mathrm{d}s^2]_{\text{M}} = -2$. However, there remains an ambiguity in assigning the dimensionality to $g_{\mu\nu}$ or to x^μ. In the following, we will choose a dimensionless metric $[g_{\mu\nu}]_{\text{M}} = [g^{\mu\nu}]_{\text{M}} = 0$ and shift the dimensionality in the coordinates $[x^\mu]_{\text{M}} = [x_\mu]_{\text{M}} = -1$. Within our choice the coordinates x^μ are dimensionful and so are the derivatives

$$[\partial_\mu]_{\text{M}} = \left[\frac{\partial}{\partial x^\mu}\right]_{\text{M}} = [\partial^\mu]_{\text{M}} = \left[\frac{\partial}{\partial x_\mu}\right]_{\text{M}} = 1. \tag{4.5}$$

Thus, each space-time derivative increases the mass dimension by one. We can convince ourselves that the kinetic term in (4.1) has the correct dimensionality

$$[g^{\mu\nu}\,\partial_\mu\,\phi\partial_\nu\,\phi]_{\text{M}} = 2\frac{(d-2)}{2} + 2 = d. \tag{4.6}$$

It remains to determine the mass dimension of the coupling constants g_{n}. From (4.3) and (4.1), we can simply read off the dimensionality of g_n

$$[g_n]_{\text{M}} = d - n\,(d-2) \overset{d=4}{=} 4 - 2\,n. \tag{4.7}$$

We obtain the result that the mass dimension of the coupling constants depends on the space-time dimension. This dimensionality will become crucial when taking into account quantum corrections. We will establish the concept of renormalization in the next sections. But at this stage, we can already anticipate that for a theory in $d = 4$, a necessary condition for its renormalizability is that no coupling constant has a negative mass dimension $[g_n]_{\text{M}} < 0$. In this way, the criterion of renormalizability reduces the infinitely many polynomial interactions of (4.1) in $d = 4$ to one single term $\propto g_2\,\phi^4$. Only for $n = 2$, the coupling constant $[g_2]_{\text{M}} = 0$ is not negative. Later, we will relax the strict condition of renormalizability and consider non-renormalizable theories. In such a more general setup, the only way to reduce the number of allowed interaction terms induced by quantum corrections is to invoke a symmetry that prevents certain terms to appear [like the reflection symmetry which only allowed even powers of ϕ in (4.1)].

4.2 Divergences, Regularization and Renormalization

In a quantum field theory, we usually encounter infinities in terms of divergent integrals. In general, it is only possible to formulate a quantum field theory perturbatively, which means to perform an expansion in a small parameter. The most efficient way to perform such a perturbative expansion is an expansion in $\hbar$, which is therefore at the same time a semi-classical expansion. In this perturbative treatment we have to evaluate divergent integrals at each order of $\hbar$. It was Richard Feynman who invented a diagrammatic technique as a tool to express the physical processes described by a quantum field theory [7]. In this pictorial language a divergent integral arises due to closed loops. An expansion in loops is equivalent to an expansion in $\hbar$. Physically, a loop corresponds to the creation of a virtual particle-antiparticle pair, which is allowed by Heisenberg's indeterminacy relation. The fundamental superposition principle of quantum theory requires that we have to integrate over all possible momenta running in such a loop, including infinite momenta. This is the origin of the ultraviolet[1] divergent integrals in a quantum field theory. Every Feynman diagram corresponds to the amplitude of the depicted physical process to happen. The probability for such a process to take place is proportional to the amplitude squared. An infinite integral would lead to an infinite probability. No sensible physical theory can contain infinities, since we do not (and cannot) measure infinite quantities. We can only measure scattering probabilities or decay rates for certain processes, and the result is definite not infinite. Thus, the infinities are only allowed to appear during the intermediate stages of calculations. We need some calculational tools to manipulate these divergences so that they cancel out somehow in the final results. There is a systematical procedure to get rid of the infinities and to obtain finite answers which are in agreement with the experiment. The procedure roughly consists of two steps. The first step is called *regularization* and comprises the isolation of the infinite parts by the introduction of a *regulator* R. Then, the infinities in the divergent integrals are parametrized as singularities in this regulator. The second step concerns the removal of the isolated divergences which is equivalent to the removal of the regulator R and is called *renormalization*. In order to illustrate the basic idea behind this algorithm, we schematically consider a divergent integral over the momentum p_μ in four space-time dimensions

$$I = \int_0^\infty \mathrm{d}^4 p\, F(p). \tag{4.8}$$

[1] For massless fields, also infrared divergences might occur for small momenta.

There are multiple regularization schemes which differ in the choice of the regulator R. In a *cut-off* or *Pauli–Villars*[2] [15] regularization scheme, we consider (4.8) as a function of the upper integration bound, so that R becomes the dimensionful quantity $R = \Lambda$

$$I(\infty) \to I(\Lambda) = \int_0^{\Lambda} \mathrm{d}^4 p\, F(p) = A(\Lambda) + B + O(1/\Lambda). \tag{4.9}$$

The integral splits into a divergent part parametrized by Λ, into a finite part independent of Λ and into higher orders in $1/\Lambda$. The original integral I is recovered in the limit $\Lambda \to \infty$. In this limit, the $O(1/\Lambda)$ terms vanish and only the finite part and the isolated divergences parametrized by Λ remain.

There is also a regularization scheme in which the dimension of space-time is used as a regulator [13, 23]. We can first treat (4.8) as a function of some continuous dimension d and perform a Wick rotation to Euclidean space. Since the integrand $F(p)$ only depends on $p^2 = p_\mu p^\mu$, it has a rotational symmetry and we can introduce polar coordinates to separate the angular integration. The remaining integral can then be evaluated in terms of the Gamma function $\Gamma(\epsilon) := \int_0^\infty \mathrm{d}t t^{\epsilon-1} e^{-t}, \quad \mathfrak{Re}(\epsilon) > 0$. The Gamma function can be analytically continued to the *entire* complex plane with poles at $\epsilon = 0$ and the negative integers. For the complex dimension we write $d \to 2\omega := 4 - 2\epsilon$. Using integration by parts, we obtain a basic property of $\Gamma(\epsilon)$, the functional identity $\Gamma(\epsilon + 1) = \epsilon\, \Gamma(\epsilon)$. The regularized integral yields

$$I(4) \to I(\epsilon) = \mu^{2\epsilon} \int_0^\infty \mathrm{d}^{2\omega} p F(p) = A(1/\epsilon) + B(\mu) + O(\epsilon). \tag{4.10}$$

Although the left-hand side of (4.10) may not exist, the right-hand side $I(\epsilon)$ can be expressed as an analytic function of 2ω (or ϵ) that can be expanded as a Laurent series around the true dimensionality $\omega = 2$ (or $\epsilon = 0$). Thus $I(\epsilon)$ is convergent except for the limit $\epsilon \to 0$, where the divergences appear as poles in dimension [13]. After removing these poles, we can analytically continue the integral back to four dimensions $2\omega \to 4$. We note that in order to maintain the correct dimensionality, we had to introduce a dimensionful but otherwise arbitrary constant μ in the finite part $B(\mu)$.

So far, we have only parametrized or isolated the divergences and separated them from the finite parts of (4.8). In a second step, we have to get rid of the dependence on the regulator R in order to obtain finite results. In general, the integrand $F(p)$ of (4.8) will in addition depend parametrically on the coupling constants, masses and fields that are involved in the action. These are the free parameters of the theory which have to be determined by experiments. Therefore, we should rather write $F(p;\, \lambda, m, \Phi)$, where λ represents a set of coupling constants, m a set of masses and Φ a set of fields. During the regularization procedure we have evaluated the integral in terms of a regulator R so that $I = I(R, \lambda, m, \Phi)$. Since λ, m and Φ

[2] Here we use Λ also for the Pauli–Villars regularization scheme, although in this scheme, usually a heavy ghost field with mass M is added to render the integrals finite such that $R = M$.

are just parameters which are not fixed by any theoretical principle, but have to be determined by experiments, we can promote them to functions of the regulator

$$\lambda \to \tilde{\lambda}(R), \quad m \to \tilde{m}(R), \quad \Phi \to \tilde{\Phi}(R) \tag{4.11}$$

and in this way absorb the dependence of I on the regulator:

$$I(R, \lambda, m, \Phi) \to I(\tilde{\lambda}(R), \tilde{m}(R), \tilde{\Phi}(R)). \tag{4.12}$$

The resulting expression (4.12) is now *finite* in terms of the renormalized quantities. This "two-step algorithm" works order by order in perturbation theory. If we increase the order of perturbation theory, the parameters have to be re-adjusted again.

The procedure of renormalization has originally been defined only for renormalizable theories in the strict sense. In such theories, just a finite number of different structures in the action are induced to an arbitrary order of perturbation theory. Thus, we only have a *finite* number of free parameters and the theory preserves its predictive power. But the procedure of renormalization also makes sense in a field theory which is not renormalizable in the strict sense. In fact, the modern viewpoint is, that most of the physical theories are *effective* field theories which are only valid up to a certain energy scale Λ. In a non-renormalizable theory, there will, in general, appear new structures of field operators at each perturbative order which do not share the structure of those terms that are already present in the original action. Thus, the infinities cannot be hidden by absorbing them just in a re-definition of the free parameters. However, these higher dimensional operators are usually suppressed by a large mass scale M. This mass scale $M \sim \Lambda$ limits the theory's scope of application.

We have not yet clarified the meaning of the arbitrary mass scale μ, we had to introduce during the regularization procedure.[3] Only the finite part of the renormalized integral I does depend on the arbitrary parameter μ: $I(\lambda(R), m(R), \mu)$. In order to fix these finite parts, we have to state renormalization conditions at a specific scale μ. Since physical results should be independent of this choice, we have to investigate how the physical parameters change under a variation of μ. This leads to the renormalization group, to be discussed in the next section. In addition to the ambiguity of the regularization scheme and the ambiguity connected with the renormalization point μ, there is a certain arbitrariness regarding the treatment of the finite parts. In dimensional regularization, we must subtract the pole parts in order to render the result finite. However, there are different possibilities how to treat the finite part. The most common choice is the so-called minimal subtraction (MS) scheme in which only the pole part gets subtracted. Another popular choice is the $\overline{\text{MS}}$ scheme in which, beside the pole parts, also some constant factors are subtracted. Again, physics should not depend on this choice. Therefore, it is necessary to mention which scheme has

---

[3] From now on, we will focus on dimensional regularization, but the features connected to the RG are similar for other regularizations. Dimensional regularization respects the Lorentz and gauge symmetries. Moreover, calculations using dimensional regularization are much simpler since the structure of divergences shows up as simple poles in dimension.

been used when presenting a result. Results calculated within different schemes can then be converted into each other.

4.3 Renormalization Group

Before investigating the origin of the RG on the basis of our simple scalar field model of Sect. 4.1, we want to demonstrate the basic underlying features of "screening", "coarse-graining" and "self-similar reproduction" by means of two illustrative examples.

4.3.1 Illustrative Examples

4.3.1.1 Screening Charges

First, we consider a particle with a positive (renormalized) electric charge q_r^+, observed at a distance r.

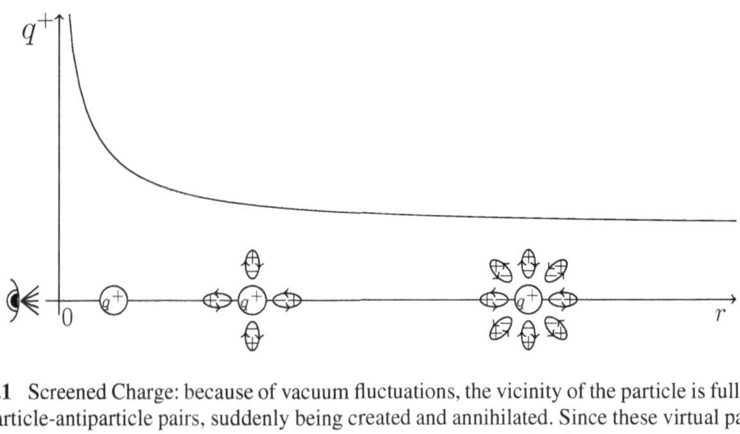

Fig. 4.1 Screened Charge: because of vacuum fluctuations, the vicinity of the particle is full of virtual particle-antiparticle pairs, suddenly being created and annihilated. Since these virtual particles also carry an electric charge, the cloud around the *positive charge* will align in such a way that they screen the (infinite) bare *positive charge*, leading to a polarization of the vacuum

The closer we approach (the smaller r) the bare charge q^+, the weaker becomes the screening effect and the stronger becomes the positive charge. In fact, the charge becomes infinite in the limit $r \to 0$, when no virtual cloud screens the bare charge anymore. This is the origin of the notation of "bare" quantities and it suggests that the coupling in QED depends on the scale $r \sim 1/E$.

4.3.1.2 Block Spin Renormalization

The second example is borrowed from statistical physics and is denoted Kadanoff block spin renormalization [12]. It emphasizes the self-similar reproduction process (the group structure) of the RG. We can think of the black dots as spins on a lattice labelled by coordinates $i, j, \ldots$ with a local pairwise interaction $\propto -J_0 \vec{s}_i \vec{s}_j$. The interaction strength is described by J_0. In general, we do not have full information about all microscopic degrees of freedom.

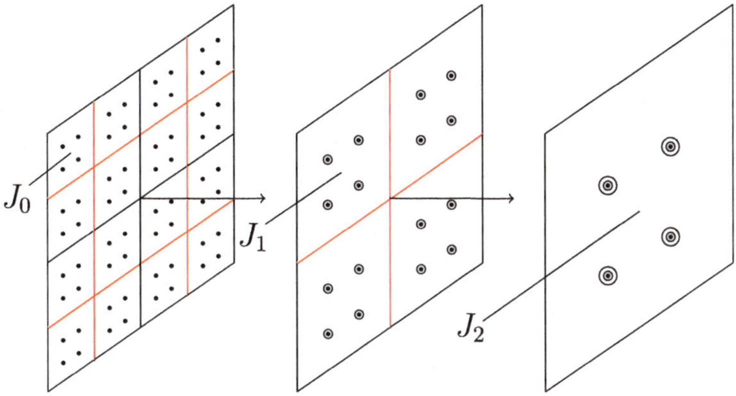

Fig. 4.2 Block spin renormalization

However, in most situations it is sufficient to describe the behaviour at some macroscopic scale. Therefore, we can imagine an averaging procedure which only collects the information of the interaction with the nearest neighbours. This information is effectively encoded in a modified coupling strength $J_0 \to J_1$. Due to the symmetry of the system, the averaged structure is the same as the original one but with a renormalized coupling strength J_1. We have created a self-similar copy of the system. We can repeat this procedure, increasing to higher scales (lower energies) and obtain in each step a new coupling constant $J_k \to J_{k+1}$. In the field theoretical context, we can obtain a low energy effective action by integrating out the heavy degrees of freedom (large momentum modes) [25, 26].

4.3.2 Gell–Mann-Low Equations

Let us consider again our renormalizable scalar field model of Sect. 4.1 with the interaction term $\propto \lambda \phi^4$. Let us furthermore assume that we have performed the calculations of quantum corrections and regularized the divergences by dimensional regularization. The renormalizability of the $\lambda \phi^4$ theory ensures that only those structures appear in the quantum corrections which were already present in the original

Lagrangian (4.1). We can therefore summarize the result of the loop calculations after dimensional regularization as the subtraction of the counterterms from the original Lagrangian $\mathcal{L}_\mathrm{r} = \mathcal{L} + \mathcal{L}_\mathrm{c}$ with the original Lagrangian in $d = 2\omega$ dimensions

$$\mathcal{L} = -\frac{1}{2}\,\partial^\mu\,\phi\,\partial_\mu\,\phi - \frac{1}{2}\,m^2\,\phi^2 - \mu^{2\epsilon}\,\frac{\lambda}{4}\,\phi^4. \tag{4.13}$$

The free parameters of this theory are the field configuration variable ϕ, the mass m and the self-interaction constant λ. The counterterm Lagrangian density shares the same structure as (4.13), but with coefficients A, B and C

$$\mathcal{L}_\mathrm{c} = -\frac{1}{2}\,A\,\partial^\mu\,\phi\,\partial_\mu\,\phi - \frac{1}{2}\,B\,\phi^2 - \mu^{2\epsilon}\,\frac{C}{4}\,\phi^4. \tag{4.14}$$

The regularized Green's functions[4] G^n of the unrenormalized theory (4.13), calculated based on $\mathcal{L}$ are divergent in the limit $\epsilon \to 0$. These divergences must be cancelled by the coefficients A, B and C of the counterterms contained in $\mathcal{L}_\mathrm{c}$. In order to obtain finite answers for the Green's functions, we have to subtract these divergences form the original Lagrangian and calculate the Green's functions on the basis of the renormalized Lagrangian $\mathcal{L}_\mathrm{r}$. Since both, $\mathcal{L}$ and $\mathcal{L}_\mathrm{c}$, share the same structure we can simply renormalize the free parameters by absorbing the infinities contained in A, B and C in the definition of the bare parameters

$$\begin{aligned}
\phi_\mathrm{b} &:= (1+A)^{1/2}\phi_\mathrm{r} := Z_\phi^{1/2}\,\phi_\mathrm{r},\\
m_\mathrm{b}^2 &:= (1+B)\,Z_\phi^{-1}\,m_\mathrm{r}^2,\\
\lambda_\mathrm{b} &:= \mu^{2\epsilon}(1+C)\,Z_\phi^{-2}\,\lambda_\mathrm{r}
\end{aligned} \tag{4.15}$$

and obtain a Lagrangian similar to (4.13), but with all parameters substituted by their bare (divergent) values

$$\mathcal{L}_\mathrm{r} = -\frac{1}{2}\,\partial^\mu\,\phi_\mathrm{b}\,\partial_\mu\,\phi_\mathrm{b} - \frac{1}{2}m_\mathrm{b}^2\,\phi_\mathrm{b}^2 - \mu^{2\epsilon}\,\frac{\lambda_\mathrm{b}}{4}\,\phi_\mathrm{b}^4. \tag{4.16}$$

Calculating the Green's functions on the basis of $\mathcal{L}_\mathrm{r}$ from the beginning and taking the limit $\epsilon \to 0$, the divergences exactly cancel and we obtain the desired finite answers. It is the property of renormalizability which allows to simply absorb these infinities in a redefinition of the free parameters of the original theory thereby creating a self-similar copy of the theory with modified coupling constants. It is clear by construction that the bare quantities (4.15) all diverge in the limit $\epsilon \to 0$, while the renormalized ϕ_r, m_r, λ_r give finite but *arbitrary* values. This arbitrariness is partly due to the regularization scheme, but also due to the dependence on the arbitrary mass scale μ that was necessarily introduced during the process of regularization. In

[4] In the next chapter we will give an explicit definition of the n-point Green's function in the path integral approach.

dimensional regularization, the coefficients A, B and C of the counterterms perturbatively calculated at loop order L have the general structure [17]

$$A = \sum_{i=-\infty}^{L} \frac{a_i(\lambda, \frac{m}{\mu})}{\epsilon^i}, \quad B = \sum_{i=-\infty}^{L} \frac{b_i(\lambda, \frac{m}{\mu})}{\epsilon^i}, C = \sum_{i=-\infty}^{L} \frac{c_i(\lambda, \frac{m}{\mu})}{\epsilon^i}. \tag{4.17}$$

In the limit $\epsilon \to 0$, only the finite parts $a_0(\mu), b_0(\mu), c_0(\mu)$ and the poles in dimensions remain in the series (4.17). First, we calculate the bare Green's functions in terms of the bare fields ϕ_b

$$\langle \phi_b(x_1)...\phi_b(x_n)\rangle = G_b^n(x_1, ..., x_n; m_\text{b}, \lambda_\text{b}; \epsilon). \tag{4.18}$$

These only depend on the bare quantities ϕ_b, m_b and λ_b and therefore not on μ. Then, we can express (4.18) via (4.15) in terms of the renormalized quantities

$$\begin{aligned} G_b^n(x_1, ..., x_n; m_\text{b}, \lambda_\text{b}; \epsilon) &= \langle Z_\phi{}^{1/2}(\mu)\, \phi_r(x_1)...Z_\phi^{1/2}(\mu)\, \phi_r(x_n)\rangle \\ &= Z_\phi^{n/2}(\mu)\, \langle \phi_r(x_1)...\phi_r(x_n)\rangle \\ &= Z_\phi^{n/2}(\mu)\, G_r^n(x_1, ..., x_n; \mu, m_\text{r}(\mu), \lambda_\text{r}(\mu); \epsilon). \end{aligned} \tag{4.19}$$

In this way, the renormalized Green's functions depend on the arbitrary scale μ

$$G_\text{r}^n(x_1, ..., x_n; \mu, m_\text{r}(\mu), \lambda_\text{r}(\mu); \epsilon) = Z_\phi^{-n/2}(\mu)\, G_b^n(x_1, ..., x_n; m_\text{b}, \lambda_\text{b}; \epsilon). \tag{4.20}$$

We can use the renormalized n-point Green's functions to fix this arbitrariness in the counterterms by imposing renormalization conditions at a specific momentum scale $p = \mu$. However, since μ is arbitrary, we could have fixed the values of $m_\text{r}(\mu)$ and $\lambda_\text{r}(\mu)$ at a different scale μ' as well. Nevertheless, we still describe the *same theory* and thus a change in μ has to be compensated by a change in $Z_\phi(\mu)$, $m_\text{r}(\mu)$ and $\lambda_\text{r}(\mu)$. A relation similar to (4.19) holds for the vertex functions in momentum space

$$\Gamma_\text{b}^n(p_1, ..., p_n; m_\text{b}, \lambda_\text{b}; \epsilon) = Z_\phi^{-n/2}(\mu)\, \Gamma_\text{r}^n(p_1, ..., p_n; \mu, m_\text{r}(\mu), \lambda_\text{r}(\mu); \epsilon), \tag{4.21}$$

where only $n-1$ momenta p are independent due to momentum conservation. We also note that we have to multiply with a factor Z_ϕ^n when switching from the 1PI Green's functions to the vertex functions because we have to remove n external legs (propagators) to obtain the latter. According to (4.19), each propagator $G^{n=2}$ is accompanied by a factor of Z_ϕ. Thus, we have to multiply the n-point vertex function $\Gamma_\text{r}^n(p_1, ..., p_n)$ by a total factor of Z_ϕ^n. The consequences of a change in μ are obtained by differentiating relation (4.21) with respect to μ and using the independence of Γ_b^n on μ

$$\left[\mu \frac{\partial}{\partial \mu} + \mu \frac{\partial m_{\rm r}}{\partial \mu} \frac{\partial}{\partial m_{\rm r}} + \mu \frac{\partial \lambda_{\rm r}}{\partial \mu} \frac{\partial}{\partial \lambda_{\rm r}} - \frac{n}{2} \mu \frac{\partial \ln Z_\phi}{\partial \mu}\right] \Gamma_r^n = 0 \tag{4.22}$$

By defining the anomalous dimensions γ_m and γ_ϕ and the beta function β_λ

$$\gamma_m(\mu, m_{\rm r}/\mu, \epsilon) := -\frac{\mu}{2} \frac{\partial \ln m_{\rm r}^2}{\partial \mu}, \quad \gamma_\phi(\mu, m_{\rm r}/\mu, \epsilon) := \frac{\mu}{2} \frac{\partial \ln Z_\phi}{\partial \mu},$$
$$\beta_\lambda(\mu, m_{\rm r}/\mu, \epsilon) := \mu \frac{\partial \lambda_{\rm r}}{\partial \mu}, \tag{4.23}$$

we can write the differential equation (4.22) in a more compact way

$$\left[\mu \frac{\partial}{\partial \mu} + \beta_\lambda \frac{\partial}{\partial \lambda_{\rm r}} - \gamma_m m_{\rm r} \frac{\partial}{\partial m_{\rm r}} - n \gamma_\phi\right] \Gamma_r^n = 0. \tag{4.24}$$

These are the Gell-Mann–Low equations which were first obtained in the context of QED [10].

4.3.3 Relation Between Counterterms and Beta Functions

Now, we will show that the beta functions are determined by the lowest logarithmic divergences in each order of perturbation theory for a particular renormalization scheme. In the previous section, we have mentioned that beside the arbitrariness of μ, there is an ambiguity in the renormalization prescription that fixes the finite parts of the counterterms. In view of the structure of the counterterms in dimensional regularization (4.17), there is a choice which drastically simplifies the calculations [22]. In this choice, the finite parts are simply set to zero. Thus, in the limit $\epsilon \to 0$, the counterterms (4.17) only consist of the pure pole structure $\propto \sum_{i=1\ldots L} 1/\epsilon^i$, necessary to cancel the divergences. It follows that we can neglect the mass dependence which entered in (4.21) and (4.23) only through the finite parts of the counterterms. This choice can be physically motivated by the following argument: The ultraviolet divergences occur at very high energies $p \to \infty$. Any dependence on a fixed mass scale should therefore be negligible in the limit $p \to \infty$ [17]. In fact, for dimensional reasons each mass m can only appear in the dimensionless combination m/μ. Thus, the massless renormalization prescription corresponds to fixing the renormalization conditions at $p = \mu \to \infty$. Without a mass dependence, the system of beta functions is easier to solve. We will derive this dependence for a general dimensionless coupling λ. Thus, the infinities using dimensional regularization and the mass independent renormalization prescription come as a set of general counterterms A which are simple poles in dimension and only depend on the coupling constant λ

$$A = \sum_{n=1}^{\infty} \frac{a_n(\lambda)}{\epsilon^n}. \tag{4.25}$$

These infinitely many counterterms must be subtracted in order to render the theory finite. Following [22], we can therefore express the bare charge in terms of the renormalized charge λ_r

$$\lambda_{\mathrm{b}} := (\mu^2)^{\epsilon} \left[\lambda_{\mathrm{r}} + \sum_{n=1}^{\infty} \frac{a_n(\lambda_{\mathrm{r}})}{\epsilon^n}\right]. \tag{4.26}$$

This is analogous to the definition in (4.15). We define the derivative of λ_{r} by keeping the bare charge λ_{b} constant while taking the derivative

$$\mu^2 \left.\frac{\mathrm{d}\,\lambda_{\mathrm{r}}}{\mathrm{d}\mu^2}\right|_{\lambda_{\mathrm{b}}} := -\epsilon\,\lambda_{\mathrm{r}} + \beta_{\lambda}(\lambda_{\mathrm{r}}). \tag{4.27}$$

Since λ_{b} is independent of the regularization procedure, it cannot depend on μ. Differentiation of (4.26) with respect to μ leads to

$$\sum_{n=1}^{\infty} \frac{1}{\epsilon^{n-1}} \left[\lambda_{\mathrm{r}} \frac{\partial a_n(\lambda_{\mathrm{r}})}{\partial \lambda_{\mathrm{r}}} - a_n(\lambda_{\mathrm{r}}) - \frac{1}{\epsilon}\beta(\lambda_{\mathrm{r}}) \frac{\partial a_n(\lambda_{\mathrm{r}})}{\partial \lambda_{\mathrm{r}}}\right] = \beta_{\lambda}(\lambda_{\mathrm{r}}). \tag{4.28}$$

Starting with $n = 1$ and comparing equal powers of ϵ we can deduce

$$\beta_{\lambda}(\lambda_{\mathrm{r}}) = \left(\lambda_{\mathrm{r}} \frac{\partial}{\partial \lambda_{\mathrm{r}}} - 1\right) a_1. \tag{4.29}$$

This shows explicitly that the function $\beta_{\lambda}(\lambda_{\mathrm{r}})$ is completely determined by the lowest order a_1. Comparing equal powers of ϵ for general n by shifting the summation index in (4.28), we obtain the recursion relation

$$\left(\lambda_{\mathrm{r}} \frac{\partial}{\partial \lambda_{\mathrm{r}}} - 1\right) a_n(\lambda_{\mathrm{r}}) = \beta(\lambda_{\mathrm{r}}) \frac{\partial\, a_{n-1}(\lambda_{\mathrm{r}})}{\partial\, \lambda_{\mathrm{r}}}. \tag{4.30}$$

In particular, this relation implies that knowing $a_1(\lambda_{\mathrm{r}})$ at the pole $1/\epsilon$ is sufficient to know *all* coefficients a_n at higher poles $1/\epsilon^n$ via (4.30). The beta function $\beta(\lambda_{\mathrm{r}})$ that determines the RG flow is given by the differential equation

$$\mu^2 \frac{\mathrm{d}\,\lambda_{\mathrm{r}}}{\mathrm{d}\mu^2} = \beta(\lambda_{\mathrm{r}}) \tag{4.31}$$

and simply states that a change in the parameter μ is compensated by a change in λ_{r}. The treatment for one coupling can easily be generalized for a set of different couplings $g_i, i = 1, ..., m$. In order to keep the notation clear, we will suppress the subscript r from now on. The counterterms $a_{i,n}(g_1, ..., g_m)$ will then in general depend on all couplings g_i and (4.26), (4.27) become

$$g_{\mathrm{b},\mathrm{i}} = (\mu^2)^{\epsilon}\Big[g_i + \sum_{n=1}^{\infty} \frac{a_{i,n}(g_1, ..., g_m)}{\epsilon^n}\Big], \tag{4.32}$$

$$\mu^2 \left.\frac{\mathrm{d}g_i}{\mathrm{d}\mu^2}\right|_{g_{\mathrm{b},\mathrm{i}}} = -\epsilon\, g_i + \beta_i(g_1, ..., g_m). \tag{4.33}$$

Following the same steps as in (4.29) and (4.30), we obtain a set of *coupled ordinary differential equations*, instead of the single RG equation (4.31)

$$\mu^2 \frac{\mathrm{d}g_i(\mu^2)}{\mathrm{d}\,\mu^2} = \beta_i(g_1, ..., g_m). \tag{4.34}$$

4.4 Path Integral and Effective Action

In the previous sections we have explained the origin of divergences only on a *formal* level. We have outlined how to regularize and renormalize them and how the dependence of the energy scale arises from the calculated quantum corrections. However, we need a theoretical tool for a *practical* calculation of the counterterms order by order. In this section, we will derive the concept of the effective action within a path integral approach.

The principle of stationary action is so important because it allows to derive the dynamical laws of a system just from a scalar functional, the *classical* action. An extremum of the classical action functional selects *a trajectory* which represents a solution of the classical equations of motion. Now, we want to extend this concept to the quantum level and the appropriate tool for such a discussion will be the *effective* action. In quantum theory, the concept of a trajectory does not make sense anymore. The linearity of quantum theory requires that each superposition of a solution to the Schrödinger equation is again a solution. We can only predict a transition amplitude for some initial state to end in some final state. We have to take into account *all* possible transitions from the initial state to the final state, or the sum over all possible paths between these states. However, not all these transitions are realized with equal probability. We have to weight each path with a phase factor exp $(i\, S/\hbar)$. This weighting has the effect that those paths "far from the classical solution" cancel out due to fast oscillations. The transition amplitude for a single particle moving in direction x can be expressed in terms of the Feynman path integral

$$\langle x_{\mathrm{out}} | x_{\mathrm{in}} \rangle = \int_{x_{\mathrm{in}}}^{x_{\mathrm{out}}} \mathcal{D}x\, e^{\frac{i}{\hbar} S[x]}. \tag{4.35}$$

Now, we have to generalize this concept to a field theory $x \to \phi(x)$. For that purpose, it is useful to introduce an idea due to Schwinger [18]. Using the Dirac bra-ket notation, we can write the transition amplitude from the vacuum in-state $|\text{in},\ 0\rangle$ to the vacuum out-state $\langle\text{out},\ 0|$ by $\langle\text{out},\ 0|\text{in},\ 0\rangle$. We artificially introduce an external source, a current $J(x)$, which couples linearly to the field. We can think of a very intuitive physical situation, a scattering experiment between two free particles. We have complete knowledge about the free asymptotic in- and out-states (the *prepared* initial state and the *measured* final state). The interaction only takes place in the intermediate state of the scattering event. The source is switched off in the initial state, switched on during the interaction and finally switched off again in the final state. The vacuum transition amplitude in the presence of a source can be described in terms of the Feynman path integral [6]

$$\langle\text{out},\ 0|\text{in},\ 0\rangle_J := \int_{\phi_{\text{in},\,0}}^{\phi_{\text{out},\,0}} \mathcal{D}\phi\ e^{\frac{i}{\hbar}S[\phi(x)]+\frac{i}{\hbar}\int \mathrm{d}^4x\ J_A(x)\phi^A(x)} \tag{4.36}$$

We note that in the context of curved space-time there is no unique definition of the vacuum anymore and $|\text{in},\ 0\rangle \neq |\text{out},\ 0\rangle$ in general, see e.g. the discussion in Sect. 2.4.3. The linear coupling between the source $J_A(x)$ and the field $\phi^A(x)$ should be understood as a useful mathematical tool to define the generating functional by a physically intuitive concept. The generalized field $\phi^A(x)$ collects all possible fields involved in the action. These can be scalar, spinor, vector or tensor fields. A capital index $A,\ B, ..$ labels the discrete internal and space-time indices of the components of ϕ^A. Since we have to work with functional identities, it is useful to adopt an even more condensed notation, which is due to DeWitt [4]. We write $\phi^i := \phi^A(x)$ where, beside the discrete indices $A, B, ..$, the DeWitt indices $i, j, ...$ also include the space-time point x as an continuous index. Furthermore, we also apply the Einstein summation convention to this continuous indices. If two equal indices $i, j, ...$ appear, we extend the summation over the discrete indices $A, B, ..$ to also include an integration $\int \mathrm{d}^4x$, $\int \mathrm{d}^4y, ...$ over the continuous space-time points $x,\ y, ...$ With this convention, the expression for the source coupling term becomes

$$\int \mathrm{d}^4x\ J_A(x)\phi^A(x) = J_i\phi^i. \tag{4.37}$$

The transition amplitude in the presence of a source is denoted by

$$Z[J] := \langle\text{out},\ 0|\text{in},\ 0\rangle_J. \tag{4.38}$$

$Z[J]$ is the *generating functional*. The name stems from the fact that by functionally differentiating $Z[J]$ with respect to the source J, we can generate all n-point Green's functions. The *connected* Green's functions are generated by another functional $W[J]$ which is defined by

$$Z[J] =: e^{\frac{i}{\hbar}W[J]}, \tag{4.39}$$

so that repeated differentiation of $W[J]$ with respect to J yields

$$\langle \phi_{k_1}, \ldots, \phi_{k_n} \rangle_J = e^{-\frac{i}{\hbar} W[J]} \left(\frac{\hbar}{i} \right)^n \frac{\delta^n}{\delta J^{k_1}(x_1) \ldots \delta J^{k_n}(x_n)} e^{\frac{i}{\hbar} W[J]}, \tag{4.40}$$

where the factor exp $(-i\, W[J]/\hbar)$ takes care of the proper normalization. The expectation value of a general operator $O(\phi)$ is defined as

$$\langle O(\phi) \rangle_J := \frac{\langle \text{out}, 0 | \mathcal{T}\, O(\phi) | \text{in}, 0 \rangle_J}{\langle \text{out}, 0 | \text{in}, 0 \rangle_J} \tag{4.41}$$

with the time-ordering operator $\mathcal{T}$. For $O(\phi) = \phi$, we can use the definitions (4.38) and (4.36) and obtain in accordance with (4.40) for $n = 1$

$$\langle \phi_k \rangle_J = \frac{\int \mathcal{D}\phi \phi_k \, e^{\frac{i}{\hbar}(S[\phi] + J_l \phi^l)}}{\int \mathcal{D}\phi \, e^{\frac{i}{\hbar}(S[\phi] + J_l \phi^l)}} = -i\,\hbar Z^{-1} \frac{\delta Z[J]}{\delta J^k} = \frac{\delta\, W[J]}{\delta\, J^k}. \tag{4.42}$$

The vacuum expectation value $\langle \phi_k \rangle_J$ is also denoted the *mean field*

$$\overline{\phi}_k := \langle \phi_k \rangle_J , \tag{4.43}$$

since the expectation value corresponds to a quantum-averaging procedure. By inverting relation (4.42), we can express the source $J_k[\overline{\phi}]$ as a functional of the mean field $\overline{\phi}_k$ and define another quantity which is a functional of the mean field $\overline{\phi}_k$, the effective action

$$\Gamma[\overline{\phi}] := W[J] - J_i \, \overline{\phi}^i . \tag{4.44}$$

Equation (4.44) corresponds to a functional Legendre transformation between the pair of variables $\overline{\phi}_k \longleftrightarrow J_k$. Differentiating (4.44) with respect to $\overline{\phi}_k$, we obtain

$$\frac{\delta \Gamma[\overline{\phi}]}{\delta\, \overline{\phi}^k} = -J_k, \quad \text{and} \quad \left. \frac{\delta \Gamma[\overline{\phi}]}{\delta\, \overline{\phi}^k} \right|_{J=0} = 0 \,. \tag{4.45}$$

Equation (4.45) generalizes the classical extremal condition $S_{,i}[\phi] = 0$ for a vanishing source $J_i = 0$ and yields the equation of motion for the mean field $\overline{\phi}_i$ including all quantum corrections. We introduce the additional abbreviation and denote functional derivatives like $\delta\, S[\phi]/\delta \phi^i := S_{,i}[\phi]$ by a comma. The origin of the name effective action $\Gamma[\phi]$ can be ascribed to the relation

$$\begin{aligned} e^{\frac{i}{\hbar}\Gamma[\overline{\phi}]} &= e^{\frac{i}{\hbar}\left(W[J] - J_k \overline{\phi}^k\right)} = \int \mathcal{D}\phi \, e^{\frac{i}{\hbar}\left(S[\phi] + J_j \phi^j - J_k \overline{\phi}^k \right)} \\ &= \int \mathcal{D}\phi \, e^{\frac{i}{\hbar}\left[S[\phi] + J_j \left(\phi^j - \overline{\phi}^j \right) \right]} . \end{aligned} \tag{4.46}$$

The effective action is the generating functional for the one-particle irreducible[5] n-point vertex-functions. Substituting J_i in (4.46) by relation (4.45), we obtain the following *exact* equation for the effective action

$$\exp \frac{i}{\hbar}\,\Gamma[\overline{\phi}] = \int \mathcal{D}\phi \exp \frac{i}{\hbar}\left(S[\phi] - \frac{\delta\Gamma[\overline{\phi}]}{\delta\overline{\phi}^k}\,(\phi^k - \overline{\phi}^k)\right). \tag{4.47}$$

In order to solve (4.47) perturbatively, we expand $S[\phi]$ around the mean field $\overline{\phi}^k$

$$\begin{aligned} S[\phi] &= S[\overline{\phi}] + \sum_{n=1}^{\infty} \frac{1}{n!} S_{,k_1\dots k_n}[\phi]\Big|_{\phi=\overline{\phi}} (\phi^{k_1} - \overline{\phi}^{k_1})\dots(\phi^{k_n} - \overline{\phi}^{k_n}) \\ &= S[\overline{\phi}] + \hbar^{1/2}\, S_{,i}[\overline{\phi}]\,\Delta^i + \frac{\hbar}{2}\, S_{,ij}[\overline{\phi}]\,\Delta^i\Delta^j + \sum_{n=3}^{\infty} \frac{\hbar^{n/2}}{n!}\, S_{,k_1\dots k_n}[\overline{\phi}]\Delta^{k_1}\dots\Delta^{k_n}, \end{aligned} \tag{4.48}$$

where we have shifted the variable of integration $\phi^k \to \hbar^{1/2}\,\Delta^k + \overline{\phi}^k$ in the last step [3]. The rescaling with $\hbar^{1/2}$ will become useful when performing an expansion in $\hbar$. We separate the effective action into a classical and a quantum part $\tilde{\Gamma}$ and expand the quantum part in powers of $\hbar$

$$\Gamma[\overline{\phi}] := S[\overline{\phi}] + \tilde{\Gamma}[\overline{\phi}] := S[\overline{\phi}] + \sum_{n=1}^{\infty} \hbar^n \Gamma^{(n)}[\overline{\phi}] \tag{4.49}$$

Inserting the expansions (4.48) and (4.49) in (4.47), we obtain

$$\exp i \left(\sum_{n=1}^{\infty} \hbar^{n-1}\,\Gamma^{(n)}[\overline{\phi}] \right) = \int \mathcal{D}\Delta \exp i \left(\frac{1}{2}\,\Delta^i\, S_{,ij}[\overline{\phi}]\,\Delta^j + S^{\text{eff}}_{\text{int}}[\overline{\phi}] \right). \tag{4.50}$$

We can think of (4.50) as the exponentiated version of an action where the right-hand side describes the splitting into a free part quadratic in Δ^i and an interaction term

$$S^{\text{eff}}_{\text{int}}[\overline{\phi}] = \sum_{n=3}^{\infty} \frac{\hbar^{n/2-1}}{n!}\, S_{,k_1\dots k_n}[\overline{\phi}]\Delta^{k_1}\dots\Delta^{k_n} - \sum_{n=1}^{\infty} \hbar^{n-1/2}\,\Gamma^{(n)}_{,j}[\overline{\phi}]\,\Delta^j. \tag{4.51}$$

[5] Each term of a n-point Green's-function graphically corresponds to a Feynman diagram. A *connected* diagram cannot be represented as a product of individual diagrams. A *one-particle-irreducible* (1PI) diagram cannot be divided into two connected diagrams by cutting one internal propagator line. Since the 1PI diagrams that are generated by the effective action form a "basis" which can be used to construct all general Feynman diagrams, it is convenient to reduce the analysis of Feynman diagrams to that of the 1PI diagrams.

We emphasize that the terms proportional to $S_{,i}$ cancel out exactly *without* making use of the classical equations of motion $S_{,i} = 0$. Now we recognize the usefulness of the rescaling with $\hbar^{1/2}$, since the propagators of the free theory do not carry any factors of $\hbar$ [3]. By collecting terms of equal powers in $\hbar$ on the left- and right-hand side of (4.50), we can evaluate this expression iteratively order by order. For $n = 1$, we obtain

$$\exp i\left(\Gamma^{(1)}[\bar{\phi}]\right) = \int \mathcal{D}\Delta \exp i\left(\frac{1}{2}\,\Delta^i\, S_{,ij}[\bar{\phi}]\,\Delta^j\right), \tag{4.52}$$

since the lowest order of $S^{\text{eff}}_{\text{int}}[\bar{\phi}]$ already starts with $\hbar^{1/2}$. We note that no non-integer powers of $\hbar$ for odd n can appear. In order to show this, we expand $e^{S^{\text{eff}}_{\text{int}}}$ in powers of Δ

$$\int \mathcal{D}\Delta\; e^{\left\{\frac{i}{2}\,\Delta^j\, S_{,jk}[\bar{\phi}]\,\Delta^k + S^{\text{eff}}_{\text{int}}[\bar{\phi}]\right\}} = \int \mathcal{D}\Delta\; e^{\frac{i}{2}\,\Delta^j\, S_{,jk}[\bar{\phi}]\,\Delta^k}\;\left(1 + \alpha_3\Delta^3 + \alpha_4\Delta^4 + \ldots\right) \tag{4.53}$$

and notice that the Gaussian average $\langle \ldots \rangle_{\mathrm{g}}$ of an odd number of fields Δ^i vanishes [3]

$$\langle (\Delta^k)^n \rangle_{\mathrm{g}} = \frac{\int \mathcal{D}\Delta\;(\Delta^k)^n\; e^{\frac{i}{2}\,\Delta^i\, S_{,ij}[\bar{\phi}]\,\Delta^j}}{\int \mathcal{D}\Delta\; e^{\frac{i}{2}\,\Delta^i\, S_{,ij}[\bar{\phi}]\,\Delta^j}} \overset{n\ \text{odd}}{=} 0\,. \tag{4.54}$$

We have used an abbreviation of the form $\alpha_3\Delta^3 = \alpha_{ijk}\Delta^i\Delta^j\Delta^k$ with some expansion coefficient α_{ijk} for the terms $\alpha_3\Delta^3$, $\alpha_4\Delta^4$, etc. Thus, we only have to consider even powers of n or integer powers of $\hbar$. The Gaussian integral (4.52) for the one-loop contribution quadratic in Δ^k can be evaluated exactly

$$\exp\left(i\,\Gamma^{(1)}[\bar{\phi}]\right) = \left(\det S_{,ij}\,\big|_{\phi=\bar{\phi}}\right)^{-1/2} \tag{4.55}$$

The one-loop effective action can thus be formulated with (4.49) as

$$\begin{aligned}\Gamma_{1-\text{loop}}[\bar{\phi}] &= S[\bar{\phi}] + \hbar\,\Gamma^{(1)}[\bar{\phi}] = S[\bar{\phi}] - i\,\hbar \ln\left(\det S_{,ij}\right)^{-1/2} \\ &= S[\bar{\phi}] + \frac{i\,\hbar}{2}\ln\det S_{,ij} = S[\bar{\phi}] + \frac{i\,\hbar}{2}\operatorname{Tr}\ln S_{,ij}.\end{aligned} \tag{4.56}$$

In the last step, we have used the operator identity $\ln(\det O) = \operatorname{Tr}(\ln O)$.

4.4.1 Faddeev–Popov-Method

In the case of gauge theories and gravity, there is an additional complication in the path integral approach. The sum over all paths does "over count" field configurations that are connected by a gauge transformation or, in the case of General Relativity, by a general coordinate transformation. These infinitely many non-physical configurations have to be factored out in order to avoid divergences. In [5], the authors have developed a method to handle this problem. As part of this method, we will break the gauge invariance by adding an additional term that fixes the gauge in the action. In order to restore gauge invariance, we have to introduce a compensating contribution which can be written as "ghost-contribution" in terms of Grassmannian vector fields. We will discuss this method for the case of General Relativity with the path integral

$$Z[g_{\mu\nu}] = N \int \mathcal{D}g_{\mu\nu}\, e^{\frac{i}{\hbar}\, S[g_{\mu\nu}]} = N \int \mathcal{D}h_{\mu\nu}\, e^{\frac{i}{\hbar}\, S[\bar{g}_{\mu\nu}+h_{\mu\nu}]}. \tag{4.57}$$

We use the background field method [4] and split the metric field $g_{\mu\nu} := \bar{g}_{\mu\nu} + h_{\mu\nu}$ into a background part $\bar{g}_{\mu\nu}$ and perturbations $h_{\mu\nu}$. The factor N corresponds to a normalization factor yet to be determined. In General Relativity, the gauge transformations are the four dimensional diffeomorphisms of the metric space-time manifold $(\bar{g}_{\mu\nu}, \mathcal{M})$ which can be described by a flow $\phi(x) : \mathcal{M} \to \mathcal{M}$, moving points $p \in \mathcal{M}$. Following the treatment in [24], we know that $(\bar{g}_{\mu\nu},\ \mathcal{M})$ represents the same space-time as $(\phi^* \bar{g}_{\mu\nu},\ \mathcal{M})$ for the pullback $\phi^* : \mathrm{T}\mathcal{M}_p \to \mathrm{T}\mathcal{M}_{\phi(p)}$. Studying "small" perturbations $h_{\mu\nu} = g_{\mu\nu} - \bar{g}_{\mu\nu}$ around a solution $\bar{g}_{\mu\nu}$ of the Einstein equations, we need a measure of this smallness. Supposing that we have a one parameter family of background space-times $(\bar{g}_{\mu\nu}(\epsilon),\ \mathcal{M})$, labelled by ϵ, then $(\phi_\epsilon^*\, \bar{g}_{\mu\nu}(\epsilon),\ \mathcal{M})$ represents again the same one parameter family of space-times, provided ϕ_ϵ is a one parameter group of diffeomorphisms. Small perturbations mean small $\epsilon \ll 1$, i.e. $\bar{g}_{\mu\nu}(\epsilon)$ does not "flow" too far away. A small ϵ, in turn, justifies the linear approximation

$$\left.\frac{\mathrm{d}g_{\mu\nu}(\epsilon)}{\mathrm{d}\epsilon}\right|_{\epsilon=0} = h_{\mu\nu}. \tag{4.58}$$

The corresponding perturbation for the pulled back metric is

$$h'_{\mu\nu} = \left.\frac{\mathrm{d}\,\phi_\epsilon^* g_{\mu\nu}(\epsilon)}{\mathrm{d}\epsilon}\right|_{\epsilon=0} \tag{4.59}$$

It is clear that (4.58) and (4.59) differ by the Lie derivative $\mathcal{L}_\xi$, see (B.12)

$$h'_{\mu\nu} = h_{\mu\nu} - \mathcal{L}_\xi\, \bar{g}_{\mu\nu} \tag{4.60}$$

with ξ^ρ being the infinitesimal generator of the flow ϕ. This means that the remaining gauge freedom of the perturbations $h_{\mu\nu}$ is given by

$$\begin{aligned} \mathcal{L}_\xi\, \bar{g}_{\mu\nu} &= \xi^\rho \bar{\nabla}_\rho\, \bar{g}_{\mu\nu} + \bar{g}_{\rho\nu} \bar{\nabla}_\mu\, \xi^\rho + \bar{g}_{\mu\rho} \bar{\nabla}_\nu\, \xi^\rho \\ &= \bar{\nabla}_\mu\, \xi_\nu + \bar{\nabla}_\nu\, \xi_\mu, \end{aligned} \tag{4.61}$$

since the covariant derivative $\bar{\nabla}_\rho$ is defined with respect to $\bar{g}_{\mu\nu}$. Finally, we find

$$h'_{\mu\nu} := h^{\xi}_{\mu\nu} = h_{\mu\nu} - \bar{\nabla}_{(\mu}\,\xi_{\nu)} =: h_{\mu\nu} - \Delta^{\xi} h_{ik}. \tag{4.62}$$

As a specific gauge fixing $\chi^{\rho}(h_{\mu\nu}) = 0$, we choose the common condition

$$\chi^{\rho}(h_{\mu\nu}) = \bar{\nabla}_{\sigma} h^{\sigma\rho} - \frac{1}{2}\bar{\nabla}^{\rho} h = 0 \quad \text{with} \quad h := \bar{g}^{\mu\nu} h_{\mu\nu}, \tag{4.63}$$

known as the harmonic or de Donder gauge.[6] The gauge condition χ^{ρ} for the gauge transformed perturbations $h^{\xi}_{\mu\nu}$ behaves as

$$\chi^{\rho}(h^{\xi}_{\mu\nu}) = \chi^{\rho} + (\bar{\Box}\,\delta^{\rho}_{\sigma} + \bar{R}^{\rho}_{\sigma})\xi^{\sigma} = \chi^{\rho} + Q^{\rho}_{\sigma}\xi^{\sigma}. \tag{4.64}$$

Thus, the change of the gauge condition $\chi^{\rho} = 0$ due to the gauge transformation of the perturbations $h^{\xi}_{\mu\nu}$ can be described by the ghost operator Q^{ρ}_{σ}

$$Q^{\rho}_{\sigma} := \bar{\Box}\,\delta^{\rho}_{\sigma} + \bar{R}^{\rho}_{\sigma}. \tag{4.65}$$

In order to implement the gauge condition (4.63), we insert a functional **1** in (4.57)

$$\mathbf{1} = \int \mathcal{D}\xi^{\rho}\, \delta[\chi^{\sigma}(h^{\xi}_{\mu\nu})]\, \det\left(\frac{\delta\,\chi^{\sigma}(h^{\xi}_{\mu\nu})}{\delta\xi^{\rho}}\right), \tag{4.66}$$

with the functional delta function $\delta[\chi^{\sigma}(h^{\xi}_{\mu\nu})]$. We can think of the determinant

$$\det\left(\frac{\delta\chi^{\rho}(h^{\xi}_{\mu\nu})}{\delta\xi^{\sigma}}\right) = \det\left(\frac{\delta(\chi^{\rho} + Q^{\rho}_{\nu}\xi^{\nu})}{\delta\xi^{\sigma}}\right) = \det(Q^{\rho}_{\sigma}) \tag{4.67}$$

as of the Jacobian which transforms from the"coordinate" χ^{ρ} to the "coordinate" ξ^{σ} and since (4.67) does neither depend on ξ^{ρ} nor on $h_{\mu\nu}$, we can pull this factor out of the functional integral (4.57) and the path integral (4.57) acquires the form

$$Z[g_{\mu\nu}] = N \det(Q^{\rho}_{\sigma}) \int \mathcal{D}\xi^{\alpha} \int \mathcal{D}h_{\mu\nu}\delta[\chi^{\alpha}(h^{\xi}_{\mu\nu})] \exp\left(\frac{i}{\hbar}\, S[h_{\mu\nu}]\right). \tag{4.68}$$

The functional delta function ensures that the integration over $h_{\mu\nu}$ is constrained to the gauge-hypersurface, defined by the gauge (4.63). The important point is that the measure $\mathcal{D}h_{\mu\nu} \to \mathcal{D}h^{\xi}_{\mu\nu} = \mathcal{D}h_{\mu\nu}$ and the action $S[h_{\mu\nu}] \to S[h^{\xi}_{\mu\nu}] = S[h_{\mu\nu}]$ are invariant under the gauge transformation $h_{\mu\nu} \to h^{\xi}_{\mu\nu}$, so that we can write

$$Z[g_{\mu\nu}] = N \det(Q^{\rho}_{\sigma}) \int \mathcal{D}\xi^{\alpha} \int \mathcal{D}h_{\mu\nu}\delta\left[\chi^{\alpha}(h_{\mu\nu})\right] \exp\left(i S[h_{\mu\nu}]\right). \tag{4.69}$$

[6] The treatment is similar to that of linearised gravity $\bar{g}_{\mu\nu} = \eta_{\mu\nu}$ in the context of gravitational waves.

We have re-labelled the "dummy" integration variable $h^{\xi}_{\mu\nu}$ to $h_{\mu\nu}$ again in the last step. We note that a modification of the gauge condition (4.63) of the form $\chi'^{\rho} = \chi^{\rho} - \omega^{\rho}$ for an arbitrary constant ω^{ρ}, leaves the ghost operator unchanged

$$\frac{\delta \chi'^{\rho}}{\delta \xi^{\sigma}} = \frac{\delta \chi^{\rho}}{\delta \xi^{\sigma}}. \tag{4.70}$$

With the modified gauge condition the path integral can be written as

$$Z[g_{\mu\nu}] = N \det(Q^{\rho}_{\sigma}) \int \mathcal{D}\xi^{\alpha} \int \mathcal{D}h_{\mu\nu} \delta[\chi^{\alpha} - \omega^{\alpha}] \exp\left(\frac{i}{\hbar} S[h_{\mu\nu}]\right). \tag{4.71}$$

Since the integrand (4.71) is independent of ξ^{μ}, the integration over the gauge group $\int \mathcal{D}\xi^{\rho}\, 1$ just yields an unimportant contribution which can be absorbed by a proper normalization $N \to \tilde{N}$. Therefore, we will omit this integration in the following. Moreover, since the value of the path integral (4.71) is completely independent of the choice for ω^{α}, we can integrate over all values of ω^{α} together with an arbitrary weight function. We choose this weight function to be a Gaussian distribution, see Appendix A of [21]

$$g(k^2) \propto \exp\left(-\frac{i}{\hbar} \int \mathrm{d}^4x \, \frac{\sqrt{-\bar{g}}}{2k^2} \, \bar{g}_{\alpha\beta} \, \omega^{\alpha} \omega^{\beta}\right). \tag{4.72}$$

Then, the path integral (4.71) becomes

$$Z[g_{\mu\nu}] = \tilde{N} \det(Q^{\rho}_{\sigma}) \int \mathcal{D}h_{\mu\nu} \int \mathcal{D}\omega^{\alpha} \, e^{\frac{i}{\hbar} S[h_{\mu\nu}]} \, \delta(\chi^{\alpha} - \omega^{\alpha}) \, e^{-\frac{i}{\hbar} \int \mathrm{d}^4x \, \frac{\sqrt{-\bar{g}}}{2k^2} \, \bar{g}_{\alpha\beta} \, \omega^{\alpha} \omega^{\beta}}. \tag{4.73}$$

Performing the integration over ω^{α}, absorbing all inessential parts by re-defining $\tilde{N}$ once again and setting $\tilde{N} = 1$ at the end, we can write the path integral as

$$Z[g_{\mu\nu}] = \det(Q^{\rho}_{\sigma}) \int \mathcal{D}h_{\mu\nu} \, e^{\frac{i}{\hbar} \left(S[\bar{g}_{\mu\nu}, h_{\mu\nu}] + S_{\mathrm{gb}}[\bar{g}_{\mu\nu}, h_{\mu\nu}]\right)}. \tag{4.74}$$

We have displayed the dependence on $\bar{g}_{\mu\nu}$ and defined the gauge breaking term

$$S_{\mathrm{gb}} := -\frac{1}{2\,k^2} \int \mathrm{d}^4x \, \sqrt{-\bar{g}} \, \bar{g}_{\alpha\beta} \, \chi^{\alpha} \chi^{\beta}. \tag{4.75}$$

We can write the Faddeev–Popov determinant $\det(Q^{\rho}_{\sigma})$ as a Gaussian integral over the *independent* anti-commutative Grassmann fields $\eta^{\sigma}(x)$ and $\eta^{*\,\rho}(x)$

$$\det(Q^{\rho}_{\sigma}) = \int \mathcal{D}\eta^{\alpha} \int \mathcal{D}\eta^{*\,\alpha} \, e^{\frac{i}{\hbar} S_{\mathrm{ghost}}[\eta, \eta^*, \bar{g}_{\mu\nu}]}. \tag{4.76}$$

Now, the meaning of the term "ghost operator" becomes clear. It refers to the fact that Q^{ρ}_{σ} corresponds to the inverse propagator of the Grassmannian ghost-fields

$$S_{\text{ghost}}[\eta, \eta^{*}, \bar{g}_{\mu\nu}] := \int \mathrm{d}^4 x \eta^{\rho}(x)\, Q_{\rho\sigma}(\bar{\nabla})\, \eta^{*\sigma}(x). \tag{4.77}$$

The Faddeev–Popov ghost fields do not correspond to physical particles, since vector fields which obey fermionic statistics $\{\eta^{*\sigma}, \eta^{\rho}\} = 0$ do violate the spin-statistic theorem. In a diagrammatic language this means that they cannot be associated with external legs in Feynman diagrams. However, as virtual particles they run in quantum loops and thus give contributions. Finally, we can write the path integral (4.74) as

$$Z[g_{\mu\nu}] = \int \mathcal{D}h_{\mu\nu} \int \mathcal{D}\eta^{\alpha} \int \mathcal{D}\eta^{*\alpha}\, e^{\frac{i}{\hbar} S_{\text{tot}}[\bar{g}_{\mu\nu}, h_{\mu\nu}, \eta, \eta^{*}]}, \tag{4.78}$$

with a total action defined as the sum of the original, gauge fixing and ghost action

$$S_{\text{tot}}[\bar{g}_{\mu\nu}, h_{\mu\nu}, \eta, \eta^{*}] := S[\bar{g}_{\mu\nu}, h_{\mu\nu}] + S_{\text{gb}}[\bar{g}_{\mu\nu}, h_{\mu\nu}] + S_{\text{ghost}}[\bar{g}_{\mu\nu}, \eta, \eta^{*}]. \tag{4.79}$$

We have managed to avoid the over counting of gauge equivalent configurations in the naive definition of the path integral. The same procedure works for additional fields with an internal gauge symmetry.

From now on, we will assume that the gauge breaking term is always included in the definition of the effective action. In other words, we assume that we perform a second functional derivative of S_{tot} (instead of S) in order to obtain the operator $F_{ij'}$. In addition, the ghost operator Q^{ν}_{μ} has to be considered separately and gives a similar contribution. However due to its fermionic statistics, the one-loop contribution contains a factor of two and a negative sign. Setting $\hbar = 1$ again, we can write the one-loop effective action for gravity as

$$\Gamma_{1-\text{loop}} = S + \frac{i}{2} \operatorname{Tr} \ln S_{,ij} - i \operatorname{Tr} \ln Q^{\nu}_{\mu}. \tag{4.80}$$

4.5 Generalized Schwinger–DeWitt Algorithm

In order to further evaluate the one-loop effective action (4.80), we have to regularize the expression $\operatorname{Tr} \ln S_{,ij}$. In Sect. 4.5.1, we will first consider the general structure of the second variation operator $F_{ij} \sim S_{,ij}$ and follow the technique developed in [2], to bring this operator into a minimal form $\hat{F}$ suitable for the application of the Schwinger–DeWitt algorithm [4]. This algorithm consists of several steps which will be discussed in Sect. 4.5.1. First we use an integral representation of Green's functions developed by Schwinger [19] to rewrite the minimal operator. Then we exploit the fact that the kernel $K(x, x')$ of the so defined operator satisfies the heat equation. Finally, in order to solve the heat equation, we adopt the proposal of DeWitt [4] and use a geometrically motivated ansatz for the kernel $K(x, x';\, s) \sim \sum_n \hat{a}_n(x, x')\,(is)^n$ which includes a series expansion in proper time s. The kernel $K(x, x';\, s)$ defined in this way diverges for small proper times s and we will use dimensional regularization to isolate the divergent part in Sect. 4.5.2. This result can be inserted into the definition of the one-loop effective action. The functional trace in (4.80), requires to find the coincidence limits $x \to x'$ of the coefficients $\hat{a}_n(x, x')$. In Sect. 4.5.3, we obtain the coincidence limits of these coefficients with the help of the recursion equation mentioned above and by using the universal results for the coincidence limits of other geometrical quantities contained in the DeWitt ansatz. We have collected the necessary definitions and coincidence limits of these quantities in Appendix D.

4.5.1 Reduction Algorithm for Minimal Operators

The second variation has the structure of the kernel of an a differential operator

$$S_{,ij} = \frac{\delta^2 S[\phi]}{\delta\phi^A(x)\,\delta\phi^B(x')} = F_{AB}(x, x') = F_{AB}(\bar{\nabla}^x)\delta(x, x')\,. \tag{4.81}$$

The bar denotes again background quantities and the superscript x in $\bar{\nabla}^x_\mu$ indicates that the derivatives act on x only. It will turn out to be convenient to consider the operator F_{AB} itself rather than its kernel $F_{AB}(x, x')$. If the original action $S[\phi]$ does contain the conventional kinetic terms, quadratic in the first derivatives $\nabla_\mu\phi^i$, with ∇_μ being the covariant derivative with respect to some general affine connection,[7] the operator F_{AB} will contain at most second derivatives. It can be divided into three parts according to the number of derivatives

$$F_{AB} := C^{\mu\nu}_{AB}\bar{\nabla}_\mu\bar{\nabla}_\nu + 2\,\Gamma^\mu_{AB}\bar{\nabla}_\mu + W_{AB}. \tag{4.82}$$

The authors of [2] extended the treatment to even more general operators, containing higher derivatives. However, for the discussion in this thesis, the structures 4.82, will be sufficient. Assuming that the Faddeev–Popov procedure has been taken into account (i.e. $S = S_{\text{tot}} = S + S_{\text{gb}}$), we use the reduction algorithm of [2] in order to bring the differential operator $F_{AB}(\bar{\nabla})$ into a so-called *minimal* form that is suitable for an application of the general Schwinger–DeWitt method. If the operator has a minimal structure, we can write the part with second derivatives as

$$C^{\mu\nu}_{AB}\bar{\nabla}_\mu\bar{\nabla}_\nu = C_{AB}\bar{g}^{\mu\nu}\bar{\nabla}_\mu\bar{\nabla}_\nu =: C_{AB}\bar{\Box}, \tag{4.83}$$

[7] In the case of pure Einstein gravity, the connection would correspond to the Christoffel symbols. However, the treatment here is more general. If ϕ^i does include fields with internal gauge symmetries, it is understood that ∇_μ acts with respect to the corresponding gauge field.

The freedom of choosing S_{gb} can be used to cancel those terms that are not proportional to the d'Alembertian $\bar{\Box}$ when performing the second functional derivative of S. We assume that S_{gb} has been chosen in such a way. Thus, the operator F_{AB} has the structure

$$F_{AB} := C_{AB}\,\bar{\Box} + 2\,\Gamma^{\mu}_{AB}\bar{\nabla}_{\mu} + W_{AB}. \tag{4.84}$$

Next, we would like to normalize the "kinetic part" $C_{AB}\,\bar{\Box}$, i.e. $C_{AB}\,\bar{\Box} \to \hat{I}\bar{\Box}$, with $\hat{I} := \delta^{B}_{A}$. We can achieve this by multiplying F_{AB} with the inverse $(C^{-1})^{AC}$.

However, since the operator F_{AB} enters the definition of the one-loop effective action, we must investigate whether the multiplication with $(C^{-1})^{AC}$ can affect the final result. It can be shown [2] that this operation does not change the divergent part of (4.80), since it only leads to additional terms $\mathrm{Tr}\ln\delta(x,x') \propto \delta(0)\,(...)$ which are cancelled by the local measure [2, 8, 9]. Thus, after multiplication with $(C^{-1})^{AC}$, the operator (4.84) takes the form

$$\hat{F} := (C^{-1})^{BC}F_{AC} = \hat{I}\bar{\Box} + 2\,\hat{\Gamma}^{\mu}\bar{\nabla}_{\mu} + \hat{W}, \tag{4.85}$$

We have used $\delta^{C}_{A} = C_{AB}(C^{-1})^{BC}$ and defined $\hat{\Gamma}^{\mu} := (C^{-1})^{BC}\,\Gamma^{\mu}_{CA}$ and $\hat{W} := (C^{-1})^{BC}\,W_{CA}$. In general, we will follow the abbreviation used in [2, 4] and denote operators with one index raised by a hat $\hat{O} = O^{B}_{A}$ and write the generalized field ϕ^{A} as ϕ.

The next step of the reduction algorithm is to further simplify the structure by absorbing the terms linear in the derivative $\sim \hat{\Gamma}^{\mu}\,\bar{\nabla}_{\mu}$ by a redefinition of the covariant derivative

$$\hat{\bar{\nabla}}^{\mu} \to \hat{\mathcal{D}}^{\mu} := \hat{\bar{\nabla}}^{\mu} + \hat{\Gamma}^{\mu}. \tag{4.86}$$

The redefinition of the covariant derivative entails a redefinition of the potential term $\hat{W}$ and the field curvature defined in Appendix D.4, which has to be redefined in terms of the new covariant derivative

$$[\hat{\mathcal{D}}_{\mu},\,\hat{\mathcal{D}}_{\nu}]\,\phi = \hat{\mathcal{R}}_{\mu\nu}\phi. \tag{4.87}$$

Re-expressing these contributions in terms of the original derivative $\hat{\bar{\nabla}}_{\mu}$, we find

$$\hat{P} = P^{B}_{A} := W^{B}_{A} + \frac{1}{6}\,\bar{R}\,\delta^{B}_{A} - (\bar{\nabla}_{\mu}\Gamma_{A}^{\;B\,\mu}) - g_{\mu\nu}\Gamma_{A}^{\;B\,\mu}\Gamma_{B}^{\;C\,\nu}, \tag{4.88}$$

$$\hat{\mathcal{R}}_{\mu\nu} = \mathcal{R}_{A\;\;\mu\nu}^{\;\;B} := \mathcal{R}^{0\,B}_{A\;\;\mu\nu} + 2\,\bar{\nabla}_{[\mu}g_{\nu]\lambda}\,\Gamma^{B\;\lambda}_{A} + 2\,g_{\sigma[\mu}g_{\nu]\lambda}\,\Gamma^{C\;\sigma}_{A}\,\Gamma^{B\;\lambda}_{C}, \tag{4.89}$$

with the background field curvature $\hat{\mathcal{R}}^{0}_{\mu\nu}$ being defined with respect to $\hat{\bar{\nabla}}_{\mu}$

$$[\hat{\bar{\nabla}}_{\mu},\,\hat{\bar{\nabla}}_{\nu}]\,\phi = \hat{\mathcal{R}}^{0}_{\mu\nu}\,\phi. \tag{4.90}$$

After all these manipulations, we have managed to bring the operator (4.82) into its final form

$$\hat{F} := \hat{\bar{\Box}} + \hat{P} - \frac{1}{6}\,\bar{R}\,\hat{I}. \tag{4.91}$$

From $\hat{P}$, we have extracted a term $1/6\,\bar{R}\,\hat{I}$ in the definition (4.91). This is just a convention and will simplify some further expressions.

4.5.1.1 Heat Kernel Technique

Now the operator (4.91) is ready for the Schwinger–DeWitt algorithm. The Green's function $G_{AB}(x, x')$ of the operator F_{AB} has to satisfy the equation

$$-\hat{I}\,\delta(x, x'') = \int \mathrm{d}^4x'\,\hat{F}(x, x')\,\hat{G}(x', x'') = \hat{F}(\bar{\nabla}^x)\,\hat{G}(x, x''). \tag{4.92}$$

If we just work with the operators instead of the kernels, this equation reads

$$\hat{F}\,\hat{G} = -\hat{I} \quad \text{or} \quad \hat{G} = -\frac{1}{\hat{F}}. \tag{4.93}$$

We can use an integral representation for the Feynman Green's function [19]

$$\frac{1}{\hat{F}} = i\int_0^\infty \mathrm{d}s\exp\left(-i\,\hat{F}\,s\right), \tag{4.94}$$

where we have implicitly assumed that a small imaginary part $+i\epsilon$ has been added to $\hat{F}$ in order to choose the correct integration contour for the Feynman propagator. The parameter s is called proper time for reasons that will become clear in a moment. By a formal integration of (4.94), we obtain

$$\ln\hat{F} = -\int_0^\infty \frac{\mathrm{d}s}{s}\exp\left(-i\,\hat{F}\,s\right). \tag{4.95}$$

We can now define the kernel of the exponentiated operator

$$\hat{K}(x, x';\, s) := e^{-i\,\hat{F}\,s}\,\delta(x, x'). \tag{4.96}$$

It satisfies the heat equation with the initial condition $\hat{K}(x, x'|0) = \delta(x, x')$

$$i\,\frac{\partial}{\partial s}\,\hat{K}(x, x'|s) = \hat{F}\,\hat{K}(x, x'|s). \tag{4.97}$$

We can use (4.95) in (4.56) to calculate the one-loop effective action

$$\Gamma^{(1)} = \frac{i}{2}\,\mathrm{Tr}\ln\Big[\hat{F}(\bar{\nabla}^x)\,\delta(x,x')\Big] = \frac{i}{2}\,\mathrm{Tr}\left[-\int\frac{\mathrm{d}s}{s}\,\hat{K}(x,x'|s)\right] \tag{4.98}$$

$$= -\frac{i}{2}\int \mathrm{d}^4x \int\frac{\mathrm{d}s}{s}\,\mathrm{tr}\,\hat{K}(x,x|s). \tag{4.99}$$

The functional trace of a general two-point object $O_A^B(x,x')$ is defined as

$$\mathrm{Tr}\,O_A^B(x,x') := \int \mathrm{d}^4x\,\mathrm{tr}\,O_A^B(x,x) := \int \mathrm{d}^4x\,O_A^A(x,x) = \int \mathrm{d}^4x\,\hat{O}(x,x). \tag{4.100}$$

Thus, we have to calculate the coincidence limit $x \to x'$ for the kernel defined in (4.96). DeWitt proposed the following ansatz for the kernel $\hat{K}(x,x'|s)$ [4].

$$\hat{K}(x,x'|s) = \frac{i}{(4\,\pi\,i)^{\omega}}\,\frac{D^{1/2}(x,x')}{s^{\omega}}\,\exp\left[i\,\frac{\sigma(x,x')}{2s}\right]\hat{\Omega}(x,x'|s). \tag{4.101}$$

The geometrical objects $\sigma(x,\ x')$, $D(x,\ x')$, $\Omega(x,\ x'|s)$ contained in this ansatz correspond to the bi-scalar world function, the Van-Vleck determinant and a general two point configuration space matrix, that contains information about the background curvature. These objects are defined and explained in Appendix D. A detailed treatment can be found in [4, 16, 20]. We have generalized the space-time dimension to 2ω, since we intend to use dimensional regularization. Comparing this with the solution of the heat kernel for the propagator of a massive scalar field in flat space-time, it becomes obvious that this ansatz is strongly motivated by

$$G(x,x') = \int_0^{\infty}\mathrm{d}s\,K(x,x'|s), \quad K(x,x'|s) = \frac{1}{(4\,\pi)^2}e^{-i\left[m^2\,s-\frac{(x-x')^2}{4\,s}\right]}. \tag{4.102}$$

The aim of DeWitt's ansatz is the generalization to curved space-times for general fields ϕ^i. Before proceeding, it is useful to compare and relate the new symbols and objects for a curved space-time in (4.101) with those contained in (4.102). In flat space, the expansion of the world function near the coincidence limit $x \to x'$ is equivalent to an expansion near the light cone. The Van-Vleck determinant $\mathcal{D}_{\mathrm{flat}}(x,x') = 1$ simply reduces to 1 and from (4.102) it is clear that the general bi-tensor $\hat{\Omega}_{\mathrm{flat}}(x,x'|s) = \hat{I}$ is trivial, since it contains information about the curvature of space-time. Our task is to find an expression for $\hat{\Omega}(x,x'|s)$ in a generally curved space-time. In order to do so, we expand $\hat{\Omega}(x,x'|s)$ in powers of the proper time s. Thus, we separate space-time and proper time arguments by expanding $\hat{\Omega}(x,x'|s)$ in the series

$$\hat{\Omega}(x,x'|s) = \sum_{n=0}^{\infty}(is)^n\,\hat{a}_n(x,x'). \tag{4.103}$$

This turns out to be especially useful in order to isolate the UV divergences of a theory. Inserting the ansatz (4.101) with the expansion (4.103) into the expression for the one-loop effective action (4.99) and using the coincidence limits $[D^{1/2}(x, x')]$ and $[\sigma(x, x')]$, derived in Appendix D.5 yields

$$\Gamma^{(1)} = \frac{(i)^{n-\omega}}{2\,(4\pi)^{\omega}} \int_0^{\infty} \mathrm{d}s \int \mathrm{d}^{2\omega}x g^{1/2} \sum_{n=0}^{\infty} \mathrm{tr}\left[(a_n)_A^{\ B}(x, x')\right] s^{n-\omega-1}. \tag{4.104}$$

This integral diverges at the *lower* limit for $s = 0$ for $n \leq \omega$.

4.5.2 Dimensional Regularization and the $\hat{a}_2$ Coefficient

Following the treatment of [14], we use dimensional regularization in order to extract the UV divergent parts in the limit $s \to 0$, $\omega \to 2$. From the condition $n \leq \omega$, it follows that only the three integrals with $n = 0, 1, 2$ can contribute to the divergences in the limit $\omega \to 2$. Since we are looking for the small s limit, we can fix an arbitrary *upper* integration bound for s. Consider the integral

$$I(\omega, n) := \int_0^{s_{\max}} \mathrm{d}\, s^{n-\omega-1} \tag{4.105}$$

For $\mathfrak{Re}(2\,\omega) < 0$, this integral becomes

$$I(\omega, n) = \frac{s_{\max}^{(n-\omega)}}{(n-\omega)} \tag{4.106}$$

and has only simple poles at $n = \omega$. Thus we can analytically continue 2ω to $\mathfrak{Re}(2\,\omega) \geq 0$. The divergent part only results from the $n = 2$ coefficient. Introducing an effective cut-off Λ by

$$\lim_{\omega \to 2} \frac{1}{(2-\omega)} := \lim_{\Lambda \to \infty} \ln \Lambda^2, \tag{4.107}$$

we finally find for the divergent part of the one-loop effective action

$$\begin{aligned}\Gamma^{(1)}_{\mathrm{div}} &= \lim_{\omega \to 2} \left[\frac{1}{2\,(2-\omega)\,(4\pi)^{\omega}} \int \mathrm{d}^{2\omega}x\, g^{1/2}\, \mathrm{tr}\,[(a_2)_A^{\ B}(x,\, x')]\right] \\ &= \lim_{\Lambda \to \infty} \frac{\ln \Lambda^2}{32\,\pi^2} \int \mathrm{d}^4x\, g^{1/2}\, (a_2)_A^{\ A}(x, x).\end{aligned} \tag{4.108}$$

It is remarkable that in four space-time dimensions $\Gamma^{(1)}_{\mathrm{div}}$ only depends on the coincidence limit $x \to x'$ of the trace of the single $(a_2)_A^{\ B}(x, x')$ coefficient.

4.5.3 Recursion Relation and Coincidence Limits

We have already managed to reduce the divergent part of the one-loop effective action to the calculation of one single quantity $\mathrm{tr}\,[\hat{a}_2(x,x')]$. However, in order to extract physical results, we have to find an expression for $\hat{a}_2(x,x')$ in terms of the background quantities $\{\bar{R}_{\mu\nu\sigma\rho}\,\hat{I},\ \hat{\mathcal{R}}_{\mu\nu},\ \hat{P}\}$.

Inserting the ansatz (4.101) into the heat equation (4.97) with the minimal form (4.91) of $\hat{F}$, yields a recursion relation for the coefficients $\hat{a}_n(x,x)$

$$(n+1)\,\hat{a}_{n+1} + \sigma^{\mu}\,\bar{\nabla}_{\mu}\,\hat{a}_{n+1} = \Delta^{-1/2}\,\bar{\Box}\,(\Delta^{1/2}\,\hat{a}_n) + (\hat{P} - \frac{1}{6}\,\bar{R}\,\hat{I})\,\hat{a}_n, \tag{4.109}$$

which can be solved iteratively with the initial conditions

$$\sigma^{\mu}\bar{\nabla}_{\mu}\,\hat{a}_0 = 0 \quad \text{and} \quad \lim_{x\to x'}\hat{a}_0(x,x') = 1. \tag{4.110}$$

We only have to evaluate this recursion relation up to $n=2$. With (4.109), we can calculate the coincidence limits for $\hat{a}_1$ and $\hat{a}_2$ in terms of the coincidence limits of the world function and the Van-Vleck determinant and their higher derivatives. Those coincidence limits were calculated in [2, 4] and can be found in Appendix D.5. Inserting the expressions for the coincidence limits in the recursive expressions for $\hat{a}_1$ and $\hat{a}_2$ obtained by (4.109), we finally obtain the coincidence limits for the minimal operator (4.91)

$$[\,\hat{a}_1\,] = -\frac{\hat{\mathbf{1}}}{6}\bar{R}, \tag{4.111}$$

$$[\,\hat{a}_2\,] = \frac{\hat{\mathbf{1}}}{180}\left(\bar{R}^2_{\alpha\beta\mu\nu} - \bar{R}^2_{\mu\nu} + \bar{\Box}\bar{R}\right) + \frac{1}{2}\,\hat{P}^2 + \frac{1}{12}\,\hat{\mathcal{R}}^2_{\mu\nu} + \frac{1}{6}\,\bar{\Box}\hat{P}. \tag{4.112}$$

The higher coefficients $\hat{a}_3$ and $\hat{a}_4$ were calculated by Gilkey [11] and by Avramidi [1]. For our purpose it is sufficient to consider the coincidence limit of the $\hat{a}_2$ coefficient, since all information of the divergent part of the one-loop effective action in four space-time dimensions is contained in $[\hat{a}_2]$. Now we can write down the important formula for (4.108) only in terms of the background quantities

$$\Gamma^{(1)}_{\text{div}} = \lim_{\omega\to 2}\int \frac{\mathrm{d}^{2\omega}x\,\bar{g}^{1/2}}{32\pi^2(2-\omega)}\,\mathrm{tr}\left[\frac{\hat{\mathbf{1}}\left(\bar{R}^2_{\alpha\beta\mu\nu} - \bar{R}^2_{\mu\nu} + \bar{\Box}\bar{R}\right)}{180} + \frac{\hat{P}^2}{2} + \frac{\hat{\mathcal{R}}^2_{\mu\nu}}{12} + \frac{\bar{\Box}\hat{P}}{6}\right]. \tag{4.113}$$

The same algorithm works for the ghost operator, as defined for the case of pure gravity in (4.65). The final result for the complete one-loop contribution can then be expressed as

$$\Gamma^{(1)}_{\rm div} = \lim_{\omega\to 2} \frac{1}{32\,\pi^2\,(2-\omega)} \left[\int d^{2\omega}x\, \bar{g}^{1/2}\, {\rm tr}\,[\hat{a}_2] - 2\int d^{2\omega}x\, \bar{g}^{1/2}\, {\rm tr}\,[(a_2^{(Q)})^{\nu}_{\mu}] \right]. \tag{4.114}$$

References

1. Avramidi, I.G.: The covariant technique for calculation of one loop effective action. Nucl. Phys. B **355**, 712 (1991)
2. Barvinsky, A.O., Vilkovisky, G.A.: The generalized Schwinger-DeWitt technique in gauge theories and quantum gravity. Phys. Rep. **119**, 1 (1985)
3. Buchbinder, I.L., Odintsov, S.D., Shapiro, I.L.: Effective Action in Quantum Gravity. Taylor and Francis, New York (1992)
4. DeWitt, B.S.: Dynamical Theory of Groups and Fields. Blackie and Son, London (1965)
5. Faddeev, L.D., Popov, V.N.: Feynman diagrams for the Yang-Mills field. Phys. Lett. B **25**, 29 (1967)
6. Feynman, R.P.: Space-time approach to non-relativistic quantum mechanics. Rev. Mod. Phys. **20**, 367 (1948)
7. Feynman, R.P.: Space-time approach to quantum electrodynamics. Phys. Rev. **76**, 769 (1949)
8. Fradkin, E.S. Vilkovisky, G.A.: S Matrix for gravitational field. II. Local measure; general relations; elements of renormalization theory. Phys. Rev. D **8**, 4241 (1973)
9. Fradkin, E.S., Vilkovisky, G.A.: On the renormalization of quantum field theory in curved space-time. Lett. Nuovo Cimento **19**, 47 (1977)
10. Gell-Mann, M., Low, F.: Quantum electrodynamics at small distances. Phys. Rev. **95**, 1300 (1954)
11. Gilkey, P.B.: The spectral geometry of a Riemannian manifold. J. Diff. Geom. **10**, 601 (1975)
12. Kadanoff, L.P., Gotze, W., Hamblen, D., Hecht, R., Lewis, E.A.S., et al.: Static phenomena near critical points: theory and experiment. Rev. Mod. Phys. **39**, 395 (1967)
13. Leibbrandt, G.: Introduction to the technique of dimensional regularization. Rev. Mod. Phys. **47**, 849 (1975)
14. Parker, L.E., Toms, D.J.: Quantum Field Theory in Curved Spactime. Cambridge University Press, Cambridge (2009)
15. Pauli, W., Villars, F.: On the invariant regularization in relativistic quantum theory. Rev. Mod. Phys. **21**, 434 (1949)
16. Poisson, E., Pound, A. and Vega, I.: The motion of point particles in curved spacetime. Living Rev. Rel. **7**, 6 (2004). Accessed 19 December 2011. http://relativity.livingreviews.org/Articles/lrr-2011-7
17. Ramond, P.: Field Theory A Modern Primer. Benjamin, Massachusetts (1981)
18. Schwinger, J.: Particles and Sources. Gordon and Breach, New York (1969)
19. Schwinger, J.S.: On gauge invariance and vacuum polarization. Phys. Rev. **82**, 664 (1951)
20. Synge, J.L.: Relativity: The General Theory. North-Holland, Amsterdam (1960)
21. 't Hooft, G.: Renormalizable Lagrangians for massive Yang-Mills fields. Nucl. Phys. B. **35, 167 (1971)**
22. 't Hooft, G. : Dimensional regularization and the renormalization group. Nucl. Phys. B **61**, 455 (1973)
23. 't Hooft, G. and Veltman, M.J.G.: Regularization and renormalization of gauge fields. Nucl. Phys. B **44**, 189 (1972)
24. Wald, R.M.: General Relativity. The University of Chicago Press, Chicago (1984)
25. Wilson, K.G.: Renormalization group and critical phenomena. I. Renormalization group and the Kadanoff scaling picture. Phys. Rev. B **4**, 3174 (1971)
26. Wilson, K.G.: Renormalization group and critical phenomena. II. Phase-space cell analysis of critical behavior. Phys. Rev. B **4**, 3184 (1971)

Chapter 5
One-Loop Cosmology and Frame Dependence

We will start with the most technical chapter of this thesis because the results obtained in this part will serve as a basis for the cosmological investigations of the Higgs inflation scenario we will discuss in Chap. 6. We will make use of the generalized Schwinger–DeWitt algorithm discussed in Sect. 4.5 in order to calculate the one-loop effective action for gravity coupled to scalar fields—a setup especially interesting in the cosmological context of inflation.

In Sect. 5.1 we will demonstrate the practical details of the algorithm, when applying it to the model of a *single* scalar field *minimally*[1] coupled to gravity. In order to keep the calculations as general as possible, we will allow generalized couplings which can depend on the scalar field. A similar calculation has already been performed in [2], but here we will repeat and correct this calculation. This will serve as a guiding line for the kind of calculations to be performed in the more complicated setup of a $O(N)$-symmetric *multiplet of scalar fields non-minimally coupled to gravity*, we will discuss in Sect. 5.2. During these calculations, the number of the single tensorial expressions that have to be properly contracted is at least $\mathcal{O}(10^4)$, which is why we have to resort on the help of a computer algebra system, capable of performing *analytical* tensor manipulations—the `Mathematica` package `MathTensor`. In view of this computer assistance, it is of crucial importance to be aware of each individual step in the algorithm.

Furthermore, in Sect. 5.3.2 of this part, we will investigate the frame dependence of quantum corrections by explicitly calculating the difference of the divergent part of the one-loop off-shell effective action in the two cosmological field parametrizations, denoted by Jordan frame and Einstein frame. As part of these investigations we will show that while the explicit transformation rules between Jordan frame and Einstein frame for a $O(N)$ symmetric scalar multiplet non-minimally coupled to gravity can be constructed, this leads to a non-trivial integrability condition. Finally, in Sect. 5.3.3, we will identify the source of this frame dependence and show how these results can be naturally embedded into a more general formalism, developed in [26].

[1] A minimal coupling to gravity means no coupling with derivative terms of the metric $R \sim \partial^2 g$.

C. F. Steinwachs, *Non-minimal Higgs Inflation and Frame Dependence in Cosmology*, Springer Theses, DOI: 10.1007/978-3-319-01842-3_5,
© Springer International Publishing Switzerland 2014

5.1 Minimally Coupled Single Field in the Einstein Frame

In this section we will explicitly calculate the divergent part of the one-loop effective action for a single scalar field minimally coupled to gravity. This calculation was already performed in [2] as an important application of the general Schwinger–DeWitt formalism described in Sect. 4.5, but here we will repeat it in detail. This setup is much easier than the model we will discuss in Sect. 5.2 so that we can concentrate on the transparent presentation of the practical implementation of each individual step of the general Schwinger–DeWitt algorithm. It will serve us as a guideline, when resorting on the assistance of the computer algebra program `MathTensor` in Sect. 5.2. Moreover, the results can considered as an independent confirmation (correction) of the results obtained in [2]. In addition, since they represent a particular limit of the more general model to be investigated in Sect. 5.2, they will serve as an important cross-check.

5.1.1 Calculation of the Fluctuation Operator

We will use the background field method and therefore split the basic fields of the model (metric and scalar field) into a background part, denoted by a bar, and a fluctuation part. In addition, we will combine these two fields in a generalized field

$$\Psi^A := \bar{\Psi}^A + \delta\Psi^A := \begin{pmatrix} \bar{g}_{\mu\nu} \\ \bar{\varphi} \end{pmatrix} + \begin{pmatrix} h_{\mu\nu} \\ \sigma \end{pmatrix}, \tag{5.1}$$

with the fluctuations $\delta g_{\mu\nu} =: h_{\mu\nu}$ and $\delta\varphi =: \sigma$.

We consider the action of a single scalar field minimally coupled to gravity

$$S[\varphi, g_{\mu\nu}] = \int d^4x\, g^{1/2} \left(\frac{R}{k^2} - \frac{1}{2} g^{\mu\nu} (\partial_\mu\varphi)(\partial_\nu\varphi) - V(\varphi) \right). \tag{5.2}$$

Following the algorithm outlined in Sect. 4.5, we have to calculate the second variation of S with respect to the metric $g_{\mu\nu}$ and the scalar field φ. In order to simplify the calculations, we will fix the constant k^2 by choosing $k^2 \equiv 1$ from the very beginning.

5.1.1.1 Gauge-Breaking Term

According to the prescriptions in Sect. 4.4.1, we have to choose a gauge-breaking term. This choice is arbitrary and therefore we will search for a form that simplifies the calculations. As outlined in Sect. 4.5, the further Schwinger–DeWitt algorithm requires that the part of the operator F_{AB} containing second order derivatives has a Laplacian structure. In the language of [26], this means that F_{AB} is a *minimal* operator, i.e. terms like $h\,\bar{\nabla}^\mu\bar{\nabla}^\nu h_{\mu\nu}$ should be absent. This can be achieved by the so-called "harmonic" (or "de Donder") gauge condition, cf. Sect. 4.4.1.

$$\chi_\nu = \bar{\nabla}^\mu h_{\mu\nu} - \frac{1}{2}\bar{\nabla}_\nu h. \tag{5.3}$$

The gauge-breaking term then takes the form

$$\begin{aligned} S_{\rm gb} &= -\frac{1}{2}\int \mathrm{d}^4x\,\bar{g}^{1/2}\left(\bar{\nabla}^\mu h_{\mu\nu} - \frac{1}{2}\bar{\nabla}_\nu h\right)\bar{g}^{\nu\sigma}\left(\bar{\nabla}^\rho h_{\rho\sigma} - \frac{1}{2}\bar{\nabla}_\sigma h\right) \\ &= \frac{1}{2}\int \mathrm{d}^4x\,\bar{g}^{1/2}\,h_{\alpha\beta}\left(\bar{\nabla}^\alpha\bar{\nabla}^\delta\bar{g}^{\beta\gamma} - \frac{1}{2}\bar{\nabla}^\gamma\bar{\nabla}^\delta\bar{g}^{\alpha\beta} - \frac{1}{2}\bar{\nabla}^\alpha\bar{\nabla}^\beta\bar{g}^{\gamma\delta} + \frac{1}{4}\bar{g}^{\alpha\beta}\bar{g}^{\gamma\delta}\bar{\Box}\right)h_{\gamma\delta}, \end{aligned} \tag{5.4}$$

where we have performed an integration by parts in the last step. The gauge-breaking term only consists of background quantities $\bar{\Psi}^A$ and therefore is insensitive to functional differentiation with respect to the field Ψ^A. Thus, we can add the gauge-breaking term $S_{\rm gb}$ to the action S already *before* the variation. However, since (5.4) only depends on gravitational degrees of freedom, we will add it to the graviton-graviton sector *after* variation of (5.2) for the sake of a clear presentation. The fluctuation operator F_{AB} for this model corresponds to a (2×2) matrix in field space. It can be subdivided into four sectors associated with the components of the (2×2) matrix: The graviton-graviton sector, the mixed graviton-scalar and scalar-graviton sectors and the scalar-scalar sector. In order to figure out the contributions of $S_{\rm gb}$ to the corresponding sectors we can perform the second functional derivative of $S_{\rm gb}$ with respect to the *perturbations* $h_{\mu\nu}$ and σ.[2]

5.1.1.2 First Variations

Using the expressions collected in Appendix B.4, the first variation with respect to the scalar field yields

$$\delta_\varphi S = \int \mathrm{d}^4x\,g^{1/2}\,\left(-g^{\mu\nu}(\partial_\mu\varphi)(\partial_\nu\sigma) - V'(\varphi)\sigma\right). \tag{5.5}$$

In order to free the perturbation σ from the derivative we integrate by parts

$$\delta_\varphi S = \int \mathrm{d}^4x\,\sigma\,\left[\partial_\nu\left(g^{1/2}\,g^{\mu\nu}(\partial_\mu\varphi)\right) - g^{1/2}\,V'(\varphi)\right]. \tag{5.6}$$

---

[2] In general, this is important in order to obtain the correct symmetrization structure and—especially in the mixed sectors—to obtain the correct contributions to the structures with derivatives acting on different perturbations. However, in this case it does not matter because $S_{\rm gb}$ only contributes to the graviton-graviton sector.

We note that in the case of the functional derivative (instead of the variation we have used here) we had to deal with a delta distribution instead of the perturbation σ, which would "eat" the integral over space-time. We have introduced the abbreviation that a prime represents a partial derivative with respect to the argument

$$\frac{\partial V(\varphi)}{\partial \varphi} = V'(\varphi), \quad \frac{\partial^2 V(\varphi)}{\partial \varphi^2} = V''(\varphi), \quad \left.\frac{\partial V(\varphi)}{\partial \varphi}\right|_{\varphi=\bar{\varphi}} = \bar{V}'(\bar{\varphi}), \text{ etc.} \tag{5.7}$$

In the context of metric variations it is convenient to introduce the abbreviations

$$G^{\alpha\beta\gamma\delta} := \frac{1}{4}\left(g^{\alpha\gamma}g^{\beta\delta} + g^{\alpha\delta}g^{\beta\gamma} - g^{\alpha\beta}g^{\gamma\delta}\right), \tag{5.8}$$

$$G_{\alpha\beta\gamma\delta} := g_{\alpha\gamma}g_{\beta\delta} + g_{\alpha\delta}g_{\beta\gamma} - g_{\alpha\beta}g_{\gamma\delta}, \tag{5.9}$$

which satisfy $G^{\alpha\beta\gamma\delta}G_{\alpha\beta\gamma\delta} = 10$. We note that although the partial derivative ∂_μ commutes with the variation δ_g, this is not true for the covariant derivative ∇_μ since it involves the variation of the Christoffel symbol. The first variation with respect to the metric yields

$$\delta_g S = \int d^4x\, g^{1/2}\Big[-2G^{\alpha\beta\mu\nu}R_{\mu\nu} + G^{\alpha\beta\mu\nu}(\partial_\mu\varphi)(\partial_\nu\varphi) - \frac{1}{2}g^{\alpha\beta}\,V(\varphi)$$
$$+\frac{1}{2}g^{\mu\nu}g^{\alpha\beta}\Big(\nabla_\beta\nabla_\nu\, h_{\mu\alpha} + \nabla_\alpha\nabla_\mu\, h_{\nu\beta} - \nabla_\beta\nabla_\alpha\, h_{\mu\nu} - \nabla_\mu\nabla_\nu\, h_{\alpha\beta}\Big)\Big]. \tag{5.10}$$

The last term is a total divergence and can be omitted. Thus, the first variation yields

$$\delta_g S = \int d^4x\, g^{1/2}\, h_{\alpha\beta}\left(-2G^{\alpha\beta\mu\nu}R_{\mu\nu} + G^{\alpha\beta\mu\nu}(\partial_\mu\varphi)(\partial_\nu\varphi) - \frac{1}{2}g^{\alpha\beta}\,V(\varphi)\right). \tag{5.11}$$

By the extremal conditions $\delta_g S = 0$ and $\delta_\varphi S = 0$, we recover the Einstein field equations and the Klein–Gordon equation for the scalar field .

In order to perform the second variations of the corresponding sectors, we will vary the results of the first variations (5.6) and (5.10) once again and evaluate the fields at their background values $g_{\mu\nu} \to \bar{g}_{\mu\nu}$, $\nabla_\mu \to \bar{\nabla}_\mu$, $\varphi \to \bar{\varphi}$ afterwards.

5.1.1.3 Second Variation: Scalar-Scalar Part

We use the results of the first variation and perform the second variation with respect to the scalar field. Using the abbreviation $\bar{\Box} := \bar{g}^{\mu\nu}\bar{\nabla}_\mu\bar{\nabla}_\nu$, we find

$$\delta_\varphi^2 S = \frac{1}{2}\int d^4x\, \bar{g}^{1/2}\sigma\left[\bar{\Box} - \bar{V}''(\bar{\varphi})\right]\sigma. \tag{5.12}$$

5.1.1.4 Second Variation: Graviton-Graviton Part

We perform the second variation with respect to the metric and arrange the result in a potential and a kinetic part

$$\begin{aligned}\delta_g^2 S = \frac{1}{2}\int \mathrm{d}^4x\, g^{1/2}\Big\{h_{\alpha\beta}\Big[2\Big(g^{\beta\nu}G^{\gamma\delta\alpha\mu} + g^{\nu\delta}G^{\alpha\beta\mu\gamma}\Big)R_{\mu\nu} - G^{\alpha\beta\gamma\delta}R \\ + G^{\alpha\beta\gamma\delta}\,V(\varphi) + \frac{1}{2}g^{\alpha\beta}(\nabla^\gamma\varphi)(\nabla^\delta\varphi) - g^{\alpha\gamma}(\nabla^\beta\varphi)(\nabla^\delta\varphi) \\ - \frac{1}{8}\,g^{\alpha\beta}g^{\gamma\delta}(\nabla^\nu\varphi)(\nabla_\nu\varphi) + \frac{1}{4}g^{\alpha\gamma}g^{\beta\delta}(\nabla^\nu\varphi)(\nabla_\nu\varphi)\Big]\,h_{\gamma\delta} \\ - h^{\mu\nu}h_{\mu\rho\,;\,\nu}{}^{\rho} + \frac{1}{2}h^{\mu\nu}h_{\mu\nu\,;\,\rho}{}^{\rho} + \frac{1}{2}h^{\mu\nu}h_{;\,\mu\nu} + \frac{1}{2}h\,h^{\mu\nu}{}_{;\,\mu\nu} - \frac{1}{2}h\,h_{;\rho}{}^{\rho}\Big\}.\end{aligned} \tag{5.13}$$

We have made use of the symmetry in $\alpha \leftrightarrow \beta, \gamma \leftrightarrow \delta$ and $\mu \leftrightarrow \nu$ following from the symmetry of the perturbation $h_{\rho\sigma}$. In general, when performing the functional derivative rather than the variation, we should pay attention to the correct symmetrization which otherwise can get lost in the fluctuation operator F_{AB}. In other words, we have to use the symmetry of the perturbations $h_{\mu\nu}$ when deriving the explicit form of F_{AB}, i.e. using (B.14). However, by keeping this in mind, we can proceed with the variations as derived above. Now we have to add the gauge breaking term

$$S_{\mathrm{gb}} = \frac{1}{2}\int \mathrm{d}^4x\, g^{1/2}\left(h^{\alpha\gamma}h_{\gamma\delta}{}^{;\delta}{}_{\alpha} - \frac{1}{2}h\,h_{\gamma\delta}{}^{;\,\gamma\delta} - \frac{1}{2}h_{\alpha\beta}h^{;\,\alpha\beta} + \frac{1}{4}h\,h_{;\,\rho}{}^{\rho}\right), \tag{5.14}$$

which only contributes to the kinetic part of the gravitational fluctuations $\propto h\,\bar{\nabla}\bar{\nabla}h$

$$\delta_g^2 S^{\mathrm{kin}} + S_{\mathrm{gb}} = \frac{1}{2}\int \mathrm{d}^4x\,(g)^{1/2}\left(h_{\alpha\gamma}G^{\alpha\beta\gamma\delta}\bar{\Box}h_{\gamma\delta} + h^{\alpha\gamma}h_{\gamma\delta}{}^{;\delta}{}_{\alpha} - h^{\alpha\gamma}h_{\gamma\delta\,;\,\alpha}{}^{\delta}\right). \tag{5.15}$$

The last two terms give rise to additional curvature terms, which can be calculated by

$$h^{\beta\epsilon}h^{\lambda}{}_{\beta;\,\epsilon\lambda} - h^{\beta\epsilon}h^{\lambda}{}_{\beta;\,\lambda\epsilon} = h_{\alpha\beta}g^{\alpha\mu}g^{\beta\nu}R^{\gamma}{}_{\mu}h_{\gamma\nu} - h_{\alpha\beta}g^{\alpha\mu}g^{\beta\nu}g^{\gamma\delta}R^{\lambda}{}_{\nu\mu\delta}h_{\lambda\gamma}. \tag{5.16}$$

By substituting this result and evaluating at $g = \bar{g},\ \varphi = \bar{\varphi}$, we find

$$\frac{\delta_g^2 S}{2} + S_{\rm gb} = \frac{1}{2}\int \mathrm{d}^4x\,\bar{g}^{1/2}\,h_{\alpha\beta}\Big[\bar{G}^{\alpha\beta\gamma\delta}\bar{\Box} + \bar{G}^{\alpha\beta\gamma\delta}\,\bar{V}(\bar{\varphi}) - \bar{G}^{\alpha\beta\gamma\delta}\,\bar{R} + 2\bar{g}^{\beta\nu}\bar{G}^{\gamma\delta\alpha\mu}\,\bar{R}_{\mu\nu}$$
$$+ 2\bar{g}^{\nu\delta}\bar{G}^{\alpha\beta\mu\gamma}\,\bar{R}_{\mu\nu} - \bar{g}^{\alpha\mu}\bar{g}^{\beta\delta}\bar{g}^{\gamma\nu}\,\bar{R}_{\mu\nu} + \bar{g}^{\alpha\mu}\bar{g}^{\beta\nu}\bar{g}^{\gamma\lambda}\,\bar{R}^{\delta}_{\ \nu\mu\lambda} + \frac{1}{2}\bar{g}^{\alpha\beta}(\bar{\nabla}^{\gamma}\bar{\varphi})(\bar{\nabla}^{\delta}\bar{\varphi})$$
$$- \bar{g}^{\alpha\gamma}(\bar{\nabla}^{\beta}\bar{\varphi})(\bar{\nabla}^{\delta}\bar{\varphi}) - \frac{1}{8}\bar{g}^{\alpha\beta}\bar{g}^{\gamma\delta}(\bar{\nabla}^{\nu}\bar{\varphi})(\bar{\nabla}_{\nu}\bar{\varphi}) + \frac{1}{4}\bar{g}^{\alpha\gamma}\bar{g}^{\beta\delta}(\bar{\nabla}^{\nu}\bar{\varphi})(\bar{\nabla}_{\nu}\bar{\varphi})\Big]\,h_{\gamma\delta}. \tag{5.17}$$

We can further simplify this result by arranging parts involving the curvature and the scalar field in a total symmetric form $G^{\alpha\beta\mu\nu}\,O^{\gamma\delta}_{\mu\nu}$. This can be achieved by making an ansatz for the object $O^{\gamma\delta}_{\mu\nu}$, so that it contains all possible contractions. By inverting and comparing coefficients, we find for the curvature terms

$$\bar{K}^{\gamma\delta}_{\lambda\sigma} := 2\,\bar{R}_{(\lambda\ \ \sigma)}^{\ \ \gamma\ \ \delta} + 2\delta^{(\gamma}_{(\lambda}\,\bar{R}^{\delta)}_{\sigma)} - \delta^{\gamma\delta}_{\lambda\sigma}\,\bar{R} - \bar{g}^{\gamma\delta}\,\bar{R}_{\lambda\sigma} - \bar{R}^{\gamma\delta}\,\bar{g}_{\lambda\sigma} + \frac{1}{2}\bar{g}^{\gamma\delta}\bar{g}_{\lambda\sigma}\,\bar{R}, \tag{5.18}$$

and for the scalar field terms

$$\bar{S}^{\gamma\delta}_{\lambda\sigma} := \frac{1}{2}\,\bar{\varphi}_{,\mu}\bar{\varphi}_{,\nu}\bar{g}^{\mu\nu}\,\delta^{\gamma\delta}_{\lambda\sigma} - 2\,\delta^{(\gamma}_{(\lambda}\bar{\varphi}_{,\sigma)}\bar{\varphi}^{,\delta)} + \frac{1}{2}\,\bar{g}_{\lambda\sigma}\,\bar{\varphi}^{,\gamma}\bar{\varphi}^{,\delta}$$
$$+ \frac{1}{2}\,\bar{g}^{\gamma\delta}\,\bar{\varphi}_{,\lambda}\bar{\varphi}_{,\sigma} - \frac{1}{4}\,\bar{g}^{\gamma\delta}\bar{g}_{\lambda\sigma}\bar{\varphi}^{,\nu}\bar{\varphi}_{,\nu}. \tag{5.19}$$

Finally, we can write the graviton-graviton contribution including the gauge-breaking term in the compact form

$$\frac{\delta_g^2 S}{2} + S_{\rm gb} = \frac{1}{2}\int \mathrm{d}^4x\,\bar{g}^{1/2}\,h_{\alpha\beta}\left[\bar{G}^{\alpha\beta\mu\nu}\left(\delta^{\gamma\delta}_{\mu\nu}\bar{\Box} + \bar{K}^{\gamma\delta}_{\mu\nu} + \bar{S}^{\gamma\delta}_{\mu\nu} + \delta^{\gamma\delta}_{\mu\nu}\,\bar{V}\right)\right]\,h_{\gamma\delta}. \tag{5.20}$$

In order to have a better comparison with [2], we define in addition

$$\bar{P}^{\gamma\delta}_{\mu\nu} := \bar{K}^{\gamma\delta}_{\mu\nu} + \bar{S}^{\gamma\delta}_{\mu\nu} + \delta^{\gamma\delta}_{\mu\nu}\,\bar{V}. \tag{5.21}$$

5.1.1.5 Second Variation: Mixed Parts

The mixed second variation $\delta_g\delta_\varphi S$ is given by

$$\delta_g\delta_\varphi S = \frac{1}{2}\int \mathrm{d}^4x\,\sigma\,\left[-\partial_\nu\left(g^{1/2}\,2\,G^{\alpha\beta\mu\nu}(\partial_\mu\varphi)\,h_{\alpha\beta}\right) - \frac{1}{2}\,g^{1/2}\,g^{\alpha\beta}\,V'(\varphi)\,h_{\alpha\beta}\right]$$
$$= \frac{1}{2}\int \mathrm{d}^4x\,\sigma\,\left[-g^{1/2}\,\nabla_\nu\left(\,2\,G^{\alpha\beta\mu\nu}(\partial_\mu\varphi)\,h_{\alpha\beta}\right) - \frac{1}{2}\,g^{1/2}\,g^{\alpha\beta}\,V'(\varphi)\,h_{\alpha\beta}\right]$$
$$= \frac{1}{2}\int \mathrm{d}^4x\,\bar{g}^{1/2}\,\sigma\,\left[-2\,\bar{G}^{\alpha\beta\mu\nu}\left(\bar{\nabla}_\nu\bar{\nabla}_\mu\bar{\varphi} + (\bar{\nabla}_\mu\bar{\varphi})\,\bar{\nabla}_\nu\right) - \frac{1}{2}\,\bar{g}^{\alpha\beta}\,\bar{V}'(\bar{\varphi})\right]\,h_{\alpha\beta}. \tag{5.22}$$

The remaining mixed variation with opposite order $\delta_\varphi \delta_g S$ yields

$$\begin{aligned}\delta_\varphi \delta_g S &= \frac{1}{2} \int d^4 x \, g^{1/2} \, h_{\alpha\beta} \left[2G^{\alpha\beta\mu\nu} (\partial_\mu \varphi) \, \partial_\nu - \frac{1}{2} g^{\alpha\beta} \, V'(\varphi) \right] \sigma \\ &= \frac{1}{2} \int d^4 x \, \bar{g}^{1/2} \, h_{\alpha\beta} \left[2\bar{G}^{\alpha\beta\mu\nu} (\bar{\nabla}_\mu \bar{\varphi}) \, \bar{\nabla}_\nu - \frac{1}{2} \bar{g}^{\alpha\beta} \, \bar{V}'(\bar{\varphi}) \right] \sigma . \end{aligned} \tag{5.23}$$

5.1.1.6 Fluctuation Operator

We use the definition of F_{AB} in terms of the variations $\delta\Phi^A$

$$\frac{1}{2} \delta^2 S + S_{\text{gb}} =: \frac{1}{2} \int \bar{g}^{1/2} \, \delta\Psi^A \, F_{AB} \, \delta\Psi^B . \tag{5.24}$$

We write the operator in the form as outlined in the general algorithm of Sect. 4.5.1 as

$$F_{AB} = C_{AB} \, \bar{\Box} + 2 \, \Gamma^\sigma_{AB} \bar{\nabla}_\sigma + W_{AB} . \tag{5.25}$$

Thus, we can simply read off the entries of the matrices C_{AB}, Γ_{AB} and W_{AB} from the results of the second variations (5.12), (5.20), (5.22) and (5.23):

$$C_{AB} = \begin{pmatrix} \bar{G}^{\alpha\beta\gamma\delta} & 0 \\ 0 & 1 \end{pmatrix} , \quad \Gamma^\sigma_{AB} = \begin{pmatrix} 0 & \bar{G}^{\alpha\beta\mu\sigma} \, \bar{\varphi}_{,\mu} \\ -\bar{G}^{\alpha\beta\mu\sigma} \, \bar{\varphi}_{,\mu} & 0 \end{pmatrix} , \tag{5.26}$$

and

$$W_{AB} = \begin{pmatrix} \bar{G}^{\alpha\beta\mu\nu} \, \bar{P}^{\gamma\delta}_{\mu\nu} & -\frac{1}{2} \bar{g}^{\alpha\beta} \, \bar{V}' \\ -2\bar{G}^{\gamma\delta\mu\nu} (\bar{\nabla}_\nu \bar{\nabla}_\mu \bar{\varphi}) - \frac{1}{2} \bar{g}^{\gamma\delta} \, \bar{V}' & -\bar{V}'' \end{pmatrix} . \tag{5.27}$$

5.1.2 Calculation of the Minimal Operator

5.1.2.1 Normalization and Inverse

The next step in the algorithm is to normalize the coefficient matrix C_{AB} in front of the d'Alembertian to unity. Thus, we need the inverse of the matrix C_{AB} which is given by

$$\left(C^{-1} \right)^{AB} = \begin{pmatrix} \bar{G}_{\alpha\beta\gamma\delta} & 0 \\ 0 & 1 \end{pmatrix} . \tag{5.28}$$

We assure ourselves that this is indeed the correct inverse by checking the relation

$$\hat{I} := \delta_A^B = (C^{-1})^{BC} C_{CA} = \begin{pmatrix} \delta_{\mu\nu}^{\alpha\beta} & 0 \\ 0 & 1 \end{pmatrix}. \tag{5.29}$$

The remaining pieces $\Gamma_A{}^{B\,\mu}$ and W_{AB} of F_{AB} must also be multiplied by the inverse. We follow the notation of Sect. 4.5.1 and suppress the indices $A, B, \ldots$ and write a hat over the corresponding quantity. Multiplying (5.26) and (5.27) with (5.28) yields

$$\hat{\Gamma}^{\epsilon} := \Gamma_B^{A\,\epsilon} = (C^{-1})^{AC}\,\Gamma_{CB}^{\,\epsilon} = \begin{pmatrix} 0 & \delta_{\alpha\beta}^{\mu\epsilon}\,\bar{\varphi}_{,\mu} \\ -\bar{G}^{\gamma\delta\mu\epsilon}\,\bar{\varphi}_{,\mu} & 0 \end{pmatrix} \tag{5.30}$$

and

$$\hat{W} := W_B^A = (C^{-1})^{AC}\,W_{CB} = \begin{pmatrix} \bar{P}_{\alpha\beta}^{\gamma\delta} & \bar{g}_{\alpha\beta}\,\bar{V}' \\ -2\bar{G}^{\gamma\delta\mu\nu}(\bar{\nabla}_\nu\bar{\nabla}_\mu\bar{\varphi}) - \frac{1}{2}\bar{g}^{\gamma\delta}\,\bar{V}' & -\bar{V}'' \end{pmatrix}. \tag{5.31}$$

5.1.2.2 Re-definition of the Covariant Derivative

As explained in Sect. 4.5.1, we can absorb the term linear in the derivative by defining

$$\hat{\mathcal{D}}_\mu := \hat{\bar{\bar{\nabla}}}_\mu + \bar{g}_{\mu\nu}\,\hat{\Gamma}^\nu \tag{5.32}$$

and then write the operator in the compact form

$$\hat{F} = \hat{\bar{\Box}} + \hat{P} - \frac{1}{6}\,\bar{R}\,\hat{I}. \tag{5.33}$$

In this form the operator $\hat{F}$ is suited for the application of the general algorithm of Sect. 4.5and we only have to calculate the expressions which enter the second DeWitt coefficient $\hat{a}_2$ defined in (4.113). Re-expressed in terms of the original derivative $\bar{\nabla}_\mu$, the quantities $\hat{P}$ and $\hat{\mathcal{R}}_{\mu\nu}$ are given by (4.88), (4.89) with the background curvature (4.90). We can now compose these expressions explicitly from the expressions found for $\hat{W}$ and $\hat{\Gamma}^\nu$. Using (D.39), we find for the action of $\hat{\mathcal{R}}^0_{\mu\nu}$ on the covariant field $h_{\gamma\delta}$

$$\hat{\mathcal{R}}^0_{\mu\nu} = -2\delta_{(\alpha}^{(\gamma}\,\bar{R}^{\delta)}{}_{\beta)\mu\nu}. \tag{5.34}$$

We note that it makes a difference in sign whether the commutator acts on a covariant or a contravariant index. For the terms appearing in $\hat{P}$, we find the explicit expressions

$$\bar{\nabla}_\epsilon \hat{\Gamma}^\epsilon = \begin{pmatrix} 0 & \bar{\varphi}_{;\,\alpha\beta} \\ -\bar{G}^{\mu\epsilon\gamma\delta}\,\bar{\varphi}_{;\,\mu\epsilon} & 0 \end{pmatrix}, \tag{5.35}$$

$$\bar{g}_{\rho\epsilon}\hat{\Gamma}^\epsilon\hat{\Gamma}^\rho = \begin{pmatrix} -\bar{g}_{\epsilon\rho}\,\delta^{\mu\epsilon}_{\alpha\beta}\,\bar{G}^{\gamma\delta\nu\rho}\,\bar{\varphi}_{,\,\mu}\bar{\varphi}_{,\nu} & 0 \\ 0 & -(\bar{\varphi}_{,\,\nu}\bar{\varphi}^{,\nu}) \end{pmatrix}. \tag{5.36}$$

Thus, the potential can be written as the (2×2) matrix

$$\begin{aligned}\hat{P} = P^B_A &= W^B_A + \frac{1}{6}\,\bar{R}\,\delta^B_A - (\bar{\nabla}_\mu \Gamma_A^{\;B\,\mu}) - \bar{g}_{\rho\epsilon}\Gamma_A^{\;C\,\epsilon}\Gamma_C^{\;B\,\rho} \\ &= \begin{pmatrix} \bar{P}^{\gamma\delta}_{\alpha\beta} + \frac{1}{6}\,\bar{R}\,\delta^{\gamma\delta}_{\alpha\beta} + \bar{g}_{\epsilon\rho}\,\delta^{\mu\epsilon}_{\alpha\beta}\,\bar{G}^{\gamma\delta\nu\rho}\,\bar{\varphi}_{,\,\mu}\bar{\varphi}_{,\nu} & \bar{g}_{\alpha\beta}\,\bar{V}' - (\bar{\nabla}_\alpha\bar{\nabla}_\beta\bar{\varphi}) \\ -\bar{G}^{\gamma\delta\mu\nu}(\bar{\nabla}_\nu\bar{\nabla}_\mu\bar{\varphi}) - \frac{1}{2}\bar{g}^{\gamma\delta}\,\bar{V}' & -\bar{V}'' + \frac{1}{6}\bar{R} + (\bar{\varphi}_{,\,\nu}\bar{\varphi}^{,\nu}) \end{pmatrix}.\end{aligned} \tag{5.37}$$

For the terms appearing in $\hat{\mathcal{R}}_{\mu\nu}$, we find the explicit expressions

$$2\bar{\nabla}_{[\mu}\bar{g}_{\nu]\epsilon}\hat{\Gamma}^\epsilon = \begin{pmatrix} 0 & 2\bar{\nabla}_{[\mu}\bar{g}_{\nu]\epsilon}\delta^{\lambda\epsilon}_{\alpha\beta}\bar{\varphi}_{;\,\lambda} \\ -2\bar{\nabla}_{[\mu}\bar{g}_{\nu]\epsilon}\,\bar{G}^{\gamma\delta\lambda\epsilon}\bar{\varphi}_{;\,\lambda} & 0 \end{pmatrix}, \tag{5.38}$$

$$2\bar{g}_{\epsilon[\mu}\bar{g}_{\nu]\rho}\hat{\Gamma}^\epsilon\hat{\Gamma}^\rho = \begin{pmatrix} -2\bar{g}_{\epsilon[\mu}\bar{g}_{\nu]\rho}\delta^{\tau\epsilon}_{\alpha\beta}\bar{G}^{\gamma\delta\eta\rho}\,\bar{\varphi}_{,\,\tau}\,\bar{\varphi}_{,\,\eta} & 0 \\ 0 & 0 \end{pmatrix}, \tag{5.39}$$

so that the field curvature can also be written as (2×2) matrix

$$\begin{aligned}\hat{\mathcal{R}}_{\mu\nu} = \mathcal{R}_A^{\;B}{}_{\mu\nu} &= \mathcal{R}^{0}{}_A^{\;B}{}_{\mu\nu} + 2\,\bar{\nabla}_{[\mu}\bar{g}_{\nu]\epsilon}\,\Gamma_A^{B\;\epsilon} + 2\,\bar{g}_{\epsilon[\mu}\bar{g}_{\nu]\rho}\,\Gamma_A^{C\;\epsilon}\,\Gamma_C^{B\;\rho} \\ &= \begin{pmatrix} -2\delta^{(\gamma}_{(\alpha}\,R^{\delta)}_{\beta)\mu\nu} - 2\bar{g}_{\epsilon[\mu}\bar{g}_{\nu]\rho}\delta^{\tau\epsilon}_{\alpha\beta}\bar{G}^{\gamma\delta\eta\rho}\,\bar{\varphi}_{,\,\tau}\,\bar{\varphi}_{,\,\eta} & 2\bar{\nabla}_{[\mu}\bar{g}_{\nu]\epsilon}\delta^{\lambda\epsilon}_{\alpha\beta}\bar{\varphi}_{;\,\lambda} \\ -2\bar{\nabla}_{[\mu}\bar{g}_{\nu]\epsilon}\,\bar{G}^{\gamma\delta\lambda\epsilon}\bar{\varphi}_{;\,\lambda} & 0 \end{pmatrix}.\end{aligned} \tag{5.40}$$

5.1.3 Results

5.1.3.1 DeWitt-Coefficient of the Main Operator

For the $\hat{a}_2$ coefficient we need the trace of the squared operators $\hat{P}^2$ and $\hat{\mathcal{R}}^2$:

$$
\begin{aligned}
\mathrm{tr}\,(\hat{P}\,\hat{P}) = & \left(\bar{P}^{\gamma\delta}_{\alpha\beta} + \frac{1}{6}\,\bar{R}\,\delta^{\gamma\delta}_{\alpha\beta} + \bar{g}_{\epsilon\rho}\,\delta^{\mu\epsilon}_{\alpha\beta}\,\bar{G}^{\gamma\delta\nu\rho}\,\bar{\varphi}_{,\mu}\bar{\varphi}_{,\nu}\right) \\
& \times \left(\bar{P}^{\alpha\beta}_{\gamma\delta} + \frac{1}{6}\,\bar{R}\,\delta^{\alpha\beta}_{\gamma\delta} + \bar{g}_{\lambda\sigma}\,\delta^{\tau\lambda}_{\gamma\delta}\,\bar{G}^{\alpha\beta\eta\sigma}\,\bar{\varphi}_{,\tau}\bar{\varphi}_{,\eta}\right) \\
& + 2\left[\bar{g}_{\alpha\beta}\,\bar{V}' - (\bar{\nabla}_\alpha\bar{\nabla}_\beta\bar{\varphi})\right]\left[-\bar{G}^{\alpha\beta\mu\nu}(\bar{\nabla}_\nu\bar{\nabla}_\mu\bar{\varphi}) - \frac{1}{2}\bar{g}^{\alpha\beta}\,\bar{V}'\right] \\
& + \left[-\bar{V}'' + \frac{1}{6}\bar{R} + (\bar{\varphi}_{,\nu}\bar{\varphi}^{,\nu})\right]\left[-\bar{V}'' + \frac{1}{6}\bar{R} + (\bar{\varphi}_{,\nu}\bar{\varphi}^{,\nu})\right] \\
= & \,\bar{P}^{\gamma\delta}_{\alpha\beta}\,\bar{P}^{\alpha\beta}_{\gamma\delta} + \frac{1}{36}\delta^{\gamma\delta}_{\alpha\beta}\delta^{\alpha\beta}_{\gamma\delta}\,\bar{R}^2 + \bar{g}_{\epsilon\rho}\delta^{\mu\epsilon}_{\alpha\beta}\bar{G}^{\gamma\delta\nu\rho}\bar{g}_{\lambda\sigma}\delta^{\tau\lambda}_{\gamma\delta}\bar{G}^{\alpha\beta\eta\sigma}\,\bar{\varphi}_{,\tau}\bar{\varphi}_{,\eta}\bar{\varphi}_{,\mu}\bar{\varphi}_{,\nu} \\
& + \frac{1}{3}\delta^{\gamma\delta}_{\alpha\beta}\,\bar{P}^{\alpha\beta}_{\gamma\delta}\,\bar{R} + 2\bar{P}^{\gamma\delta}_{\alpha\beta}\,\bar{g}_{\lambda\sigma}\delta^{\tau\lambda}_{\gamma\delta}\bar{G}^{\alpha\beta\eta\sigma}\,\bar{\varphi}_{,\tau}\bar{\varphi}_{,\eta} + \frac{1}{3}\bar{R}\delta^{\gamma\delta}_{\alpha\beta}\bar{g}_{\lambda\sigma}\delta^{\tau\lambda}_{\gamma\delta}\bar{G}^{\alpha\beta\eta\sigma}\,\bar{\varphi}_{,\tau}\bar{\varphi}_{,\eta} \\
& - 2\,\bar{G}^{\alpha\beta\mu\nu}\bar{g}_{\alpha\beta}\bar{\varphi}_{;\mu\nu}\,\bar{V}' - \bar{g}^{\gamma\delta}\bar{g}_{\gamma\delta}(\bar{V}')^2 + 2\,\bar{G}^{\mu\nu\alpha\beta}\bar{\varphi}_{;\mu\nu}\bar{\varphi}_{;\alpha\beta} + \bar{g}^{\alpha\beta}\bar{\varphi}_{;\alpha\beta}\,\bar{V}' \\
& + (\bar{V}'')^2 + \frac{1}{36}\bar{R}^2 + (\bar{\varphi}_{,\nu}\bar{\varphi}^{,\nu})^2 - \frac{1}{3}\bar{V}''\,\bar{R} - 2\,\bar{V}''\,(\bar{\varphi}_{,\nu}\bar{\varphi}^{,\nu}) + \frac{1}{3}\bar{R}\,(\bar{\varphi}_{,\nu}\bar{\varphi}^{,\nu}).
\end{aligned}
\tag{5.41}
$$

For the further evaluation, we decompose $\bar{P}^{\alpha\beta}_{\gamma\delta} = \bar{K}^{\alpha\beta}_{\gamma\delta} + \bar{S}^{\alpha\beta}_{\gamma\delta} + \delta^{\alpha\beta}_{\gamma\delta}\,\bar{V}$ and use $\bar{R}^{\alpha}{}_{\alpha\gamma\delta} = 0$, $\delta^{\alpha\beta}_{\gamma\delta}\delta^{\gamma\delta}_{\alpha\beta} = \delta^{\gamma\delta}_{\gamma\delta} = 10$, $\bar{G}^{\alpha\beta\gamma\delta}\bar{g}_{\gamma\delta} = -\frac{1}{2}\bar{g}^{\alpha\beta}$, $\bar{G}_{\alpha\beta\gamma\delta}\bar{g}^{\gamma\delta} = -2\bar{g}_{\alpha\beta}$, $\delta^{\gamma\delta}_{\lambda\sigma}\delta^{\lambda}_{\gamma} = \frac{5}{2}\delta^{\delta}_{\sigma}$ and

$$
\bar{g}_{\epsilon\rho}\delta^{\mu\epsilon}_{\alpha\beta}\bar{G}^{\gamma\delta\nu\rho}\bar{g}_{\lambda\sigma}\delta^{\tau\lambda}_{\gamma\delta}\bar{G}^{\alpha\beta\eta\sigma}\,\bar{\varphi}_{,\tau}\bar{\varphi}_{,\eta}\bar{\varphi}_{,\mu}\bar{\varphi}_{,\nu} = \frac{1}{4}(\bar{\varphi}_{,\nu}\bar{\varphi}^{,\nu})^2. \tag{5.42}
$$

Inserting this in (5.41), we find

$$
\begin{aligned}
\mathrm{tr}\,(\hat{P}\,\hat{P}) = & \,\bar{K}^{\alpha\beta}_{\gamma\delta}\bar{K}^{\gamma\delta}_{\alpha\beta} + \bar{S}^{\alpha\beta}_{\gamma\delta}\bar{S}^{\gamma\delta}_{\alpha\beta} + 10\,\bar{V}^2 + 2\bar{K}^{\alpha\beta}_{\gamma\delta}\bar{S}^{\gamma\delta}_{\alpha\beta} + 2\bar{K}^{\alpha\beta}_{\alpha\beta}\,\bar{V} + 2\bar{S}^{\alpha\beta}_{\alpha\beta}\,\bar{V} \\
& + \frac{10}{36}\,\bar{R}^2 + \frac{1}{4}\,(\bar{\varphi}_{,\nu}\bar{\varphi}^{,\nu})^2 + \frac{1}{3}\bar{K}^{\alpha\beta}_{\alpha\beta}\,\bar{R} + \frac{1}{3}\bar{S}^{\alpha\beta}_{\alpha\beta}\,\bar{R} + \frac{10}{3}\,\bar{R}\,\bar{V} \\
& + 2\bar{K}^{\gamma\delta}_{\alpha\beta}\bar{g}_{\lambda\sigma}\delta^{\tau\lambda}_{\gamma\delta}\bar{G}^{\alpha\beta\eta\sigma}\,\bar{\varphi}_{,\tau}\bar{\varphi}_{,\eta} + 2\bar{S}^{\gamma\delta}_{\alpha\beta}\bar{g}_{\lambda\sigma}\delta^{\tau\lambda}_{\gamma\delta}\bar{G}^{\alpha\beta\eta\sigma}\,\bar{\varphi}_{,\tau}\bar{\varphi}_{,\eta} \\
& + 2\bar{V}\,\delta^{\gamma\delta}_{\alpha\beta}\bar{g}_{\lambda\sigma}\delta^{\tau\lambda}_{\gamma\delta}\bar{G}^{\alpha\beta\eta\sigma}\,\bar{\varphi}_{,\tau}\bar{\varphi}_{,\eta} + \frac{1}{3}\,\bar{R}\,\delta^{\tau\lambda}_{\alpha\beta}\bar{g}_{\lambda\sigma}\,\bar{G}^{\alpha\beta\eta\sigma}\,\bar{\varphi}_{,\tau}\bar{\varphi}_{,\eta} \\
& - 2\,\bar{G}^{\alpha\beta\mu\nu}\bar{g}_{\alpha\beta}\bar{\varphi}_{;\mu\nu}\,\bar{V}' - 4(\bar{V}')^2 + 2\,\bar{G}^{\mu\nu\alpha\beta}\bar{\varphi}_{;\mu\nu}\bar{\varphi}_{;\alpha\beta} + (\bar{\Box}\bar{\varphi})\,\bar{V}' \\
& + (\bar{V}'')^2 + \frac{1}{36}\bar{R}^2 + (\bar{\varphi}_{,\nu}\bar{\varphi}^{,\nu})^2 - \frac{1}{3}\bar{V}''\,\bar{R} - 2\,\bar{V}''\,(\bar{\varphi}_{,\nu}\bar{\varphi}^{,\nu}) + \frac{1}{3}\bar{R}\,(\bar{\varphi}_{,\nu}\bar{\varphi}^{,\nu}).
\end{aligned}
\tag{5.43}
$$

We have made use of the symmetries of the Riemann tensor (A.4) and (A.5) in order to calculate terms like

$$
\bar{R}_{\alpha\beta\gamma\delta}\,\bar{R}^{\alpha\beta\gamma\delta} = \frac{1}{2}\,\bar{R}_{\alpha\beta\gamma\delta}\,\bar{R}^{\beta\gamma\delta\alpha}. \tag{5.44}
$$

Using these expressions, it is straightforward to calculate the quantities:

$$
\begin{aligned}
&\bar{K}^{\gamma\delta}_{\gamma\delta} = -6\,\bar{R}, \quad \bar{S}^{\gamma\delta}_{\gamma\delta} = 0, \quad \bar{S}^{\alpha\beta}_{\gamma\delta}\bar{S}^{\gamma\delta}_{\alpha\beta} = 3(\bar{\varphi}^{,\nu}\bar{\varphi}_{,\nu})^2,\\
&\bar{K}^{\alpha\beta}_{\gamma\delta}\bar{K}^{\gamma\delta}_{\alpha\beta} = 3\bar{R}_{\alpha\beta\gamma\delta}\bar{R}^{\alpha\beta\gamma\delta} - 6\bar{R}_{\alpha\beta}\bar{R}^{\alpha\beta} + 5\bar{R}^2,\\
&\bar{K}^{\alpha\beta}_{\gamma\delta}\bar{S}^{\gamma\delta}_{\alpha\beta} = 0, \quad 2\bar{K}^{\tau\lambda}_{\alpha\beta}\bar{g}_{\lambda\sigma}\bar{G}^{\alpha\beta\eta\sigma}\bar{\varphi}_{,\tau}\bar{\varphi}_{,\eta} = -\frac{3}{2}\bar{R}\,(\bar{\varphi}^{,\nu}\bar{\varphi}_{,\nu}),\\
&2\bar{S}^{\tau\lambda}_{\alpha\beta}\bar{g}_{\lambda\sigma}\bar{G}^{\alpha\beta\eta\sigma}\bar{\varphi}_{,\tau}\bar{\varphi}_{,\eta} = -\frac{3}{2}(\bar{\varphi}_{,\nu}\bar{\varphi}^{,\nu})^2, \quad \delta^{\tau\lambda}_{\alpha\beta}\bar{g}_{\lambda\sigma}\bar{G}^{\alpha\beta\eta\sigma}\bar{\varphi}_{,\tau}\bar{\varphi}_{,\eta} = (\bar{\varphi}_{,\nu}\bar{\varphi}^{,\nu}).
\end{aligned}
\tag{5.45}
$$

By substituting the expressions (5.45) in (5.43), we find

$$
\begin{aligned}
\mathrm{tr}(\hat{P}\hat{P}) = {}& 3\bar{R}_{\alpha\beta\gamma\delta}\bar{R}^{\alpha\beta\gamma\delta} - 6\,\bar{R}_{\mu\nu}\bar{R}^{\mu\nu} + \frac{119}{36}\,\bar{R}^2 - \frac{5}{6}\bar{R}(\bar{\varphi}_{,\nu}\bar{\varphi}^{,\nu}) + \frac{11}{4}\,(\bar{\varphi}_{,\nu}\bar{\varphi}^{,\nu})^2\\
&+ 10\,\bar{V}^2 - \frac{26}{3}\,\bar{R}\,\bar{V} - \frac{1}{3}\,\bar{R}\,\bar{V}'' - 4\,\bar{V}'^2 + \bar{V}''^2\\
&+ 2\bar{V}'\,\bar{\Box}\bar{\varphi} + 2\,\bar{V}\,(\bar{\varphi}_{,\nu}\bar{\varphi}^{,\nu}) - 2\bar{V}''\,(\bar{\varphi}_{,\nu}\bar{\varphi}^{,\nu}) + \bar{\varphi}_{;\,\mu\nu}\,\bar{\varphi}^{;\mu\nu} - \frac{1}{2}\,(\bar{\Box}\bar{\varphi})^2.
\end{aligned}
\tag{5.46}
$$

For the field curvature, we obtain in an analogous way

$$
\begin{aligned}
\mathrm{tr}(\hat{\mathcal{R}}_{\mu\nu}\hat{\mathcal{R}}^{\mu\nu}) = {}& \left(-2\delta^{(\gamma}_{(\alpha}R^{\delta)}_{\beta)\mu\nu} - 2\bar{g}_{\epsilon[\mu}\bar{g}_{\nu]\rho}\delta^{\tau\epsilon}_{\alpha\beta}\bar{G}^{\gamma\delta\eta\rho}\,\bar{\varphi}_{,\tau}\,\bar{\varphi}_{,\eta}\right)\\
&\times\left(-2\delta^{(\alpha}_{(\gamma}R^{\beta)}_{\delta)}{}^{\mu\nu} - 2\delta^{[\mu}_{\pi}\delta^{\nu]}_{\iota}\delta^{\sigma\pi}_{\gamma\delta}\bar{G}^{\alpha\beta\lambda\iota}\,\bar{\varphi}_{,\sigma}\,\bar{\varphi}_{,\lambda}\right)\\
&- 8\left(\bar{\nabla}_{[\mu}\bar{g}_{\nu]\epsilon}\,\bar{G}^{\gamma\delta\lambda\epsilon}\bar{\varphi}_{;\lambda}\right)\left(\bar{\nabla}^{[\mu}\delta^{\nu]}_{\eta}\delta^{\sigma\eta}_{\gamma\delta}\bar{\varphi}_{;\sigma}\right)\\
= {}& 4\,\delta^{(\gamma}_{(\alpha}R^{\delta)}_{\beta)\mu\nu}\delta^{(\alpha}_{(\gamma}R^{\beta)}_{\delta)}{}^{\mu\nu} + 8\,\delta^{(\gamma}_{(\alpha}R^{\delta)}_{\beta)\mu\nu}\,\delta^{[\mu}_{\pi}\delta^{\nu]}_{\iota}\delta^{\sigma\pi}_{\gamma\delta}\,\bar{G}^{\alpha\beta\lambda\iota}\,\bar{\varphi}_{,\sigma}\,\bar{\varphi}_{,\lambda}\\
&+ 4\bar{g}_{\epsilon[\mu}\bar{g}_{\nu]\rho}\delta^{\tau\epsilon}_{\alpha\beta}\bar{G}^{\gamma\delta\eta\rho}\,\delta^{\mu}_{\pi}\delta^{\nu}_{\iota}\delta^{\sigma\pi}_{\gamma\delta}\,\bar{G}^{\alpha\beta\lambda\iota}\,\bar{\varphi}_{,\tau}\,\bar{\varphi}_{,\eta}\,\bar{\varphi}_{,\sigma}\,\bar{\varphi}_{,\lambda}\\
&- 8\bar{G}^{\gamma\delta\lambda\epsilon}\delta^{\sigma\eta}_{\gamma\delta}(\bar{\nabla}_{[\mu}\bar{g}_{\nu]\epsilon}\,\bar{\varphi}_{;\lambda})(\bar{\nabla}^{\mu}\delta^{\nu}_{\eta}\bar{\varphi}_{;\sigma}).
\end{aligned}
\tag{5.47}
$$

Evaluating the individual pieces is again straightforward and yields

$$
\begin{aligned}
&4\,\delta^{(\gamma}_{(\alpha}\bar{R}^{\delta)}_{\beta)\mu\nu}\delta^{(\alpha}_{(\gamma}\bar{R}^{\beta)}_{\delta)}{}^{\mu\nu} = -6\bar{R}_{\alpha\beta\gamma\delta}R^{\alpha\beta\gamma\delta},\\
&8\,\delta^{(\gamma}_{(\alpha}\bar{R}^{\delta)}_{\beta)\mu\nu}\,\delta^{[\mu}_{\pi}\delta^{\nu]}_{\iota}\delta^{\sigma\pi}_{\gamma\delta}\,\bar{G}^{\alpha\beta\lambda\iota}\,\bar{\varphi}_{,\sigma}\,\bar{\varphi}_{,\lambda} = \bar{R}(\bar{\varphi}_{,\nu}\bar{\varphi}^{,\nu}) + 2\bar{R}_{\mu\nu}(\bar{\varphi}^{,\mu}\bar{\varphi}^{,\nu}),\\
&4\bar{g}_{\epsilon[\mu}\bar{g}_{\nu]\rho}\delta^{\tau\epsilon}_{\alpha\beta}\bar{G}^{\gamma\delta\eta\rho}\,\delta^{\mu}_{\pi}\delta^{\nu}_{\iota}\delta^{\sigma\pi}_{\gamma\delta}\,\bar{G}^{\alpha\beta\lambda\iota}\,\bar{\varphi}_{,\tau}\,\bar{\varphi}_{,\eta}\,\bar{\varphi}_{,\sigma}\,\bar{\varphi}_{,\lambda} = -\frac{3}{2}(\bar{\varphi}_{,\nu}\bar{\varphi}^{,\nu})^2,\\
&-8\bar{G}^{\gamma\delta\lambda\epsilon}\delta^{\sigma\eta}_{\gamma\delta}(\bar{\nabla}_{[\mu}\bar{g}_{\nu]\epsilon}\,\bar{\varphi}_{;\lambda})(\bar{\nabla}^{\mu}\delta^{\nu}_{\eta}\bar{\varphi}_{;\sigma}) = -4\bar{\varphi}_{;\,\mu\nu}\bar{\varphi}^{;\,\mu\nu} + (\bar{\Box}\varphi)^2.
\end{aligned}
\tag{5.48}
$$

The result for the trace of the squared field curvature is therefore simply

$$\mathrm{tr}(\hat{\mathcal{R}}_{\mu\nu}\hat{\mathcal{R}}^{\mu\nu}) = -6\bar{R}_{\alpha\beta\gamma\delta}\bar{R}^{\alpha\beta\gamma\delta} + \bar{R}(\bar{\varphi}_{,\nu}\bar{\varphi}^{,\nu}) + 2\bar{R}_{\mu\nu}(\bar{\varphi}^{,\mu}\bar{\varphi}^{,\nu}) - \frac{3}{2}(\bar{\varphi}_{,\nu}\bar{\varphi}^{,\nu})^2 - 4\bar{\varphi}_{;\mu\nu}\bar{\varphi}^{;\mu\nu} + (\bar{\Box}\bar{\varphi})^2. \quad (5.49)$$

In the expression for the effective action we can use integration by parts and generate curvature terms by commuting covariant derivatives to transform structures like

$$\begin{aligned}\int d^4x\, \bar{g}^{1/2}\, (\bar{\nabla}_\mu\bar{\nabla}_\nu\bar{\varphi})(\bar{\nabla}^\mu\bar{\nabla}^\nu\bar{\varphi}) &= -\int d^4x\, \bar{g}^{1/2}\, (\bar{\nabla}^\mu\bar{\nabla}_\mu\bar{\nabla}_\nu\bar{\varphi})(\bar{\nabla}^\nu\bar{\varphi}) \\ &= -\int d^4x\, \bar{g}^{1/2}\, (\bar{\nabla}_\nu\bar{\Box}\bar{\varphi} + \bar{R}_{\mu\nu}\bar{\nabla}^\mu\bar{\varphi})(\bar{\nabla}^\nu\bar{\varphi}) \\ &= \int d^4x\, \bar{g}^{1/2}\left[(\bar{\Box}\bar{\varphi})^2 - \bar{R}_{\mu\nu}(\bar{\nabla}^\mu\bar{\varphi})(\bar{\nabla}^\nu\bar{\varphi})\right]\end{aligned} \quad (5.50)$$

into the structures $(\bar{\Box}\bar{\varphi})^2$ and $\bar{R}_{\mu\nu}(\bar{\nabla}^\mu\bar{\varphi})(\bar{\nabla}^\nu\bar{\varphi})$. Similarly, we transform terms like

$$\int d^4x\, \bar{g}^{1/2}\, 2\bar{V}'\bar{\Box}\bar{\varphi} = -2\bar{V}''(\bar{\varphi}_{,\nu}\bar{\varphi}^{,\nu}). \quad (5.51)$$

The one-loop contributions of the main operator (4.114) are given by

$$\begin{aligned}W^{\mathrm{div,main}}_{1-\mathrm{loop}} &= \frac{1}{32\pi^2(\omega-2)}\int d^4x\, \bar{g}^{1/2}\, \mathrm{tr}\,[\hat{a}_2] \\ &= \frac{1}{32\pi^2(\omega-2)}\int d^4x\, \bar{g}^{1/2}\left[\frac{191}{180}\bar{R}_{\alpha\beta\gamma\delta}\bar{R}^{\alpha\beta\gamma\delta} - \frac{551}{180}\bar{R}_{\mu\nu}\bar{R}^{\mu\nu} + \frac{119}{72}\bar{R}^2\right. \\ &\quad - \frac{1}{3}\bar{R}(\bar{\varphi}_{,\nu}\bar{\varphi}^{,\nu}) + \frac{5}{4}(\bar{\varphi}_{,\nu}\bar{\varphi}^{,\nu})^2 + \frac{1}{6}\bar{R}_{\mu\nu}\bar{\varphi}^{,\mu}\bar{\varphi}^{,\nu} + \frac{1}{6}\bar{\varphi}_{;\mu\nu}\bar{\varphi}^{;\mu\nu} - \frac{1}{6}(\bar{\Box}\bar{\varphi})^2 \\ &\quad + 5\bar{V}^2 - 2\bar{V}'^2 + \frac{1}{2}\bar{V}''^2 - \frac{13}{3}\bar{R}\,\bar{V} - \frac{1}{6}\bar{R}\,\bar{V}'' + \bar{V}'(\bar{\Box}\bar{\varphi}) + \bar{V}\,(\bar{\varphi}_{,\nu}\bar{\varphi}^{,\nu}) \\ &\quad \left. - \bar{V}''\,(\bar{\varphi}_{,\nu}\bar{\varphi}^{,\nu})\right]. \end{aligned} \quad (5.52)$$

Performing the partial integration of the terms mentioned above, (5.52) takes the form

$$\begin{aligned}W^{\mathrm{div,main}}_{1-\mathrm{loop}} &= \frac{1}{32\pi^2(\omega-2)}\int d^4x\, \bar{g}^{1/2}\left[\frac{191}{180}\bar{R}_{\alpha\beta\gamma\delta}\bar{R}^{\alpha\beta\gamma\delta} - \frac{551}{180}\bar{R}_{\mu\nu}\bar{R}^{\mu\nu}\right. \\ &\quad + \frac{119}{72}\bar{R}^2 - \frac{1}{3}\bar{R}(\bar{\varphi}_{,\nu}\bar{\varphi}^{,\nu}) + \frac{5}{4}(\bar{\varphi}_{,\nu}\bar{\varphi}^{,\nu})^2 + 5\bar{V}^2 - 2\bar{V}'^2 + \frac{1}{2}\bar{V}''^2 \\ &\quad \left. - \frac{13}{3}\bar{R}\,\bar{V} - \frac{1}{6}\bar{R}\,\bar{V}'' + \bar{V}\,(\bar{\varphi}_{,\nu}\bar{\varphi}^{,\nu}) - 2\bar{V}''\,(\bar{\varphi}_{,\nu}\bar{\varphi}^{,\nu})\right]. \end{aligned} \quad (5.53)$$

5.1.3.2 Ghost Operator

The ghost operator acting on ghost vector fields η^α can easily be derived by the technique explained in Sect. 4.4.1 and it is given by

$$Q_\alpha^\beta = \bar{\Box}\delta_\alpha^\beta + \bar{R}_\alpha^\beta. \tag{5.54}$$

By expressing this operator again in the form suitable for the application of the general algorithm of [6], we find the potential and curvature operators

$$\hat{P} = P_\alpha^\beta = \bar{R}_\alpha^\beta + \frac{1}{6}\bar{R}\delta_\alpha^\beta \quad \text{and} \quad \hat{\mathcal{R}}_{\mu\nu} = \mathcal{R}_\alpha{}^\beta{}_{\mu\nu} = -\bar{R}^\beta{}_{\alpha\mu\nu}. \tag{5.55}$$

The trace of the squared potential is given by

$$\mathrm{tr}(\hat{P}^2) = \bar{R}_{\alpha\beta}\bar{R}^{\alpha\beta} + \frac{4}{9}\bar{R}^2 \tag{5.56}$$

and the trace of the squared curvature term yieldsa

$$\mathrm{tr}(\hat{\mathcal{R}}_{\mu\nu}\hat{\mathcal{R}}^{\mu\nu}) = \mathcal{R}_\alpha{}^\beta{}_{\mu\nu}\mathcal{R}_\beta{}^{\alpha\,\mu\nu} = -\bar{R}_{\alpha\beta\gamma\delta}\bar{R}^{\alpha\beta\gamma\delta}. \tag{5.57}$$

Please note the minus sign in front of (5.57) due to the different index positions when contracting the Riemann tensors. The divergent part of the ghost operator yields

$$\begin{aligned} W_{1-\text{loop}}^{\text{div,ghost}} &= -\frac{1}{16\pi^2(\omega-2)}\int \mathrm{d}^4x\,\bar{g}^{1/2}\,\mathrm{tr}\,[a_{2\alpha}^{\ \beta}] \\ &= \frac{1}{32\pi^2(\omega-2)}\int \mathrm{d}^4x\,\bar{g}^{1/2}\left(\frac{22}{180}\bar{R}_{\alpha\beta\gamma\delta}\bar{R}^{\alpha\beta\gamma\delta} - \frac{86}{90}\bar{R}_{\alpha\beta}\bar{R}^{\alpha\beta} - \frac{4}{9}\bar{R}^2\right). \end{aligned} \tag{5.58}$$

Recalling the definition (4.80), we note again that the ghost contributions enter with the opposite sign and a factor 2 compared to (5.53).

5.1.3.3 One-Loop Effective Action: Final Result

The contributions of the divergent part of the total operator can be summarized as

$$\begin{aligned} W_{1-\text{loop}}^{\text{div,tot}} &= W_{1-\text{loop}}^{\text{div,main}} + W_{1-\text{loop}}^{\text{div,ghost}} \\ &= \frac{1}{32\pi^2(\omega-2)}\int \mathrm{d}^4x\,\bar{g}^{1/2}\left[\frac{213}{180}\bar{R}_{\alpha\beta\gamma\delta}\bar{R}^{\alpha\beta\gamma\delta} - \frac{241}{60}\bar{R}_{\mu\nu}\bar{R}^{\mu\nu}\right. \end{aligned}$$

$$+\frac{29}{24}\bar{R}^2 - \frac{1}{3}\bar{R}(\bar{\varphi}_{,\nu}\bar{\varphi}^{,\nu}) + \frac{5}{4}(\bar{\varphi}_{,\nu}\bar{\varphi}^{,\nu})^2 + 5\bar{V}^2 - 2\bar{V}'^2$$
$$+\frac{1}{2}\bar{V}''^2 - \frac{13}{3}\bar{R}\,\bar{V} - \frac{1}{6}\bar{R}\,\bar{V}'' + \bar{V}\,(\bar{\varphi}_{,\nu}\bar{\varphi}^{,\nu}) - 2\bar{V}''\,(\bar{\varphi}_{,\nu}\bar{\varphi}^{,\nu})\Bigg]. \quad (5.59)$$

In addition, we eliminate the Riemannian part by using the topological invariant

$$\int d^4x\,\bar{g}^{1/2}\left(\bar{R}_{\alpha\beta\gamma\delta}\bar{R}^{\alpha\beta\gamma\delta} - 4\,\bar{R}_{\mu\nu}\bar{R}^{\mu\nu} + \bar{R}^2\right). \quad (5.60)$$

Then the final result reads

$$W^{\text{div,tot}}_{1-\text{loop}} = \frac{1}{32\pi^2(\omega-2)}\int d^4x\,\bar{g}^{1/2}\Bigg[\frac{43}{60}\bar{R}_{\mu\nu}\bar{R}^{\mu\nu} + \frac{1}{40}\bar{R}^2 - \frac{1}{3}\bar{R}(\bar{\varphi}_{,\nu}\bar{\varphi}^{,\nu})$$
$$+\frac{5}{4}(\bar{\varphi}_{,\nu}\bar{\varphi}^{,\nu})^2 + 5\bar{V}^2 - 2\bar{V}'^2 + \frac{1}{2}\bar{V}''^2 - \frac{13}{3}\bar{R}\,\bar{V} - \frac{1}{6}\bar{R}\,\bar{V}''$$
$$+\bar{V}\,(\bar{\varphi}_{,\nu}\bar{\varphi}^{,\nu}) - 2\bar{V}''\,(\bar{\varphi}_{,\nu}\bar{\varphi}^{,\nu})\Bigg]. \quad (5.61)$$

This result coincides with the result of [2] for the choice $k^2 = 1$, except for the term $5\,\bar{V}^2$—a misprint in [2], already noted and corrected by the same authors in [1].

5.1.4 Comparison with 't Hooft and Veltman

For the sake of completeness, we will show the consistency of this result with the 1974 't Hooft and Veltman result [24]. In order to do so, we have to calculate the on-shell effective action in the case of a free scalar field, i.e. $V(\varphi) = 0$. From the extremal condition of the first variations (5.6) and (5.10) we obtain the Klein–Gordon equation

$$\bar{\Box}\bar{\varphi} = 0 \quad (5.62)$$

and the Einstein field equations for the scalar field energy momentum tensor

$$\bar{R}^{\alpha\beta} - \frac{1}{2}\bar{g}^{\alpha\beta}\bar{R} = \frac{1}{2}(\bar{\nabla}^\alpha\bar{\varphi})(\bar{\nabla}^\beta\bar{\varphi}) - \frac{1}{4}\bar{g}^{\alpha\beta}(\bar{\varphi}_{,\nu}\bar{\varphi}^{,\nu}). \quad (5.63)$$

Taking the trace of (5.63) leads to the condition

$$\bar{R} = \frac{1}{2}(\bar{\varphi}_{,\nu}\bar{\varphi}^{,\nu}). \quad (5.64)$$

Substituting this back in (5.63) yields an additional relation for $\bar{R}^{\alpha\beta}$

$$\bar{R}^{\alpha\beta} = \frac{1}{2}(\bar{\nabla}^{\alpha}\bar{\varphi})(\bar{\nabla}^{\beta}\bar{\varphi}). \tag{5.65}$$

Now, we can express the structures $\bar{R}_{\alpha\beta}\bar{R}^{\alpha\beta}$, $\bar{R}(\bar{\varphi}_{,\nu}\bar{\varphi}^{,\nu})$, $\bar{R}_{\mu\nu}(\bar{\varphi}^{,\mu}\bar{\varphi}^{,\nu})$, $(\bar{\varphi}_{,\nu}\bar{\varphi}^{,\nu})^2$ appearing in the effective action (5.61) in terms of $\bar{R}^2$ only:

$$\begin{aligned}
\bar{R}_{\alpha\beta}\bar{R}^{\alpha\beta} &= \frac{1}{4}(\bar{\nabla}^{\alpha}\bar{\varphi})(\bar{\nabla}^{\beta}\bar{\varphi})(\bar{\nabla}_{\alpha}\bar{\varphi})(\bar{\nabla}_{\beta}\bar{\varphi}) = \frac{1}{4}(\bar{\varphi}_{,\nu}\bar{\varphi}^{,\nu})^2 = \bar{R}^2\\
\bar{R}\,(\bar{\varphi}_{,\nu}\bar{\varphi}^{,\nu}) &= 2\,\bar{R}^2, \quad (\bar{\varphi}_{,\nu}\bar{\varphi}^{,\nu})^2 = 4\,\bar{R}^2\\
\bar{R}_{\mu\nu}(\bar{\varphi}^{,\mu}\bar{\varphi}^{,\nu}) &= \frac{1}{2}(\bar{\nabla}^{\alpha}\bar{\varphi})(\bar{\nabla}^{\beta}\bar{\varphi})(\bar{\nabla}_{\alpha}\bar{\varphi})(\bar{\nabla}_{\beta}\bar{\varphi}) = \frac{1}{2}(\bar{\varphi}_{,\nu}\bar{\varphi}^{,\nu})^2 = 2\,\bar{R}^2
\end{aligned} \tag{5.66}$$

Substituting these expressions in (5.61), we finally obtain in agreement[3] with [24]

$$W^{\text{div,tot}}_{1-\text{loop,on-shell}} = \left(\frac{43}{60} + \frac{1}{40} + 5 - \frac{2}{3}\right)\bar{R}^2 = \frac{203}{40}\bar{R}^2. \tag{5.67}$$

5.2 Non-minimal O(N) Multiplet in the Jordan Frame

In the previous section, we have seen how to perform the explicit one-loop calculation for a single scalar field minimally coupled to gravity by the application of the generalized Schwinger–DeWitt algorithm described in Sect. 4.5. Even for this simple model the calculations within the Schwinger–DeWitt algorithm become quite involved. It is obvious that the degree of complexity and the number of terms will increase very fast by taking into account more fields or additional couplings between the fields. The main purpose of the detailed presentation in Sect. 5.1 was to accurately show the technical details and the individual steps that have to be performed during such a calculation. Now we are going to use this knowledge in order to approach a more complex setup with the assistance of the `Mathematica` package `MathTensor` [20]. By using such a computer based algorithm, it is essential to have full clarity about each individual step in the algorithm and to search for independent methods to check the results. A simpler model which contains the essential features of the more complex model, but which can be calculated without computer assistance, provides such a possibility and again motivates the calculations of the previous section.

In this section, we are interested in calculating the divergent part of the one-loop effective action for a $O(N)$ symmetric *multiplet* of scalar fields *non-minimally* coupled to gravity. The motivation for such a setup originates from cosmological models, in particular from the Higgs inflation model [3, 7, 8, 12] which will be presented in Chap. 6. We will see that the predictions of such a model are in accordance with observational data only by assuming a strong non-minimal coupling of the Higgs field to gravity. As we have explained in Sect. 3.2.4, the Standard Model Higgs field Φ^a

3 An additional factor of $1/2$ in [24] is due to a different definition of the pole in dimension ϵ.

is a doublet of complex scalar fields and can be represented as a quartet of four *real* scalar fields, obeying an internal $O(4)$ symmetry associated with the indices $a, b, \ldots$. Only one component of Φ^a, the massive one, represents the Higgs boson. The other three components correspond to the Goldstone bosons which render the massless gauge bosons of the Standard Model massive. However, at the quantum level, the Goldstone bosons will give significant contributions, while running in the quantum loops. Thus, the setup for the Higgs inflation scenario requires two additional complications compared to the simpler model of the previous section: a *non-minimal coupling* and a *multiplet* of scalar fields.

5.2.1 Second Variation and Minimal Operator

Keeping in mind the physical motivation, we can forget about it for a moment and just consider the technical task to perform the one-loop calculations for the action

$$S = \int_{\mathcal{M}} \mathrm{d}^4 x \sqrt{g} \left(U(\varphi) R - \frac{1}{2} g^{\mu\nu} G(\varphi) \nabla_\mu \Phi^a \nabla_\nu \Phi_a - V(\varphi) \right). \tag{5.68}$$

In order to keep the setup as general as possible, we have assumed a multiplet of N real scalar fields which is invariant under N dimensional internal rotations. In addition, we also consider generalized couplings $U(\varphi)$, $G(\varphi)$ and $V(\varphi)$ of the non-minimal, kinetic and potential term. They are not just constants, but functions of the modulus

$$\varphi := \sqrt{\delta_{ab} \Phi^a \Phi^b}, a = 1, \ldots, N. \tag{5.69}$$

The generalized couplings U, G and V are invariant with respect to rotations in the internal N-dimensional space and ultra-local, but otherwise arbitrary functions of the modulus φ.[4] In cosmology, such a setup containing a non-minimal coupling, is usually called *Jordan frame*, in contrast to the *Einstein frame*, in which the scalar fields are minimally coupled and the kinetic term is canonically normalized—corresponding formally to $U = G = 1$ in (5.68). From now on, we abstain from labelling a background quantity with a bar.

The results of this general setup can directly be used in the Higgs inflation scenario by setting $N = 4$ and specifying the functions $U(\varphi)$, $G(\varphi)$ and $V(\varphi)$ in the appropriate way. In this case, the modulus φ corresponds to the Higgs-inflaton. Due to its universality, the results are also applicable for a much wider class of cosmological models and of course also formally include the minimally coupled case by $U \to U_0$, where U_0 is some constant. We recover ordinary General Relativity for the choice $U_0 = M_{\mathrm{P}}^2/2$.

---

[4] A more general dependence of the functional couplings on the multiplet Φ^a would destroy the $O(N)$ invariance of the action.

However, the universality of this setup, in return, will drastically increase the complexity of the calculations. In particular, there will appear nearly twice as much different contracted scalar structures in the effective action than in the much simpler model of Sect. 5.1 and the number of individual terms generated during the calculations is immense. Thus, we will follow exactly the same calculational steps as in Sect. 5.1, but rely on the support of the `Mathematica` package `MathTensor` to perform the tensor manipulations.

We consider an Euclidean manifold $\mathcal{M}$ and use the conventions defined in Sect. 5.1. We use again the background field method and split the generalized field

$$\psi_A := \begin{pmatrix} g_{\alpha\beta} \\ \Phi_a \end{pmatrix} \tag{5.70}$$

into a background part $\bar{\psi}_A$ and fluctuations $\delta\psi_A$, with

$$\bar{\psi}_A := \begin{pmatrix} \bar{g}_{\alpha\beta} \\ \bar{\Phi}_a \end{pmatrix} \quad \text{and} \quad \delta\psi_A := \begin{pmatrix} \delta g_{\alpha\beta} \\ \delta\Phi_a \end{pmatrix} := \begin{pmatrix} h_{\alpha\beta} \\ \sigma_a \end{pmatrix}. \tag{5.71}$$

The action (5.68) should be complemented by the gauge-breaking term

$$S_{\rm gb} = -\frac{1}{2}\int_{\mathcal{M}} \mathrm{d}^4x\, \sqrt{g}\, \chi_\mu \chi^\mu, \tag{5.72}$$

with the function χ_μ representing the generalization of the well-known background covariant DeWitt condition [6, 14], cf. (5.4):

$$\chi_\mu = \sqrt{U}\left(\nabla^\alpha h_{\alpha\mu} - \frac{1}{2}\nabla_\mu h + f^a(\varphi)\nabla_\mu \sigma_a\right). \tag{5.73}$$

The explicit form of the arbitrary function $f^a(\varphi)$ will be chosen later and the trace of the metric perturbation is defined by $h := g^{\alpha\beta}h_{\alpha\beta}$.

In order to calculate the one-loop divergent part of the effective action in the model with the action (5.68) by using the background field formalism [6, 14], we need to know the second-order differential operator of the total action $S_{\rm tot} := S + S_{\rm gb}$

$$F_{AB} = \frac{\delta^2 S_{\rm tot}}{\delta\psi_A\,\delta\psi_B}. \tag{5.74}$$

We also have to take into account the contribution of the Faddeev–Popov ghost term, which is given by the determinant of the ghost operator Q^α_β defined as

$$Q^\alpha_\beta = \frac{\delta(\chi_\xi)^\alpha}{\delta\xi^\beta}. \tag{5.75}$$

Here, ξ^β is a vector field realizing the general gauge transformation of the field variables by their Lie dragging as discussed in Sect. 4.4.1 and Appendix B.12.

For the application of the Schwinger–DeWitt algorithm, the operator F_{AB} should have the so-called minimal form, i.e.

$$F_A^{\ B} = g^{\mu\nu} \mathcal{D}_\mu \mathcal{D}_\nu \delta_A^{\ B} + P_A^{\ B} - \frac{1}{6} R \delta_A^{\ B}, \tag{5.76}$$

with $\mathcal{D}_\mu$ being a new covariant derivative defined with respect to a new connection and the term $R/6$ is subtracted for convenience. In order to arrive from (5.74) at the minimal form (5.76), the matrix $C^{\mu\nu}_{AB}$ should satisfy the condition

$$C^{\mu\nu}_{AB} = \tilde{C}_{AB} \, g^{\mu\nu}, \tag{5.77}$$

The proper choice of the gauge condition (5.73) guarantees that the condition (5.77) is satisfied. This means that the second-order derivatives in the operator F_{AB} are proportional to the d'Alembertian. Afterwards, we multiply the operator F_{AB} by the inverse matrix $(\tilde{C}^{-1})^{CA}$ to eliminate the dependence of the operator (5.76) on the matrix $\tilde{C}_{AB}$. Finally, in order to remove the term linear in the derivative, we have to redefine the affine connection by adding the term proportional to Γ^μ_{AB} to the original covariant derivative .

In order to follow the steps in the Schwinger–DeWitt algorithm of Sect. 4.5 we first have to fix the function $f^a(\varphi)$ for the gauge condition (5.73). The gravity part of (5.73), which coincides with the well-known de Donder gauge, ensures that the part in the graviton-graviton block $\delta^2 S^{(2)}_{\rm tot}/\delta h_{\mu\nu}\, \delta h_{\alpha\beta}$ of the operator F_{AB}, containing second order derivatives, is proportional to the d'Alembertian [6] and $S^{(2)}_{\rm tot} := \delta^2_\psi \, S_{\rm tot}$. It can easily be seen that the second variation of the action with respect to the scalar fields is also proportional to the d'Alembertian, independently of the form of the function $f^a(\varphi)$. The non-diagonal terms arise in the mixed functional derivatives and have the following form:

$$\frac{\delta S_{\rm tot}}{\delta h_{\alpha\beta}\, \delta\sigma^b} = U' \, n_b \, \nabla^\alpha \nabla^\beta + U \, f_b \, \nabla^\alpha \nabla^\beta + \cdots . \tag{5.78}$$

A "prime" denotes the derivative with respect to φ and

$$n_a := \frac{\Phi_a}{\varphi}. \tag{5.79}$$

An immediate consequence of (5.78) is that by choosing

$$f_a = -\frac{U'}{U} n_a, \tag{5.80}$$

the second derivatives contained in F_{AB} have a d'Alembertian structure.

Dividing F_{AB} as in (5.1) into single pieces according to the number of derivatives, we can represent the explicit expressions for the coefficients $\tilde{C}_{AB}$, Γ^{μ}_{AB} and W_{AB} as (2×2) matrices. For the "kinetic part" of F_{AB}, we find

$$\tilde{C}_{AB} = \begin{pmatrix} U\, G^{\alpha\beta\gamma\delta} & -\frac{1}{2}\, U'\, g^{\alpha\beta}\, n_b \\ -\frac{1}{2}\, U'\, n_a\, g^{\gamma\delta} & G\, \delta_{ab} + \frac{(U')^2}{U}\, n_a\, n_b \end{pmatrix}. \tag{5.81}$$

The part linear in the derivative is given by the (2×2) matrix

$$\Gamma^{\nu}_{AB} = \begin{pmatrix} \Gamma^{\nu}_{11} & \Gamma^{\nu}_{12} \\ \Gamma^{\nu}_{21} & \Gamma^{\nu}_{22} \end{pmatrix}, \tag{5.82}$$

where the subscript 1 stands for the graviton perturbation $h_{\mu\nu}$, while the subscript 2 denotes the scalar field perturbation σ^a:

$$\Gamma^{\nu}_{11} = U' n_a \Phi^{a}_{\ ;\,\mu}\Big(g^{\mu(\alpha} G^{\beta)\nu\gamma\delta} - g^{\mu(\gamma} G^{\delta)\nu\alpha\beta} + \frac{1}{2}\, g^{\nu\mu} G^{\alpha\beta\gamma\delta}\Big), \tag{5.83}$$

$$\Gamma^{\nu}_{12} = G\, G^{\alpha\beta\mu\nu}\, \Phi_{a;\,\mu} + \frac{1}{2}\, U_{ab}\, \Phi^{b}_{\ ;\,\mu}\Big(g^{\mu(\alpha}\, g^{\beta)\nu} - \frac{3}{2}\, g^{\alpha\beta}\, g^{\mu\nu}\Big), \tag{5.84}$$

$$\Gamma^{\nu}_{21} = -\, G^{\gamma\delta\mu\nu}\Big(G\,\delta_{ab} + U_{ab}\Big)\,\Phi^{a}_{\ ;\,\mu}, \tag{5.85}$$

$$\begin{aligned} \Gamma^{\nu}_{22} = {} & \frac{1}{2}\, G'\, g^{\mu\nu}\, \Phi^{c}_{\ ;\,\mu}\, n^d\Big(- \delta_{bc}\delta_{ad} + \delta_{ac}\delta_{bd} + \delta_{ab}\delta_{cd}\Big) \\ & + \frac{1}{2}\,\frac{U'}{U}\Big(U_{bc}\, n_a + U_{ac}\, n_b - \frac{(U')^2}{U}\, n_a\, n_b\, n_c\Big)\, g^{\mu\nu}\, \Phi^{c}_{\ ;\,\mu}. \end{aligned} \tag{5.86}$$

In an analogous manner, we find the entries for the matrix W_{AB}:

$$W_{11} = U\, K^{\alpha\beta\gamma\delta} + G\, S^{\alpha\beta\gamma\delta} + V G^{\alpha\beta\gamma\delta} + T^{\alpha\beta\gamma\delta}, \tag{5.87}$$

$$\begin{aligned} W_{12} = {} & G^{\alpha\beta\mu\nu}\Big(V'\, g_{\mu\nu} - 2\, U'\, R_{\mu\nu} + G'\, \Phi^{c}_{\ ,\,\mu}\, \Phi_{c,\,\nu}\Big)\, n_b \\ & + \Big(g^{\mu(\alpha} g^{\beta)\nu} - g^{\alpha\beta} g^{\mu\nu}\Big)\Big(U_{bac}\, \Phi^{a}_{\ ;\,\mu}\, \Phi^{c}_{\ ;\,\nu} + U_{ba}\, \Phi^{a}_{\ ;\,\nu\mu}\Big), \end{aligned} \tag{5.88}$$

$$\begin{aligned} W_{21} = {} & n_a\, G^{\gamma\delta\mu\nu}\Big(V'\, g_{\mu\nu} - 2\, U'\, R_{\mu\nu} + G'\, \Phi^{c}_{\ ;\,\mu}\, \Phi_{c;\,\nu}\Big) \\ & - 2\, G^{\gamma\delta\mu\nu}\, G'\, n_b\, \Phi^{b}_{\ ;\,\nu}\, \Phi_{a;\,\mu} - 2\, G^{\gamma\delta\mu\nu}\, G\, \Phi_{a;\,\mu\nu}, \end{aligned} \tag{5.89}$$

$$W_{22} = U_{ab}\, R - \frac{1}{2}\, G_{ab}\, \Phi^{c}_{\ ,\,\mu}\Phi_{c}^{\ ;\,\mu} - V_{ab} + G_{cb}\, \Phi^{c}_{\ ;\,\mu}\, \Phi_{a}^{\ ;\,\mu} + G'\, n_b(\Box\, \Phi_a). \tag{5.90}$$

The abbreviations used in (5.87) have the explicit structure

$$K^{\alpha\beta\gamma\delta} := -G^{\alpha\beta\gamma\delta}R - \frac{1}{2}g^{\alpha\beta}R^{\gamma\delta} - \frac{1}{2}g^{\gamma\delta}R^{\alpha\beta} + \frac{1}{2}g^{\alpha(\gamma}R^{\delta)\beta} + \frac{1}{2}g^{\beta(\gamma}R^{\delta)\alpha} + \frac{1}{2}R^{\alpha\gamma\beta\delta} + \frac{1}{2}R^{\alpha\delta\beta\gamma}, \tag{5.91}$$

$$S^{\alpha\beta\gamma\delta} := \frac{1}{2}G^{\alpha\beta\gamma\delta}\Phi^a_{;\mu}\Phi^{;\mu}_a + \frac{1}{4}g^{\alpha\beta}\Phi^{a;\gamma}\Phi^{;\delta}_a + \frac{1}{4}g^{\gamma\delta}\Phi^{a;\alpha}\Phi^{;\beta}_a - \frac{1}{2}g^{\alpha(\gamma}\Phi^{a;\delta)}\Phi^{;\beta}_a - \frac{1}{2}g^{\beta(\gamma}\Phi^{a;\delta)}\Phi^{;\alpha}_a, \tag{5.92}$$

$$T^{\alpha\beta\gamma\delta} := 2G^{\alpha\beta\gamma\delta}\Box U + g^{\alpha\beta}U^{;\gamma\delta} + \frac{1}{2}g^{\gamma\delta}U^{;\alpha\beta} - g^{\alpha(\gamma}U^{;\delta)\beta} - g^{\beta(\gamma}U^{;\delta)\alpha}. \tag{5.93}$$

The following formulas for a general function $Z(\varphi)$ will be useful:

$$\nabla_\mu Z = Z' n_a \nabla_\mu \Phi^a, \tag{5.94}$$

$$Z_{ab} := \frac{\partial^2 Z}{\partial\Phi^a\partial\Phi^b} = Z'' n_a n_b + \frac{Z'}{\varphi}\Big(\delta_{ab} - n_a n_b\Big), \tag{5.95}$$

$$Z_{abc} := \frac{\partial^3 Z}{\partial\Phi^a\partial\Phi^b\partial\Phi^c} = Z''' n_a n_b n_c + \frac{1}{\varphi}\Big[n_a\Big(Z''\delta_{bc} - Z_{bc}\Big) + n_b\Big(Z''\delta_{ac} - Z_{ac}\Big) + n_c\Big(Z''\delta_{ab} - Z_{ab}\Big)\Big]. \tag{5.96}$$

The next step in the algorithm is the calculation of the inverse matrix $(\tilde{C}^{-1})^{AB}$

$$(\tilde{C}^{-1})^{BC} = \begin{pmatrix} \frac{1}{U}G_{\lambda\sigma\gamma\delta} + \frac{(1-Gs)}{3U}g_{\lambda\sigma}g_{\gamma\delta} & -s\frac{U'}{U}g_{\gamma\delta}n^c \\ -s\frac{U'}{U}n^b g_{\lambda\sigma} & \frac{1}{G}\Big(\delta^{bc} - n^b n^c\Big) + s\,n^b n^c \end{pmatrix}, \tag{5.97}$$

with the factor s defined by the formula

$$s := \frac{U}{GU + 3U'^2}. \tag{5.98}$$

This factor will play an essential role in Chap. 6, when considering the cosmological model of Higgs inflation. It is exactly this factor that is responsible for the suppression of the contributions of quantum loop corrections, in particular for the suppression of the Higgs propagators in models with a strong non-minimally coupled Higgs field, see e.g. [3, 4, 7, 8, 12].

We refrain from presenting the lengthy results of the multiplication of the operator F_{AB} by the inverse matrix $(\tilde{C}^{-1})^{AB}$. By redefining the covariant derivative, we finally arrive at the minimal form (5.76) of the operator $F_A^{\ B}$. Now we also have to find the Faddeev–Popov operator. Applying the definition (5.75), we obtain

$$Q_\alpha^\beta = \delta_\alpha^\beta \Box + 2 \,{}^{(Q)}\Gamma_\alpha^{\beta\mu} \nabla_\mu + {}^{(Q)}W_\alpha^\beta, \tag{5.99}$$

$${}^{(Q)}\Gamma_\alpha^{\beta\mu} = -\frac{1}{2}\frac{U'}{U} n_a \Phi^a_{,\alpha} g^{\beta\mu}, \tag{5.100}$$

$${}^{(Q)}W_\alpha^\beta = R_\alpha^\beta - \frac{U'}{U} n_a \nabla_\alpha \nabla^\beta \Phi^a. \tag{5.101}$$

The Faddeev–Popov ghost operator (5.99) also contains a term linear in the derivative. In order to absorb this term into the d'Alembertian we must again redefine the general affine connection for the ghost operator, just like it was done for the main operator F_A^B. As in the case of F_A^B, we refrain from writing down the explicit formula for the final minimal form of the Faddeev–Popov operator.

Finally, we are now in the position to apply the formulas (4.114) and (4.112). Since a vast number of terms are generated during the algorithm, we use (5.81)–(5.101) as an input and implement the Schwinger–DeWitt algorithm in the `MathTensor` package [20] in order to calculate the final result for the divergent part of the one-loop effective action. This result will be presented in the next section in a closed form. The result is rather lengthy, but we have organized it in a "physical" way to facilitate its use for concrete models. First of all, for practical purposes in a cosmological setup, the most important quantum corrections are $V_{1-\text{loop}}$ and $U_{1-\text{loop}}$ and the results for these corrections do not turn out to be very long. In addition, we have arranged the results as polynomials in the suppression function $s(\varphi)$. This choice is motivated by the feature of the Higgs inflation model. In this case, $s(\varphi)$ is approximately zero for sufficiently high energies as is the case during inflation. However, the remaining structures, which are partly very long, are nevertheless important from the viewpoint of an *effective* field theory, which is valid up to a specific energy scale determined by a cut-off Λ. The magnitude of this cut-off depends on the coefficients of the induced higher dimensional structures, which are supposed to be suppressed sufficiently by powers of this cut-off in order to guarantee the applicability of the model up to the corresponding energy scale.

5.2.2 One-Loop Divergences in the Jordan Frame: The Result

The divergent part of the one-loop effective action for a scalar multiplet non-minimally coupled to gravity in the Jordan frame can be written as the following sum of a minimal set of 21 independent structures:

$$W^{\text{div}}_{1-\text{loop}} = \frac{1}{32\pi^2(2-\omega)} \int d^4x \sqrt{g} \Big[\alpha_1 + \alpha_2 R + \frac{1}{2}\alpha_3 \Phi^a_{,\mu}\Phi_a^{;\mu} + \alpha_4 \Phi^a_{,\mu} n_a \Phi^{b,\mu} n_b$$

$$+ \alpha_5 R^{\mu\nu} R_{\mu\nu} + \alpha_6 R^2 + \alpha_7 R^{\mu\nu} \Phi^a_{,\mu} \Phi_{a,\nu} + \alpha_8 R^{\mu\nu} \Phi^a_{,\mu} n_a \Phi^b_{,\nu} n_b$$

$$
\begin{aligned}
&+\alpha_9\, R\,(\Phi^a_{\,,\mu}\Phi_a^{\,;\mu}) + \alpha_{10}\, R\,(\Phi^a_{\,,\mu}n_a\Phi^{b\,,\mu}n_b) + \alpha_{11}\, R\,(\Phi^{c\;\;\mu}_{\,;\mu}\, n_c)\\
&+\alpha_{12}\,(\Phi^a_{\,,\mu}\Phi_a^{\,;\mu})^2 + \alpha_{13}\,(\Phi^a_{\,,\mu}n_a\Phi^{b\,,\mu}n_b)^2 + \alpha_{14}\,(\Phi^a_{\,,\mu}\Phi_a^{\,;\mu})\,(\Phi^c_{\,,\nu}n_c\Phi^{d\,,\nu}n_d)\\
&+\alpha_{15}\,\Phi^a_{\,;\mu}\Phi_{a\,,\nu}\Phi^{b\,,\mu}\Phi_b^{\,;\nu} + \alpha_{16}\,\Phi^a_{\,,\mu}n_a\Phi^b_{\,,\nu}n_b\Phi^{c,\mu}\Phi_c^{\,;\nu} + \alpha_{17}\,\Phi^{a\;\;\mu}_{\,;\mu}\Phi_{a;\nu}^{\;\;\;\;\nu}\\
&+\alpha_{18}\,(\Phi^{c\;\;\mu}_{\,;\mu}\, n_c)^2 + \alpha_{19}(\Phi^a_{\,,\mu}\Phi_a^{\,;\mu})(\Phi^{b\;\;\nu}_{\,;\nu}n_b) + \alpha_{20}\,(\Phi^a_{\,,\mu}n_a\Phi^{b\,,\mu}n_b)(\Phi^{c\;\;\nu}_{\,;\nu}\, n_c)\\
&+\alpha_{21}\,\Phi^{a\;\;\mu}_{\,;\mu}\Phi_a^{\,;\nu}\Phi^b_{\,,\nu}n_b\Big]. \qquad (5.102)
\end{aligned}
$$

In order to see what structures are generally allowed in in (5.102), we notice that all different structures in (5.102) are composed of multiplicative combinations of “irreducible scalar building blocks”. These fundamental building blocks are formed by contracting the different indices of the involved fields and their derivatives. In the case of our model (5.68), there are two types of fields involved. The metric $g_{\mu\nu}$ with its derived curvature quantities (each containing two derivatives) and the scalar multiplet Φ^a with its first and second derivatives. Without calculating the effective action, we first systematically collect all structures that can appear in (5.102) at the one-loop order by simple graphical considerations. We suppress the index structure and represent the contractions by contraction brackets. Therefore, we temporarily introduce the following symbolic notation for the basic quantities and their different index structure, beginning with scalar structures like the potential $V(\varphi)$ already present in (5.68). Those “intrinsic” scalar structures are represented by a **1**. All structures in (5.102) can be formed by contraction of the following basic quantities:

$$
\mathbf{1},\ \phi := \Phi^a_{\,;\mu},\ \sigma := \Phi^a_{\,;\mu\nu},\ n := n^a,\ Ric := R_{\mu\nu},\ Riem := R_{\mu\nu\rho\sigma}. \qquad (5.103)
$$

Here, the σ should of course not be confused with the perturbation $\delta\varphi$ from the previous section. The contraction operation for the different indices with the inverse background metric $g^{\mu\nu}$ and the inverse field metric δ^{ab} of the $O(N)$ multiplet can be graphically displayed by “contraction brackets”.

$$
\sqcap := g^{\mu\nu}, \qquad \text{and} \qquad \sqcup := \delta^{ab}. \qquad (5.104)
$$

We arrange the graphical representation of all irreducible fundamental scalar building blocks in three blocks according to an increasing number of derivatives:

$\mathbf{1} =: z_0$

$$\phi\phi =: z_1, \quad \phi n \phi n := z_2, \quad \sigma\, n =: z_3, \quad Ric \equiv R := z_4,$$

$$\phi\phi\phi\phi =: z_5, \quad \phi n \phi n \phi\phi := z_6, \quad \sigma\, \sigma =: z_7, \quad \sigma\, \sigma =: z_8,$$

$$\sigma n\, \sigma n =: z_9, \quad \sigma n\, \phi\, \phi =: z_{10}, \quad \sigma \phi\, \phi n =: z_{11}, \quad \sigma \phi\, n \phi =: z_{12},$$

$$\sigma n\, \phi n\, \phi n =: z_{13}, \quad Ric\, Ric =: z_{14}, \quad Ric\, \phi\phi =: z_{15}, \quad Ric\, \phi n\, \phi n =: z_{16},$$

$$Ric\, \sigma n =: z_{17}, \quad Riem\, Riem =: z_{18}. \tag{5.105}$$

Since only terms containing up to four derivatives can appear in (5.102), we have at most 29 possible structures which can be built out of the z_i. The only building block not containing any derivatives is z_0. There are four building blocks $z_1,\ z_2,\ z_3,\ z_4$ which contain two derivatives. All the remaining 14 building blocks $z_i,\ i = 5, .., 18$ contain four derivatives. Considering all possible combinations containing at most four derivatives, we can easily calculate the total number of allowed composed structures. Counting 1 for z_0 plus 4 for $z_1,\ z_2,\ z_5,\ z_{18}$, plus $4 + 3 + 2 + 1$, due to the combinations among these four building blocks, plus 14 for the remaining ones gives the total number of 29 different structures.

The *structures* in (5.102) are organized with respect to an increasing number of derivatives and furthermore sorted by pure curvature terms, mixed curvature-gradient terms and pure gradient terms. Not all of the the 29 structures are independent. Taking into account integration by parts and the Gauss–Bonnet identity (5.146), we can eliminate eight dependent structures and obtain the minimal set of 21 *independent* structures. The explicit transition formulas (F.1)–(F.8), reducing the numbers of structures can be found in Appendix F.

To further collect the terms contained in each *coefficient* α_i of (5.102) in a systematical way, we adopt the following sorting pattern: First, the contributions are arranged according to their powers of the potential V and its derivatives V', V'', V'''. Secondly, we further rewrite the coefficients as a polynomial in decreasing powers of $s(\varphi)$. Expressed in this form, we can deduce direct physical information for cosmological applications as e.g. in the context of the Higgs inflation scenario, we will discuss in the next chapter. In this model, the tree-level values of $U,\ G,\ V$ have the concrete form (5.128)–(5.130). Anticipating some results of the applications in the next parts, we can take into account the following considerations: For inflationary energy scales, corresponding to field values $\varphi \gg M_{\rm P}/\sqrt{\xi}$, the "suppression function" s scales as $s = 1/6\xi$. Since in this kind of models the non-minimal coupling constant ξ must be $\mathcal{O}(10^4)$ in order to ensure agreement with cosmological data [3, 8, 12], it follows $s \simeq 0$. Thus, in the inflationary context of a scalar multiplet non-minimally coupled to gravity most of the terms in the coefficients α_i are sufficiently suppressed and can be neglected. All structures α_i which contain more than two derivatives are supposed to be suppressed by the appropriate power of some cut-off Λ, which determines the validity of the model. All such structures appearing in (5.102) can be denoted symbolically as

$$\frac{RR}{\Lambda^2}, \qquad \frac{R\,\partial^2\Phi\Phi}{\Lambda^2}, \qquad \frac{\partial^4\Phi\Phi\Phi\Phi}{\Lambda^2}, \tag{5.106}$$

corresponding to pure curvature structures, mixed structures or pure scalar gradient structures. The structures α_i, $i = 12, ..., 21$ only contain four derivatives of the scalar fields and belong all to the pure gradient structures. In Sect. 6.6, we will show that those structures are suppressed by an additional small factor, the slow-roll parameter $\hat{\epsilon} \ll 1$, compared to the curvature structures, i.e.

$$\frac{R}{\Lambda^2} \simeq \frac{\lambda}{16\pi^2} \quad \text{compared to} \quad \frac{\partial}{\Lambda} \simeq \frac{\sqrt{\lambda}}{48\pi}\sqrt{2\hat{\epsilon}}, \tag{5.107}$$

with $\lambda \simeq 10^{-1}$ being the quartic Higgs coupling. Thus, the suppression mechanism for the gradient structures works even more efficiently. From an effective field theory point of view, these terms are less important and therefore presented in Appendix B. Below, we only present the result for the coefficients α_i, $i = 1, \ldots, 11$. The result is given in a closed form, beginning with the most important structures α_1, α_2 and α_3 corresponding to $V_{1-\text{loop}}$, $U_{1-\text{loop}}$ and $G_{1-\text{loop}}$. The factor of 1/2 in front of $G_{1-\text{loop}}$ is due to the form of the kinetic term $\frac{1}{2}\Phi^a_{\ ;\mu}\Phi_a^{\ ;\mu}$ in the action (5.68).

The one-loop correction to the potential $V(\varphi)$:

$$\begin{aligned}\alpha_1 = {} & V^2\left[\frac{2s^2\left(U'\right)^4}{U^4} - \frac{2s\left(U'\right)^2}{U^3} + \frac{5}{U^2}\right] + VV'\left[-\frac{8s^2\left(U'\right)^3}{U^3} + \frac{4sU'}{U^2}\right] \\ & + V\,V''\,\frac{2s^2\left(U'\right)^2}{U^2} + \left(V'\right)^2\left[\frac{8s^2\left(U'\right)^2}{U^2} - \frac{2s}{U} + \frac{(N-1)}{2G^2\varphi^2}\right] \\ & - V'V''\,\frac{4s^2U'}{U} + \frac{1}{2}\left(V''\right)^2 s^2.\end{aligned} \tag{5.108}$$

The one-loop correction to the non-minimal coupling $U(\varphi)$:

$$\alpha_2 = V\left[s^2\left(\frac{4\left(U'\right)^4}{U^3} - \frac{2\left(U'\right)^2U''}{U^2}\right) - \frac{7s\left(U'\right)^2}{3U^2} - \frac{13}{3U}\right]$$

$$+ V'\left[s^2\left(\frac{4U'U''}{U} - \frac{8\,(U')^3}{U^2}\right) + \frac{8sU'}{3U} - \frac{(N-1)U'}{G^2\varphi^2} - \frac{(N-1)}{6G\varphi}\right]$$
$$+ V''\left[s^2\left(\frac{2\,(U')^2}{U} - U''\right) - \frac{s}{6}\right]. \tag{5.109}$$

The one-loop correction to the kinetic term $\frac{1}{2}\,G(\varphi)$:

$$\frac{1}{2}\,\alpha_3 = V\left[s^2\left(\frac{(U')^6}{2GU^4\varphi^2} + \frac{2G'\,(U')^2}{U^2\varphi} + \frac{G''\,(U')^2}{U^2} - \frac{G'\,(U')^3}{U^3} - \frac{(G')^2\,(U')^2}{2GU^2}\right.\right.$$
$$+ \frac{6\,(U')^5}{U^4\varphi} + \frac{(U')^4}{6U^3\varphi^2} + \left.\frac{3\,(U')^6}{2U^5}\right) + s\left(-\frac{(U')^4}{3GU^3\varphi^2} - \frac{G'\,(U')^2}{2GU^2\varphi} + \frac{G'U'}{2U^2}\right.$$
$$\left.\left. - \frac{9\,(U')^3}{2U^3\varphi} - \frac{19\,(U')^2}{18U^2\varphi^2} - \frac{2\,(U')^4}{U^4}\right) + \frac{19\,(U')^2}{18GU^2\varphi^2} + \frac{G}{U^2} + \frac{3U'}{U^2\varphi} + \frac{(U')^2}{2U^3}\right]$$
$$+ V'\left[s^2\left(-\frac{3\,(U')^5}{GU^3\varphi^2} - \frac{3G'\,(U')^3}{GU^2\varphi} - \frac{5G'U'}{U\varphi} - \frac{2G''U'}{U} + \frac{2G'\,(U')^2}{U^2}\right.\right.$$
$$+ \left.\frac{(G')^2\,U'}{GU} - \frac{12\,(U')^4}{U^3\varphi} - \frac{(U')^3}{U^2\varphi^2} - \frac{3\,(U')^5}{U^4}\right) + s\left(\frac{5\,(U')^4}{4G^2U^2\varphi^3} + \frac{(U')^3}{2GU^2\varphi^2}\right.$$
$$+ \frac{3G'\,(U')^2}{2G^2U\varphi^2} + \frac{41\,(U')^2}{12GU\varphi^3} + \frac{2G'U'}{GU\varphi} - \frac{G'}{2U} - \frac{(G')^2}{4G^2\varphi} - \frac{G'}{2G\varphi^2} + \frac{25\,(U')^2}{4U^2\varphi}$$
$$\left. + \frac{11U'}{6U\varphi^2} + \frac{5\,(U')^3}{2U^3}\right) - \frac{5\,(U')^2}{12G^2U\varphi^3} - \frac{11U'}{6GU\varphi^2} - \frac{G'}{2G^2\varphi^2} - \frac{2}{U\varphi} - \frac{U'}{2U^2}$$
$$\left. + \frac{(N-1)G'}{2G^2\varphi^2}\right] + V''\left[s^2\left(\frac{3G'\,(U')^2}{2GU\varphi} - \frac{G'U'}{2U} + \frac{3G'}{2\varphi} - \frac{(G')^2}{4G} + \frac{G''}{2}\right.\right.$$
$$\left.\left. + \frac{3\,(U')^3}{U^2\varphi} + \frac{3\,(U')^4}{4U^3}\right) + s\left(-\frac{G'}{2G\varphi} - \frac{U'}{U\varphi} - \frac{(U')^2}{4U^2}\right)\right]. \tag{5.110}$$

The coefficient α_4 belonging to the structure $\Phi^a{}_{,\mu} n_a \Phi^{b\,,\,\mu} n_b$:

$$\alpha_4 = V\left[s^3\left(\frac{99\,(U')^8}{2U^6} - \frac{90U''\,(U')^6}{U^5} - \frac{9G'\,(U')^5}{U^4} - \frac{18\,(U'')^2\,(U')^4}{U^4}\right.\right.$$
$$\left. - \frac{18G'U''\,(U')^3}{U^3} - \frac{5\,(G')^2\,(U')^2}{2U^2}\right) + s^2\left(-\frac{(U')^6}{2G\varphi^2U^4} - \frac{54\,(U')^6}{U^5} - \frac{6\,(U')^5}{\varphi U^4}\right.$$
$$+ \frac{147U''\,(U')^4}{2U^4} - \frac{(U')^4}{6\varphi^2U^3} + \frac{13G'\,(U')^3}{2U^3} + \frac{(G')^2\,(U')^2}{2GU^2} + \frac{9\,(U'')^2\,(U')^2}{U^3}$$

$$
\begin{aligned}
&-\frac{2G'\left(U'\right)^2}{\varphi U^2}+\frac{5G'U''U'}{2U^2}\Bigg)+s\Bigg(\frac{\left(U'\right)^4}{3G\varphi^2U^3}+\frac{41\left(U'\right)^4}{2U^4}+\frac{9\left(U'\right)^3}{2\varphi U^3}+\frac{G'\left(U'\right)^2}{2G\varphi U^2}\\
&-\frac{17U''\left(U'\right)^2}{U^3}+\frac{19\left(U'\right)^2}{18\varphi^2U^2}-\frac{G'U'}{U^2}\Bigg)-\frac{19\left(U'\right)^2}{18G\varphi^2U^2}+\frac{33\left(U'\right)^2}{2U^3}-\frac{3U'}{\varphi U^2}+\frac{3U''}{U^2}\Bigg]\\
&+V'\Bigg[s^3\Bigg(-\frac{99\left(U'\right)^7}{U^5}+\frac{180U''\left(U'\right)^5}{U^4}+\frac{18G'\left(U'\right)^4}{U^3}+\frac{36\left(U''\right)^2\left(U'\right)^3}{U^3}\\
&+\frac{36G'U''\left(U'\right)^2}{U^2}+\frac{5\left(G'\right)^2U'}{U}\Bigg)+s^2\Bigg(\frac{3\left(U'\right)^5}{G\varphi^2U^3}+\frac{165\left(U'\right)^5}{2U^4}+\frac{12\left(U'\right)^4}{\varphi U^3}\\
&+\frac{3G'\left(U'\right)^3}{G\varphi U^2}-\frac{195U''\left(U'\right)^3}{2U^3}+\frac{\left(U'\right)^3}{\varphi^2U^2}-\frac{17G'\left(U'\right)^2}{2U^2}-\frac{\left(G'\right)^2U'}{GU}-\frac{15\left(U''\right)^2U'}{U^2}\\
&+\frac{5G'U'}{\varphi U}-\frac{5G'U''}{2U}\Bigg)+s\Bigg(-\frac{5\left(U'\right)^4}{4G^2\varphi^3U^2}-\frac{\left(U'\right)^3}{2G\varphi^2U^2}-\frac{26\left(U'\right)^3}{U^3}-\frac{3G'\left(U'\right)^2}{2G^2\varphi^2U}\\
&-\frac{41\left(U'\right)^2}{12G\varphi^3U}-\frac{25\left(U'\right)^2}{4\varphi U^2}-\frac{2G'U'}{G\varphi U}+\frac{14U''U'}{U^2}-\frac{11U'}{6\varphi^2U}+\frac{\left(G'\right)^2}{4G^2\varphi}+\frac{G'}{U}\\
&+\frac{G'}{2G\varphi^2}\Bigg)+\frac{5\left(U'\right)^2}{12G^2\varphi^3U}+\frac{G'}{2G^2\varphi^2}+\frac{11U'}{6G\varphi^2U}-\frac{5U'}{2U^2}+\frac{2}{\varphi U}+\frac{(N-1)G'}{2G^2\varphi^2}\\
&+\frac{3(N-1)\left(G'\right)^2}{4G^3\varphi}\Bigg]+V''\Bigg[s^3\Bigg(\frac{99\left(U'\right)^6}{4U^4}-\frac{45U''\left(U'\right)^4}{U^3}-\frac{9G'\left(U'\right)^3}{2U^2}-\frac{9\left(U''\right)^2\left(U'\right)^2}{U^2}\\
&-\frac{9G'U''U'}{U}-\frac{5\left(G'\right)^2}{4}\Bigg)+s^2\Bigg(-\frac{111\left(U'\right)^4}{4U^3}-\frac{3\left(U'\right)^3}{\varphi U^2}-\frac{3G'\left(U'\right)^2}{2G\varphi U}+\frac{12U''\left(U'\right)^2}{U^2}\\
&-\frac{G'U'}{2U}+\frac{\left(G'\right)^2}{4G}+\frac{3\left(U''\right)^2}{U}-\frac{3G'}{2\varphi}\Bigg)+s\Bigg(\frac{45\left(U'\right)^2}{4U^2}+\frac{U'}{\varphi U}+\frac{G'}{2G\varphi}-\frac{U''}{U}\Bigg)\\
&-\frac{2}{U}-\frac{(N-1)G'}{2G^2\varphi}\Bigg]+V'''\Bigg[s^2\Bigg(\frac{9\left(U'\right)^3}{2U^2}+\frac{G'}{2}\Bigg)-\frac{sU'}{U}\Bigg].
\end{aligned}
\tag{5.111}
$$

The coefficient α_5 belonging to the structure $R^{\mu\nu}R_{\mu\nu}$:

$$
\alpha_5=\frac{2s\left(U'\right)^2}{U}+\frac{43}{60}.
\tag{5.112}
$$

The coefficient α_6 belonging to the structure R^2 :

$$
\begin{aligned}
\alpha_6=s^2\Bigg(-\frac{2\left(U'\right)^2U''}{U}+\frac{2\left(U'\right)^4}{U^2}+\frac{\left(U''\right)^2}{2}\Bigg)+s\Bigg(\frac{U''}{6}-\frac{4\left(U'\right)^2}{3U}\Bigg)\\
+\frac{1}{40}+\frac{(N-1)}{72}+\frac{(N-1)U'}{6G\varphi}+\frac{(N-1)\left(U'\right)^2}{2G^2\varphi^2}.
\end{aligned}
\tag{5.113}
$$

The coefficient α_7 belonging to the structure $R^{\mu\nu}\Phi^a{}_{,\mu}\Phi_{a,\nu}$:

$$\alpha_7 = s\left(-\frac{3(U')^4}{GU^2\varphi^2} + \frac{G'(U')^2}{GU\varphi} - \frac{G'U'}{U} + \frac{5(U')^3}{U^2\varphi} + \frac{(U')^2}{U\varphi^2} + \frac{3(U')^4}{U^3}\right) + \frac{(U')^2}{GU\varphi^2} - \frac{2U'}{U\varphi} - \frac{(U')^2}{U^2}. \tag{5.114}$$

The coefficient α_8 belonging to the structure $R^{\mu\nu}\Phi^a{}_{,\mu}n_a\Phi^b{}_{,\nu}n_b$:

$$\alpha_8 = s^2\left(-\frac{G'U'U''}{U} - \frac{3G'(U')^3}{U^2} - \frac{15(U')^4U''}{U^3} - \frac{6(U')^2(U'')^2}{U^2} + \frac{9(U')^6}{U^4}\right) + s\left(\frac{3(U')^4}{GU^2\varphi^2} - \frac{G'(U')^2}{GU\varphi} + \frac{2G'U'}{U} - \frac{5(U')^3}{U^2\varphi} - \frac{(U')^2}{U\varphi^2} + \frac{10(U')^2U''}{U^2} - \frac{8(U')^4}{U^3} + \frac{2(U'')^2}{U}\right) - \frac{(U')^2}{GU\varphi^2} + \frac{2U'}{U\varphi} - \frac{2U''}{U}. \tag{5.115}$$

The coefficient α_9 belonging to the structure $R\,(\Phi^a{}_{,\mu}\Phi_a^{;\mu})$:

$$\alpha_9 = s^2\left(-\frac{3G'(U')^2U''}{2GU\varphi} - \frac{G'U''}{\varphi} + \frac{3(U')^6}{2GU^3\varphi^2} + \frac{3G'(U')^4}{2GU^2\varphi} + \frac{3G'(U')^2}{2U\varphi} + \frac{G''(U')^2}{U} + \frac{G'U'U''}{2U} + \frac{(G')^2U''}{4G} - \frac{G'(U')^3}{U^2} - \frac{(G')^2(U')^2}{2GU} - \frac{G''U''}{2} + \frac{3(U')^3U''}{2U^2\varphi} - \frac{3(U')^5}{U^3\varphi} + \frac{(U')^4}{2U^2\varphi^2} - \frac{3(U')^4U''}{4U^3} + \frac{3(U')^6}{2U^4}\right) + s\left(\frac{G'U''}{2G\varphi} - \frac{5(U')^5}{4G^2U^2\varphi^3} + \frac{(U')^4}{12GU^2\varphi^2} - \frac{3G'(U')^3}{2G^2U\varphi^2} - \frac{41(U')^3}{12GU\varphi^3} - \frac{2G'(U')^2}{GU\varphi} + \frac{(G')^2U'}{4G^2\varphi} + \frac{G'U'}{2G\varphi^2} + \frac{G'U'}{2U} - \frac{G'}{12\varphi} + \frac{(G')^2}{12G} - \frac{G''}{12} + \frac{(U')^3}{2U^2\varphi} - \frac{83(U')^2}{36U\varphi^2} + \frac{(U')^2U''}{4U^2} - \frac{9(U')^4}{4U^3}\right) + \frac{5(U')^3}{12G^2U\varphi^3} + \frac{47(U')^2}{36GU\varphi^2} + \frac{G'U'}{2G^2\varphi^2} - \frac{G}{3U} + \frac{G'}{6G\varphi} - \frac{2U'}{U\varphi} + \frac{7(U')^2}{12U^2} - \frac{(N-1)G'U'}{G^2\varphi^2} - \frac{(N-1)G'}{6G\varphi}. \tag{5.116}$$

The coefficient α_{10} belonging to the structure $R\,(\Phi^a{}_{,\mu}n_a\Phi^{b,\mu}n_b)$:

$$
\begin{aligned}
\alpha_{10} = s^3 \Bigg(& -\frac{9\,(U')^8}{2U^5} + \frac{81U''\,(U')^6}{4U^4} + \frac{3G'\,(U')^5}{U^3} - \frac{27\,(U'')^2\,(U')^4}{U^3} - \frac{15G'U''\,(U')^3}{2U^2} \\
& + \frac{9\,(U'')^3\,(U')^2}{U^2} - \frac{(G')^2\,(U')^2}{2U} + \frac{3G'\,(U'')^2\,U'}{U} + \frac{1}{4}\,(G')^2\,U'' \Bigg) \\
& + s^2 \Bigg(-\frac{3\,(U')^6}{2G\varphi^2U^3} + \frac{3\,(U')^6}{8U^4} + \frac{3\,(U')^5}{\varphi U^3} - \frac{3G'\,(U')^4}{2G\varphi U^2} - \frac{45U''\,(U')^4}{4U^3} \\
& - \frac{(U')^4}{2\varphi^2U^2} + \frac{9G'\,(U')^3}{4U^2} - \frac{3U''\,(U')^3}{2\varphi U^2} + \frac{(G')^2\,(U')^2}{2GU} + \frac{18\,(U'')^2\,(U')^2}{U^2} \\
& - \frac{3G'\,(U')^2}{2\varphi U} - \frac{G''\,(U')^2}{U} + \frac{3G'U''\,(U')^2}{2G\varphi U} + \frac{G'U''U'}{2U} - \frac{3\,(U'')^3}{U} + \frac{(G')^2}{24} \\
& - \frac{(G')^2\,U''}{4G} + \frac{G'U''}{\varphi} + \frac{G''U''}{2} \Bigg) + s \Bigg(\frac{5\,(U')^5}{4G^2\varphi^3U^2} - \frac{(U')^4}{12G\varphi^2U^2} + \frac{11\,(U')^4}{4U^3} \\
& + \frac{3G'\,(U')^3}{2G^2\varphi^2U} + \frac{41\,(U')^3}{12G\varphi^3U} - \frac{(U')^3}{2\varphi U^2} + \frac{2G'\,(U')^2}{G\varphi U} - \frac{2U''\,(U')^2}{U^2} + \frac{83\,(U')^2}{36\varphi^2U} \\
& - \frac{(G')^2\,U'}{4G^2\varphi} - \frac{G'U'}{U} - \frac{G'U'}{2G\varphi^2} - \frac{(G')^2}{12G} - \frac{(U'')^2}{U} + \frac{G'}{12\varphi} + \frac{G''}{12} - \frac{G'U''}{2G\varphi} \Bigg) \\
& - \frac{5\,(U')^3}{12G^2\varphi^3U} - \frac{47\,(U')^2}{36G\varphi^2U} - \frac{16\,(U')^2}{3U^2} - \frac{G'}{6G\varphi} - \frac{G'U'}{2G^2\varphi^2} + \frac{2U'}{\varphi U} - \frac{2U''}{U} \\
& + \frac{(N-1)G'}{12G\varphi} + \frac{(N-1)G'U'}{2G^2\varphi^2} + \frac{(N-1)\,(G')^2}{24G^2} + \frac{(N-1)\,(G')^2\,U'}{4G^3\varphi} \\
& - \frac{(N-1)G''}{12G} - \frac{(N-1)U'G''}{2G^2\varphi}.
\end{aligned}
\tag{5.117}
$$

The coefficient α_{11} belonging to the structure $R\,(\Phi^{c\;\;\mu}_{\;;\mu}\, n_c)$:

$$
\begin{aligned}
\alpha_{11} = s^2 \Bigg(& \frac{G'U''}{2} - \frac{G'\,(U')^2}{U} + \frac{9\,(U')^3\,U''}{2U^2} - \frac{9\,(U')^5}{U^3} \Bigg) + s \Bigg(\frac{G'}{12} \\
& - \frac{U'U''}{U} + \frac{19\,(U')^3}{4U^2} \Bigg) - \frac{3U'}{U} - \frac{(N-1)G'}{12G} - \frac{(N-1)G'U'}{2G^2\varphi}.
\end{aligned}
\tag{5.118}
$$

In this section we have presented the off-shell result for the most important terms of the divergent part of the one-loop effective action in a closed form. The remaining pure gradient structures can be found in Appendix G. Recalling that s vanishes for high energies and a large non-minimal coupling $|\xi|$, the results which are polynomials in s show, that most of the terms do vanish in the limit $s \to 0$. Likewise, in the limit $N = 1$, some terms do vanish due to the factors of $(N-1)$ and due to a cancellation resulting from the collapse of the projector $(\delta_{ab} - n_a n_b)$, see Table 5.1.

Table 5.1 Degenerated structures in the single field limit

$O(N)$ multiplet	Single field
Γ_1	Γ_1
Γ_2	Γ_2
Γ_3, Γ_4	Γ_3
Γ_5	Γ_5
Γ_6	Γ_6
Γ_7, Γ_8	Γ_7
Γ_9, Γ_{10}	Γ_9
Γ_{11}	Γ_{11}
$\Gamma_{12}, \Gamma_{13}, \Gamma_{14}, \Gamma_{15}, \Gamma_{16}$	Γ_{12}
Γ_{17}, Γ_{18}	Γ_{17}
$\Gamma_{19}, \Gamma_{20}, \Gamma_{21}$	Γ_{19}

5.2.3 Checks and Comparison with Known Results

Regarding the structure (5.118)–(5.194) of the results, it makes sense to apply all kinds of consistency checks and tests available. We can distinguish between internal consistency tests such as the correct dimensionality and the scaling behaviour of certain terms and cross-checks by comparing our general result for some limiting known cases. Fortunately, these checks can isolate certain features of our general model and test complementary aspects such as the consequences due to the non-minimal coupling, or the influence due to the presence of the $O(N)$ multiplet.

5.2.3.1 Dimensional Tests

Writing the mass dimension of some object O as $[O]_M$, as defined in (4.3), one obvious test of the one-loop results is that for all terms

$$[\alpha_i \cdot \Gamma_i]_M = 4 \tag{5.119}$$

should hold ($c = \hbar = 1$). We have introduced the symbol Γ_i for the structure corresponding to the coefficient α_i, i.e.

$$\Gamma^{\text{div}}_{1\text{-loop}} =: \frac{1}{32\,\pi^2\,(2-\omega)} \int d^4x\,\sqrt{g}\,\sum_i \alpha_i\,\Gamma_i. \tag{5.120}$$

For this check it is useful to adopt the convention that $[\varphi]_M = 1, [g_{\mu\nu}]_M = [g^{\mu\nu}]_M = 0$. With this choice, we shift the dimensionality to the coordinates and the metric remains dimensionless. This means that each space-time derivative $[\partial_\mu]_M = 1$ increases the mass dimension by one. The result of this convention is $[U(\varphi)]_M = 2$, $[G(\varphi)]_M = 0$ and $[V(\varphi)]_M = 4$. Each derivative of U, V and G with respect to φ

reduces the mass dimension by one and therefore $[s(\varphi)]_M = 0$. This check has been done successfully for all structures.

5.2.3.2 Scaling Behaviour

Another test is the scaling behaviour of the α_i. The coefficients α_i are functions of U, G, V and their derivatives $U^{(n)}$, $G^{(n)}$, $V^{(n)}$

$$\alpha_i(U^{(n)}, G^{(n)}, V^{(n)}), \; n = 0, \ldots, k. \tag{5.121}$$

The structure of the Schwinger–DeWitt algorithm requires that they are homogeneous functions of degree zero in their arguments, i.e. they have to behave as

$$\alpha_i(\lambda U^{(n)}, \lambda G^{(n)}, \lambda V^{(n)}) = \alpha_i(U^{(n)}, G^{(n)}, V^{(n)}) \tag{5.122}$$

under the simultaneous scaling

$$U^{(n)}, G^{(n)}, V^{(n)} \to \lambda U^{(n)}, \lambda G^{(n)}, \lambda V^{(n)} \tag{5.123}$$

for some constant λ. We have explicitly checked that all α_i have this property.

5.2.3.3 $O(N)$ Symmetry and Single Field Limit

A third test can be made based on the $O(N)$ symmetry. The space-time derivatives of the normal vector n_a generate powers of φ in the denominator of the α_i. In addition, we have structures which are proportional to $(N-1)$ due to the projector $(\delta_{ab} - n_a n_b)$. In the single field limit we have $\Phi^a \to \varphi$ and $\delta_{ab} = n_a = N \to 1$. This leads to a degeneracy of certain structures Γ_i listed in Table 5.1, which collapse into the same structures appearing in the effective action of [22]. We can now regard the single field case as an additional check. All terms involving φ in the denominator should either vanish due to an accompanying factor of $(N-1)$ or should exactly cancel out with other contributions from the degenerated structures. Indeed, in the single field limit there is no explicit dependence on φ at all.

One may still be concerned about the terms with powers of φ in the denominator in the multiplet case, since there is no reason why $\varphi = 0$ should be a special value and $U(0)$, $V(0)$, $G(0)$ should be perfectly regular at $\varphi = 0$. However, we can show that these inverse powers of φ are just an artefact of our notation. Instead of considering the $O(N)$-invariant quantity φ^2, we focused on the field $\varphi = \sqrt{\Phi_a \Phi^a}$, since it corresponds to the inflaton in the cosmological application of the Higgs inflation and it facilitates the comparison with the single field results. Due to the $O(N)$ symmetry, all "well-behaved" potentials U, V, G should *effectively* depend on φ^2. Defining

$$\dot{Z} := \frac{\partial Z(\varphi^2)}{\partial \varphi^2}, \tag{5.124}$$

we can express the coefficients α_i in terms of the "dot" derivatives. The expression s remains unchanged, but all derivatives Z', Z'', Z''' are replaced by $\dot{Z}$, $\ddot{Z}$, $\dddot{Z}$. The conversion formulas between the derivatives are given by

$$Z' = 2\,\varphi\,\dot{Z}, \tag{5.125}$$

$$Z'' = 4\,\varphi^2\,\ddot{Z} + 2\,\dot{Z}, \tag{5.126}$$

$$Z''' = 8\,\varphi^3\,\dddot{Z} + 12\,\varphi\,\ddot{Z}. \tag{5.127}$$

In that way, φ's are generated in the numerator which then can cancel the φ in the denominator. Moreover, the inverse powers of φ arising from the $n_a = \Phi_a/\varphi$ contained in some structures Γ_i must also at least be compensated by powers of φ in the numerator of α_i. Ultimately, only even powers of φ can remain in the numerator.

5.2.3.4 Comparison: Single Field and Einstein Frame

The comparison with [2] corresponds to the Einstein frame result of the quantum corrections in the single field case. This means that in addition to the single field limit described above, we set $U = G = 1$. Our result obtained in this limit coincides with the corrected result derived explicitly in Sect. 5.1, where we have also shown the coincidence with the well-known result of [24] in the case $V = 0$.

5.2.3.5 Comparison: Multiplet in a Cosmological Setup

In the context of the RG improved Higgs inflation scenario [3], the beta-functions for the non-minimal coupling ξ and the Higgs self-coupling λ were derived by explicit calculations of $U_{1-\text{loop}}$ and $V_{1-\text{loop}}$. The tree-level values of the couplings in this model are given by

$$U_{\text{tree}}(\varphi) = \frac{1}{2}\,(M_{\text{P}}^2 + \xi\varphi^2), \tag{5.128}$$

$$G_{\text{tree}}(\varphi) = 1, \tag{5.129}$$

$$V_{\text{tree}}(\varphi) = \frac{1}{4}\,\lambda\,(\varphi^2 - \nu^2)^2. \tag{5.130}$$

Here, ν is a symmetry breaking scale. In the calculations of [3] a multiplet Φ^a with a constant background $\nabla_\mu \bar{\Phi}^a = 0$ (leading to vanishing $\hat{\mathcal{R}}_{\mu\nu}$) was assumed. Furthermore, the contribution of graviton loops was neglected (corresponding to considering only the scalar-scalar block of $\hat{P}$). Finally, an expansion in ξ^{-1} was performed, which is justified by the assumption of a strong non-minimal coupling

$|\xi| \simeq 10^4$. Applying all these approximations to the general result (5.109)–(5.194) reproduces exactly the result derived in the appendix of [3]. We will consider these results in more detail in Chap. 6.

5.2.3.6 Comparison: Single Field and Jordan Frame

Another important source for a cross-check is the result of Shapiro and Takata [22]. They performed similar calculations for a single scalar field. In order to confront our general result in the limiting case of a single field with the result of [22] we have to bring our result into a form that is suitable for comparison. We list the following relations in order to convert between our formalism and the one used in [22]

$$\phi \leftrightarrow \varphi, \quad A \leftrightarrow -\frac{1}{2}G, \; B \leftrightarrow U, \; C \leftrightarrow -V, \tag{5.131}$$

$$A_1 \leftrightarrow -\frac{1}{2}G', \, A_2 \leftrightarrow -\frac{1}{2}G'', \text{ etc.} \tag{5.132}$$

$$X := 2AB - 3B_1^2 \leftrightarrow -(GU + 3U'^2) = -\frac{U}{s}. \tag{5.133}$$

In addition, the authors of [22] have chosen to express the R^2 and $R^{\mu\nu}R_{\mu\nu}$ contributions in terms of the Weyl tensor. Using the definition of the Weyl tensor (A.12) and the Gauss–Bonnet identity in four dimensions, we obtain for a general function $F(\phi)$

$$\begin{aligned} F\,C^{\alpha\beta\gamma\delta}C_{\alpha\beta\gamma\delta} &= F\left(\frac{1}{3}R^2 - 2R^{\alpha\beta}R_{\alpha\beta} + R^{\alpha\beta\gamma\delta}R_{\alpha\beta\gamma\delta}\right) \\ &= -\frac{2}{3}F\,R^2 + 2\,F\,R^{\alpha\beta}R_{\alpha\beta}. \end{aligned} \tag{5.134}$$

In the single-field limit, the number of different structures which do arise in the effective action reduces to eleven, see Table 5.1. Comparing our one-loop result in the single-field limit with [22], we find coincidence for the coefficients

$$\begin{aligned} &c_{12} \leftrightarrow \alpha_1, \quad c_7 \leftrightarrow \alpha_2, \quad c_{11} \leftrightarrow \frac{1}{2}\alpha_3, \\ &2c_w \leftrightarrow \alpha_5, \quad c_r - \frac{2}{3}c_w \leftrightarrow \alpha_6, \quad c_{10} \leftrightarrow \alpha_{18}. \end{aligned} \tag{5.135}$$

Differences remain for all other coefficients. Trying to trace back the origin of these discrepancies, we first reduce our input of the `MathTensor` code (the second variation) to the single field case and compare it with the input used in [22]. The authors of [22] have used the convention

$$\omega^A := \delta\Psi^A = \begin{pmatrix} \bar{h}_{\mu\nu} \\ h \\ \sigma \end{pmatrix} \tag{5.136}$$

with $\delta g_{\mu\nu} := h_{\mu\nu}$, $\delta\phi := \sigma$ and $\bar{h}_{\mu\nu} := h_{\mu\nu} - \frac{1}{4} g^{\mu\nu} h_{\mu\nu}$. We did not split $h_{\mu\nu}$ into the traceless part $\bar{h}_{\mu\nu}$ and its trace h. In order to compare their operator F_{AB} (3×3 matrix), with ours (2×2 matrix) it is useful to expand the quadratic forms of the composing parts $\omega^A \tilde{C}_{AB} \Box\, \omega^B$, $\omega^A \Gamma^{\mu}_{AB} \nabla_\mu \omega^B$, $\omega^A W_{AB}\, \omega^B$ and compare the resulting scalars. Both expressions coincide up to a factor of

$$\frac{1}{4} h\, A_1\, (\nabla\phi)^2\, \sigma \tag{5.137}$$

present in [22]. Direct recalculations starting with the action of [22] confirmed our result that this factor should indeed be absent. The origin of this is a cancellation due to contributions coming from an integration by parts. This can be seen easily by calculating the mixed variations and concentrating on the trace part of the metric perturbations. Since our input and the one of [22] coincide up to the term (5.137), we can check whether the differences in the final one-loop results all vanish when setting A_1 to zero. This does not happen, although the differences vanish e.g. for the $c_6 \leftrightarrow \alpha_7$ coefficient in this case. By programming a new `MathTensor` algorithm to calculate the single scalar field case *directly* with `MathTensor`, the result turns out to be the same as in the single field *limit* of our *general* $O(N)$ result. By further investigating the remaining differences (independent of A_1), we consider the coefficient with the biggest deviation c_8. To facilitate the calculations without computer assistance, we limit ourselves to the special case $U = B = 1$ and focus on the $G'' = -\frac{1}{2} A_2$ contributions in c_8. In our calculations (in addition repeated with `MathTensor`) these contributions are absent, but in [22] there remains a contribution G''.

Summarizing, our results are in agreement with those from paper [22] for the most important structures such as $V_{1-\text{loop}}$, $U_{1-\text{loop}}$, $G_{1-\text{loop}}$. As far as other structures are concerned, we find one source of discrepancy connected with the presence of the factor (5.137) in the input of [22], while this term is absent in our input. Some discrepancies still remain which cannot be reduced to this difference in the input. Here, our belief in the correctness of our results is based on additional cross-checks, including the calculations performed without computer assistance for some limiting cases.

5.3 Cosmological Debate: Jordan Frame Versus Einstein Frame

In cosmology there is an ongoing debate for quite a long time whether the two conformal representations in the Jordan and Einstein frame[5] are physically and mathematically equivalent, e.g. see [10, 15, 17] and references therein. There are many interesting questions related to this problem. First, it is clear that Jordan and Einstein frame correspond just to two *special parametrizations* of the basic fields $g_{\mu\nu}(x)$ and $\Phi^a(x)$ for the *same* theory. In general, there are infinitely many different field parametrizations of the same theory. However, if there is only one physical theory, then physical meaningful statements should be invariant under reparametrizations.

In cosmology, the two particular parametrizations denoted by Jordan frame and Einstein frame are distinguished among the set of all possible parametrizations in the following sense: The Jordan frame is the natural parametrization in which one usually formulates the theory according to the conventional methods, i.e. by specifying the tree-level action. The different interaction terms are modelled as scalar contractions of products between monomials of the field operators. Thus, if we want to include a non-minimal coupling, the standard method is to add a corresponding term to the action, e.g. $\xi\, R\, \varphi^2$. This parametrization is called Jordan frame. In a different parametrization, the tree-level action will look different and our physical intuition as well as the simple interpretation of the interaction terms will be lost. In this sense, the Jordan frame is the natural frame to parametrize a theory in a physically meaningful way. Likewise, the Einstein frame also has a special meaning. It formally resembles the well-known situation of General Relativity minimally coupled to a scalar field and in particular it has some mathematically appealing features, since most of the calculations are much easier to perform in the Einstein frame. However, although it might seem that the formal interpretation of the Einstein frame action is clear, it is rather subtle.

The reason is the following: In a general context, taking into account the gravitational interaction, all fields are minimally coupled to gravity via the substitution of partial derivatives by covariant ones. The only coupling between gravitational and scalar degrees of freedom in the action is due to the factors $\sqrt{g}$ and $g^{\mu\nu}$. It is again the "double role" of $g_{\mu\nu}$ causing the conceptual problems. On the one hand it determines the background manifold $(\mathcal{M}, g_{\mu\nu})$ on which the other fields propagate. On the other hand it is itself a fundamental field $g_{\mu\nu}(x)$, which is on equal footing with the other fields and whose dynamics are determined by the scalar curvature R. If all matter fields are minimally coupled to gravity, this is the mathematical expression of the equivalence principle: Gravity couples to all kind of energy universally. The non-minimal coupling in the Jordan frame violates this principle. But since the equivalence principle is a *physical* statement, it has to be independent of the formalism we use to describe it. In particular, if it is violated in the Jordan frame, it is of course still violated in the Einstein frame. Thus, the Einstein frame is *not equivalent* to ordinary General Relativity minimally coupled to a scalar field.

---

[5] We will specify *precisely* what we mean by Jordan frame and Einstein frame and give a mathematical definition in Sect. 5.3.1.

This should not be surprising, since we started from a theory different from General Relativity minimally coupled to a scalar field and thus should also end with a theory different from General Relativity after a reparametrization of fields. Thus, once we have accepted the *physical* equivalence of the different parametrizations, we should be able to freely transform between those representations in order to manipulate or simplify intermediate calculations. The physical content will be the same in all parametrizations, although the results will of course look different in different parametrizations.[6] However, to obtain a clear physical *interpretation* of the results, it seems best to transform the final results back to the "original" parametrization in which we have formulated the action.

Regardless of the *physical* equivalence of different frames, taken by itself it is an interesting purely *mathematical* question whether it is possible at all to transform between particular frames determined by certain mathematical criteria. In the context of the cosmological debate between Jordan frame and Einstein frame, the question is whether the action (5.68) formulated in the Jordan parametrization can be brought into a form which can be called Einstein parametrization, i.e. involving only minimally coupled scalar fields with a canonically normalized kinetic term. We will investigate this question in Sect. 5.3.1.

If we transform between Einstein frame and Jordan frame, we should show the mathematical equivalence between these frames to *all levels* of our formalism. It can easily be shown that this is indeed the case at the classical level. However, does this also hold at the quantum level? Coming back to the cosmological debate, the question is whether the results of quantum corrections, calculated in the Jordan frame and in the Einstein frame are equivalent. If this is the case, it would justify the strategy followed in [2] for the calculations of one-loop quantum corrections. In [2], the one-loop divergences for the effective action of a *single* scalar field *non-minimally* coupled to gravity were calculated in the following way: Starting from the action written in the Jordan frame, the authors of [2] have undertaken the transition to the Einstein frame by a conformal transformation of the metric and by a proper redefinition of the scalar field so that the interaction between the scalar field and gravity has become minimal and the kinetic term for the scalar field has acquired the standard form. Then, the one-loop divergences were calculated by using the generalized Schwinger–DeWitt technique [6, 14] for the fields in the Einstein frame. Finally, this result was undergone an inverse transformation back to the original Jordan frame. In contrast, in the preceding section we have studied the structure of divergences of the one-loop effective action for a multiplet of scalar fields non-minimally coupled to gravity [23] with the action (5.68). All calculations were carried out consistently in the *Jordan frame*. We will pursue the question of frame dependence of quantum corrections in

---

[6] This is similar to the situation of choosing an ordinary space-time coordinate system. For example, we can imagine a rotational invariant problem and try to describe it once in Cartesian coordinates and once in polar coordinates. Of course, the physical content of the problem is the same, described in both coordinate systems. But the polar coordinates will be better suited to interpret the physical results, since they are optimally adapted to the symmetries of the problem. This freedom of the choice of configuration space variables—the "generalized coordinates"—lies at the very heart of the Lagrangian formalism.

Sect. 5.3.2 by comparing the one-loop effective actions explicitly calculated in the two different cosmological parametrizations.

Finally, in Sect. 5.3.3, we will investigate the question of frame dependence in a more general context and re-investigate the conventional formalism used to calculate the effective action in Sect. 4.4. In particular, we will discuss the qualitative features of a modification of this formalism that is essentially based on a *geometrical* treatment of the effective action, introduced in [5, 26].

5.3.1 Transition Between the Frames

Before we can pursue the question of the frame dependence of quantum corrections in a cosmological context, we have to investigate whether it is possible at all to construct the transition of (5.68) from the Jordan frame to the Einstein frame and vice versa in the case of a $O(N)$ multiplet non-minimally coupled to gravity, see (5.68). By a conformal transformation with a suitable choice of the φ dependent conformal factor we can remove the non-minimal coupling that is present in the original Jordan frame. Then, the potential gets simply rescaled by the conformal factor, but the conformal transformation also leads to a modification of the kinetic term which acquires a non-standard form. By definition, the Einstein frame is that particular frame which has a minimally coupled scalar sector *and* a canonically normalized kinetic term. To bring the kinetic term in its standard form, we can make use of the freedom to perform an additional reparametrization of the scalar fields. For a single scalar field, this condition leads to a non-linear relation between the old and the new field, see e.g. [2]. For the case of a $O(N)$ symmetric multiplet of scalar fields, it was claimed in [18] that no such transformation does exist. We will explicitly construct such a transformation law for the $O(N)$ symmetric multiplet of scalar fields but then show that this leads to a non-trivial integrability relation in agreement with the statement made in [18].

5.3.1.1 Conformal Transformation

Let us consider the following conformal transformation of the metric field in four space-time dimensions:

$$g_{\mu\nu} = f\,\hat{g}_{\mu\nu}. \tag{5.138}$$

Then, the transformation for the inverse metric field is given by

$$g^{\mu\nu} = f^{-1}\,\hat{g}^{\mu\nu}. \tag{5.139}$$

The square root of the determinant transforms according to

$$\sqrt{g} = f^2 \sqrt{\hat{g}} \tag{5.140}$$

and accounts for an overall multiplicative factor of the action. In order to calculate the transformation of the Riemann tensor it is necessary to first calculate the behaviour of the Christoffel symbol and its first derivative under such a conformal transformation

$$\Gamma^{\alpha}_{\mu\nu} = \hat{\Gamma}^{\alpha}_{\mu\nu} + \frac{1}{2} f^{-1}\Big(\delta^{\alpha}_{\mu} f_{,\nu} + \delta^{\alpha}_{\nu} f_{,\mu} - \hat{g}^{\alpha\gamma} \hat{g}_{\mu\nu} f_{,\gamma}\Big), \tag{5.141}$$

$$\begin{aligned}\Gamma^{\alpha}_{\mu\nu,\beta} = \hat{\Gamma}^{\alpha}_{\mu\nu,\beta} &+ \frac{1}{2} f^{-1}\Big[\delta^{\alpha}_{\mu} f_{,\nu\mu} + \delta^{\alpha}_{\nu} f_{,\mu\beta} - \Big(\hat{g}^{\alpha\gamma}{}_{,\beta}\, \hat{g}_{\mu\nu} + \hat{g}^{\alpha\gamma} \hat{g}_{\mu\nu,\beta}\Big) f_{,\gamma} - \hat{g}^{\alpha\gamma} \hat{g}_{\mu\nu} f_{,\gamma\beta}\Big] \\ &- \frac{1}{2} f^{-2} f_{,\beta} \Big(\delta^{\alpha}_{\mu} f_{,\nu} + \delta^{\alpha}_{\nu} f_{,\mu} - \hat{g}^{\alpha\gamma} \hat{g}_{\mu\nu} f_{,\gamma}\Big).\end{aligned} \tag{5.142}$$

In order to calculate the Riemann tensor, Ricci tensor and Ricci scalar, it is convenient to use a Riemannian normal coordinate system, i.e. $\hat{\Gamma}^{\alpha}_{\mu\nu} = 0$, but $\hat{\Gamma}^{\alpha}_{\mu\nu,\beta} \neq 0$ in general. In such a coordinate system, the terms quadratic in $\hat{\Gamma}$ do not appear and we can replace each covariant derivative with a partial derivative. If the final transformation rule is a tensor equation, it will hold in all coordinate systems and we can again replace the partial derivatives by the covariant ones. The result for the conformal transformation of the Riemann tensor is

$$\begin{aligned}R^{\alpha}{}_{\beta\gamma\delta} = \hat{R}^{\alpha}{}_{\beta\gamma\delta} &+ \frac{1}{4} f^{-2} \Big(3 f_{;\gamma} f^{;\alpha} \hat{g}_{\beta\delta} - 3 f_{;\delta} f^{;\alpha} \hat{g}_{\beta\gamma} + 3 f_{;\beta} f_{;\delta} \delta^{\alpha}_{\gamma} - 3 f_{;\beta} f_{;\gamma} \delta^{\alpha}_{\delta} \\ &\qquad + f_{;\nu} f^{;\nu} \hat{g}_{\beta\gamma} \delta^{\alpha}_{\delta} - f_{;\nu} f^{;\nu} \hat{g}_{\beta\delta} \delta^{\alpha}_{\gamma}\Big) \\ &+ \frac{1}{2} f^{-1} \Big(f_{;\delta}{}^{\alpha} \hat{g}_{\beta\gamma} - f_{;\gamma}{}^{\alpha} \hat{g}_{\beta\delta} + f_{;\beta\gamma} \delta^{\alpha}_{\delta} - f_{;\beta\delta} \delta^{\alpha}_{\gamma}\Big).\end{aligned} \tag{5.143}$$

By contracting the first and the third index, we obtain the transformation law for the Ricci tensor

$$R_{\alpha\beta} = \hat{R}_{\alpha\beta} + \frac{3}{2} f^{-2} f_{;\alpha} f_{;\beta} - f^{-1} f_{;\alpha\beta} - \frac{1}{2} f^{-1} f_{;\nu}{}^{\nu} \hat{g}_{\alpha\beta}. \tag{5.144}$$

The Ricci scalar follows from contracting (5.144) with the inverse metric $g^{\mu\nu}$

$$R = g^{\alpha\beta} R_{\alpha\beta} = f^{-1} \hat{R} + \frac{3}{2} f^{-3} f_{;\nu} f^{;\nu} - 3 f^{-2} f_{;\nu}{}^{\nu}. \tag{5.145}$$

In the divergent part of the one-loop effective action, the structures $R^{\alpha\beta\gamma\delta} R_{\alpha\beta\gamma\delta}$, $R^{\alpha\beta} R_{\alpha\beta}$ and R^2 do appear. The structure $R^{\alpha\beta\gamma\delta} R_{\alpha\beta\gamma\delta}$ can be transformed into contributions to $R^{\alpha\beta} R_{\alpha\beta}$ and R^2 by making use of the Gauss–Bonnet identity, i.e.

$$\frac{\delta}{\delta g_{\mu\nu}} \Big[\int d^4x \sqrt{g} \Big(R^{\alpha\beta\gamma\delta} R_{\alpha\beta\gamma\delta} + 4 R^{\alpha\beta} R_{\alpha\beta} - R^2\Big)\Big] = 0. \tag{5.146}$$

The conformal transformation laws for the remaining two structures yield

$$R^{\alpha\beta} R_{\alpha\beta} = f^{-2} \hat{R}^{\alpha\beta} \hat{R}_{\alpha\beta} - f^{-3} \left(\hat{R} f_{;\nu}^{\ \ \nu} + 2 \hat{R}^{\alpha\beta} f_{;\alpha\beta} \right)$$
$$+ f^{-4} \left(f_{;\alpha\beta} f^{;\alpha\beta} + 3 \hat{R}^{\alpha\beta} f_{;\alpha} f_{;\beta} + 2 f_{;\nu}^{;\ \nu} f_{;\mu}^{\ \ \mu} \right)$$
$$- 3 f^{-5} \left(\frac{1}{2} f_{;\nu} f^{;\nu} f_{;\mu}^{\ \ \mu} + f^{;\alpha\beta} f_{;\alpha} f_{;\beta} \right) + \frac{9}{4} f^{-6} f^{;\alpha} f_{;\alpha} f^{;\beta} f_{;\beta}, \tag{5.147}$$

$$R^2 = f^{-2} \hat{R}^2 - 6 f^{-3} \hat{R} f_{;\nu}^{\ \ \nu} + f^{-4} \left(3 \hat{R} f_{;\nu} f^{;\nu} + 9 f_{;\nu}^{\ \ \nu} f_{;\mu}^{\ \ \mu} \right)$$
$$- 9 f^{-5} f_{;\nu} f^{;\nu} f_{;\mu}^{\ \ \mu} + \frac{9}{4} f^{-6} f_{;\nu} f^{;\nu} f_{;\mu} f^{;\mu}. \tag{5.148}$$

5.3.1.2 Field Transformation

Let us now investigate, how the non-minimal coupling term in (5.68) changes under a conformal transformation

$$U \sqrt{g} R = U \sqrt{\hat{g}} \left(f \hat{R} + \frac{3}{2} f^{-1} f_{;\nu} f^{;\nu} - 3 f_{;\nu}^{\ \ \nu} \right). \tag{5.149}$$

In order to remove the non-minimal coupling (keeping only an Einstein–Hilbert-like term associated with some constant U_0), we have to choose the conformal factor

$$f = \frac{U_0}{U}. \tag{5.150}$$

But even if we have managed to remove the non-minimal coupling by choosing the conformal factor (5.150), the other terms in the action will also be modified. The potential and kinetic term simply get rescaled by powers of the conformal factor

$$-\sqrt{g} V = \sqrt{\hat{g}} \left(\frac{U_0}{U} \right)^2 V = \sqrt{\hat{g}} \hat{V}, \tag{5.151}$$

$$-\sqrt{g} \frac{1}{2} G g^{\mu\nu} \Phi^a_{\ ;\mu} \Phi_{a\,;\nu} = -\sqrt{\hat{g}} \frac{1}{2} \left(\frac{U_0}{U} \right) G \hat{g}^{\mu\nu} \Phi^a_{\ ;\mu} \Phi_{a\,;\nu} \tag{5.152}$$

with $\hat{V} := (U_0/U)^2 V$. However, as can be seen from (5.149) after substituting (5.150) and performing an integration by parts, the transformation of the non-minimal coupling term induces an additional contribution to the kinetic term

$$U \sqrt{g} R = U_0 \sqrt{\hat{g}} \left(\hat{R} - \frac{3}{2} U^{-2} U_{;\mu} U^{;\mu} \right)$$
$$= U_0 \sqrt{\hat{g}} \left[\hat{R} - \frac{3}{2} \left(\frac{U'}{U} \right)^2 \Phi^a_{\ ;\mu} n_a \Phi^{b\,;\mu} n_b \right]. \tag{5.153}$$

Thus, after conformal transformation with (5.150), we finally obtain the kinetic term as a sum of (5.152) and the contribution from (5.153)

$$\sqrt{\hat{g}}\,\frac{1}{2}\,\hat{g}^{\mu\nu}\underbrace{\left(\frac{U_0}{U}\right)\left(\frac{G\,U\,\delta_{ab}+3\,U'^2\,n_a n_b}{U}\right)}_{=:M_{ab}}\Phi^a_{\;;\,\mu}\,\Phi^b_{\;;\nu}. \tag{5.154}$$

The Einstein frame is defined to be that particular frame in which the multiplet is coupled minimally to gravity and in addition has a kinetic term canonically normalized to one. We can exploit the fact that Φ^a is just a configuration variable and perform an additional field reparametrization

$$\Phi^a \to \hat{\Phi}^a. \tag{5.155}$$

The explicit field transformation law (5.155) for a transition to the Einstein frame is determined by the condition

$$\hat{g}^{\mu\nu} M_{ab}\Phi^a_{\;;\,\mu}\Phi^b_{\;;\,\nu} = \hat{g}^{\mu\nu}\,\delta_{ab}\,\hat{\Phi}^a_{\;;\,\mu}\hat{\Phi}^b_{\;;\,\nu}. \tag{5.156}$$

Therefore, the objective is to find a solution to (5.155), or equivalently to

$$\left(\frac{\partial\,\hat{\Phi}^a}{\partial\Phi^c}\right)\left(\frac{\partial\,\hat{\Phi}_a}{\partial\Phi^d}\right) = M_{cd} = A\,\delta_{cd} + B\,n_c n_d. \tag{5.157}$$

We have defined the abbreviations

$$A := U_0\,\frac{G}{U} \quad \text{and} \quad B := 3\,U_0\,\frac{U'^2}{U^2}. \tag{5.158}$$

There are two aspects connected to this problem. First, it was claimed in [18] that there does not even exist a field reparametrization that transforms the "non-flat field metric" M_{ab} into the "flat" form δ_{ab} and therefore a transition between Jordan and Einstein frame would be impossible for multiple scalar fields. The second aspect is connected with the way of calculating the quantum corrections corresponding to the method used in [2]. In the Einstein frame result for the one-loop effective action, the terms $\propto \hat{V}''$ and $\propto \hat{V}'$ appear, see (5.61). In order to transform these structures back into the Jordan frame, we also need the inverse solution to (5.157). In order to solve (5.157), we make the following ansatz:

$$\left(\frac{\partial\,\hat{\Phi}_a}{\partial\Phi^b}\right) := \alpha\,\delta_{ab} + \beta\,n_a n_b \quad . \tag{5.159}$$

Inserting this into (5.157) yields

$$(\alpha\,\delta_{ac}+\beta\,n_a n_b)\,(\alpha\,\delta^a_d+\beta\,n^a n_d)=\alpha^2\,\delta_{ab}+(2\alpha\beta+\beta^2)\,n_c n_d=A\,\delta_{cd}+B\,n_c n_d. \tag{5.160}$$

Comparing coefficients leads to the identification

$$\alpha^2=A \quad \text{and} \quad 2\alpha\beta+\beta^2=B. \tag{5.161}$$

With the explicit expressions

$$\alpha=\sqrt{\frac{U_0}{U}}\,\sqrt{G} \quad \text{and} \quad \beta=\sqrt{\frac{U_0}{U}}\left(-\sqrt{G}\pm\sqrt{\frac{\left(G\,U+3\,U'^2\right)}{U}}\right), \tag{5.162}$$

For the positive solution, we obtain

$$\left(\frac{\partial\,\hat{\Phi}_a}{\partial\Phi^b}\right)=\sqrt{\frac{U_0\,G}{U}}\left[\delta_{ab}+\left(\sqrt{\frac{G\,U+3\,U'^2}{G\,U}}-1\right)n_a n_b\right]. \tag{5.163}$$

We can easily calculate the inverse with a similar ansatz

$$\left(\frac{\partial\,\Phi_a}{\partial\hat{\Phi}^b}\right)=x\,\delta_{ab}+y\,n_a n_b. \tag{5.164}$$

Using the same method as above yields

$$x=\frac{1}{\alpha} \quad \text{and} \quad y=-\frac{\beta}{\alpha\,(\alpha+\beta)}. \tag{5.165}$$

Hence, the inverse transformation is given by

$$\left(\frac{\partial\,\Phi_a}{\partial\hat{\Phi}^b}\right)=\sqrt{\frac{U}{U_0\,G}}\left[\delta_{ab}+\left(\sqrt{\frac{G\,U}{G\,U+3\,U'^2}}-1\right)n_a n_b\right]. \tag{5.166}$$

We would now investigate the first aspect mentioned before. Although we have explicitly constructed the transformation formulas between the Einstein frame and the Jordan frame in the case of a $O(N)$ symmetric scalar multiplet, this does not mean that this transition will be possible in general. What is still missing, is the integrability condition connected to the differential relation (5.163). To make this point more clear, we consider for a moment the following system of (functional) differential equations

$$\left(\frac{\partial\,\hat{\Phi}_a}{\partial\Phi^b}\right)=C_{ab}(\phi_1,\ldots,\phi_N). \tag{5.167}$$

Although we treat the components Φ_a here formally as "coordinates" one should bear in mind that we are really dealing with fields $\Phi_a(x)$ as well as functional differentiation and integration, respectively. In order to find the necessary condition to make this system integrable, we consider (5.167) for some *fixed* components a and b. Formally integrating this equation with respect to Φ_b leads to

$$\hat{\Phi}_a = \int \mathrm{d}\Phi_b \, C_{ab} + \tilde{\Phi}_a. \tag{5.168}$$

No summation over equal indices was assumed. The arbitrary field $\tilde{\Phi}^a$ is an integration constant that does not depend on the component Φ_b

$$\frac{\partial \tilde{\phi}_a}{\partial \phi_b} = 0. \tag{5.169}$$

If we differentiate now (5.168) with respect to some component Φ_c with $c \neq b$, we obtain

$$\frac{\partial \hat{\phi}_a}{\partial \Phi^c} = \int \mathrm{d}\Phi_b \, \frac{\partial \, C_{ab}}{\partial \Phi^c} + \frac{\partial \tilde{\Phi}_a}{\partial \Phi^c} = C_{ac}. \tag{5.170}$$

Differentiating (5.170) once again, but this time with respect to the Φ_b component and using (5.169), we obtain the integrability relation

$$\frac{\partial C_{ab}}{\partial \Phi^c} = \frac{\partial C_{ac}}{\partial \Phi^b}. \tag{5.171}$$

Comparison with (5.163) shows that the matrix C_{ab} has the structure

$$C_{ab} := \alpha(\varphi) \, \delta_{ab} + \beta(\varphi) \, n_a n_b. \tag{5.172}$$

and therefore we find

$$\frac{\partial C_{ab}}{\partial \Phi^c} = \alpha' \, \delta_{ab} \, n_c + \beta' \, n_a n_b n_c + \frac{\beta}{\varphi} \, (n_a \delta_{bc} + n_b \delta_{ac} - 2 \, n_a n_b n_c). \tag{5.173}$$

Inserting this in the relation (5.171) yields

$$\left(\alpha' - \frac{\beta}{\varphi}\right) (\delta_{ab} n_c - \delta_{ac} n_b) = 0 \tag{5.174}$$

which in turn implies

$$\left(\alpha' - \frac{\beta}{\varphi}\right) = 0. \tag{5.175}$$

This is a non-trivial integrability condition that relates U, G and its first derivatives and thus would strongly restrict the allowed models. In general there is no transition between the two frames defined here as Einstein frame and Jordan frame for multiple fields which is in agreement with the result found in [18].

5.3.2 *Frame Dependence of Quantum Corrections*

In the previous section we have explicitly constructed the transformation laws for the transition between the Jordan frame and the Einstein frame of the particular model (5.68). Since a transition for a multiplet is not possible in general we will consider in the following discussion only the single field case $N = 1$. We will now pursue the question if the divergent parts of the off-shell one-loop effective action calculated in the different frames do coincide or not. We can rephrase this question by asking whether the following diagram commutes or not.

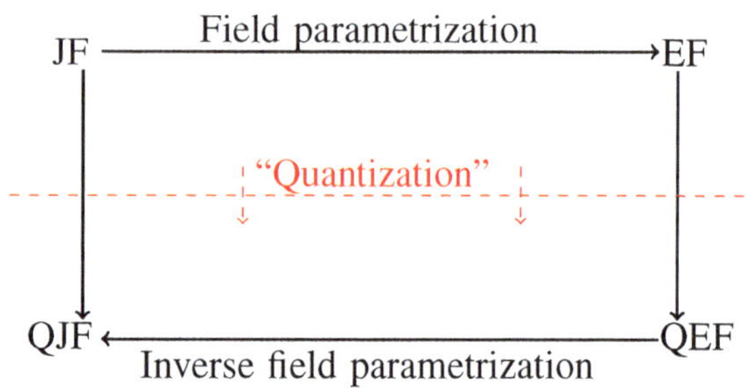

Fig. 5.1 Illustration of the two different methods: Transition between the Jordan frame and the Einstein frame and calculation of quantum corrections

In order to distinguish between the two different approaches, we will refer to the direct Jordan frame calculations as "method B" and to the detour via the Einstein frame as "method A". Technically, we will proceed as follows: The result (5.102) of Sect. 5.2 was calculated by method A. Hence, we can simply adopt it and only have to reduce it to the single field case $N = 1$, i.e. $\Phi^a \to \varphi$. As discussed in Sect. 5.2.3.3 this will lead to a degeneration of certain structures. In order to obtain the result for method B, we can use again the single field result of method A and simply set $U \equiv U_0 = \frac{1}{2}M_{\rm P}^2$ and $G \equiv 1$. Of course, this leads to the same result as if we would have calculated the effective action on the basis of the original Einstein frame action. In order to compare the result of method B with the result of method A, we have to subject the result of method A to the inverse conformal and field transformations. Then we can compare the results of both methods in the Jordan frame parametrization of the fields. By using the transformation formulas presented below and the package `MathTensor`, we can calculate the inverse conformal transformation back to the Jordan frame. The result can be compared with the single field limit of the direct Jordan frame calculations ("method B") of Sect. 5.2.

As discussed above we use the single field limit $N = 1$ of the general Jordan frame result (5.102) for the case $U = \frac{1}{2}M_{\rm P}^2$ and $G = 1$ in order to find the result for the divergent part of the off-shell effective action calculated in the Einstein frame parametrization

$$W_{1-\text{loop}}^{\text{div, A}} = \frac{1}{32\pi^2(\omega-2)} \int d^4x\, \hat{g}^{1/2} \left[\frac{43}{60}\, \hat{R}_{\mu\nu}\hat{R}^{\mu\nu} + \frac{1}{40}\, \hat{R}^2 - \frac{1}{3}\hat{R}(\hat{\varphi}_{,\nu}\hat{\varphi}^{,\nu}) \right.$$

$$+\frac{5}{4}\,(\hat{\varphi}_{,\nu}\hat{\varphi}^{,\nu})^2+5\hat{V}^2-2\hat{V}'^2+\frac{1}{2}\hat{V}''^2-\frac{13}{3}\hat{R}\,\hat{V}-\frac{1}{6}\,\hat{R}\,\hat{V}''$$
$$+\,\hat{V}\,(\hat{\varphi}_{,\nu}\hat{\varphi}^{,\nu})-2\hat{V}''\,(\hat{\varphi}_{,\nu}\hat{\varphi}^{,\nu})\Bigg]. \tag{5.176}$$

As a check of this result, we can compare (5.176) with the direct Einstein frame calculations (5.61) performed in Sect. 5.1. They both coincide.

In order to compare the Einstein frame structures (5.176) with the $N = 1$ limit of the general Jordan frame result (5.102), we have to express (5.176) in terms of the original Jordan frame parametrization. We will first collect all necessary formulas needed to express the Einstein frame result (5.176) in terms of the Jordan frame parametrizations of the fields $(g_{\mu\nu},\,\varphi)$.

When expressing the Einstein frame structures $\hat{\Gamma}_i$ in terms of the Jordan frame structures Γ_i (see Sect. 5.2.3.3 for a definition of the Γ_i associated with the α_i), it is important to note that the inverse conformal transformations of certain structures induce different structures not present in the Einstein frame effective action (5.176). Some of these structures can be eliminated again via an integration by parts. Thus, to obtain the final result for the coefficients of the minimal set of independent structures expressed in the Jordan frame parametrization, we first have to calculate the inverse transformation of all structures $\hat{\Gamma}_i \to \Gamma_i$. Then we have to perform the inverse transformation of all coefficients $\hat{\alpha}_i \to \alpha_i$. Finally, we have to use the single field limit $N = 1$ of the transfer equations listed in Appendix F to remove again certain structures via integration by parts and express the effective action in terms of the set of independent structures in the Jordan frame parametrization.

We will begin by expressing the *structures* present in (5.176) in the Jordan frame parametrization. We first perform the transformation of the scalar field $\hat{\varphi} \to \varphi$. Since the effective action in the Einstein frame parametrization (5.176) contains only structures involving first derivatives of the scalar field $\hat{\varphi}$, we can use the $N = 1$ version of (5.163) to obtain

$$\hat{g}^{\mu\nu}\,\nabla_\mu\hat{\varphi}\,\nabla_\nu\hat{\varphi}=\frac{U_0}{U}\left[G\,U+3\,(U')^2\right]\hat{g}^{\mu\nu}\,\nabla_\mu\varphi\,\nabla_\nu\varphi. \tag{5.177}$$

Later, in order to express the *coefficients* appearing in (5.176) in terms of the Jordan frame parametrization, we must also convert derivatives of the functional charges U, G and V with respect to the field $\hat{\varphi}$ in the Einstein frame parametrization into derivatives with respect to the field φ in the Jordan frame parametrization. Since (5.176) contains only powers of $\hat{V}$, $\hat{V}'$ and $\hat{V}''$, the transformation rules are easily found to be

$$\hat{V}=\frac{U_0^2}{U^2}\,V\,, \tag{5.178}$$

$$\frac{\partial\hat{V}}{\partial\hat{\varphi}}=U_0^2\,\frac{\partial\left(\frac{V}{U^2}\right)}{\partial\varphi}\,\frac{\partial\varphi}{\partial\hat{\varphi}}=\frac{U_0^{3/2}}{U^2}\,\frac{(V'\,U-2\,U'\,V)}{\sqrt{G\,U+3\,U'^2}}, \tag{5.179}$$

$$\frac{\partial^2 \hat{V}}{\partial \hat{\varphi}^2} = \frac{U_0}{U^2\left[G\,U + 3\,(U)'^2\right]^2}\Big\{12\,V\,(U')^4 - 9\,U\,(U')^3\,V' - 3\,U^2\,U'\,V'\,U'' + 3\,U^2\,(U')^2\,V'' + G\left[5\,U\,V\,(U')^2 - \frac{7}{2}U^2\,U'\,V' - 2\,U^2\,V\,U'' + U^3\,V''\right] + G'\left[U^2\,U'\,V - \frac{1}{2}U^3\,V'\right]\Big\}. \tag{5.180}$$

These expressions do coincide with the equations (2.73) and (2.74) of [2].

Finally we have to calculate the inverse conformal transformation of the metric field $\hat{g}_{\mu\nu} \rightarrow g_{\mu\nu}$. Using the general formulas (5.138)–(5.148), we obtain the following transformation rules from the Einstein frame parametrization to the Jordan frame parametrization for the metric field

$$\hat{g}_{\mu\nu} = \frac{U}{U_0}\,g_{\mu\nu}, \quad \hat{g}^{\mu\nu} = \frac{U_0}{U}\,g^{\mu\nu}, \quad \sqrt{\hat{g}} = \frac{U^2}{U_0^2}\,\sqrt{g} \tag{5.181}$$

$$\hat{R} = \frac{U_0}{U}\left\{R - 3\frac{U'}{U}(\Box\varphi) + \left[\frac{3}{2}\frac{(U')^2}{U^2} - 3\frac{U''}{U}\right](\varphi_{,\mu}\varphi^{,\mu})\right\} \tag{5.182}$$

$$\hat{R}^{\mu\nu}\,\hat{R}_{\mu\nu} = \frac{U_0^2}{U^2}\Big\{R_{\mu\nu}R^{\mu\nu} - 2\,\frac{U'}{U}\,R^{\mu\nu}\,\varphi_{;\mu\nu} + \left[3\,\frac{(U')^2}{U^2} - 2\,\frac{U''}{U}\right]R^{\mu\nu}\,\varphi_{,\mu}\varphi_{,\nu} - \frac{U'}{U}\,R\,(\Box\,\varphi) - \frac{U''}{U}\,R\,(\varphi_{,\mu}\varphi^{,\mu}) + \frac{(U')^2}{U^2}\,\varphi^{;\mu\nu}\varphi_{;\mu\nu} + \left[2\,\frac{U'\,U''}{U^2} - 3\,\frac{(U')^3}{U^3}\right]\varphi^{;\mu\nu}\,\varphi_{,\mu}\varphi_{,\nu} + 2\,\frac{(U')^2}{U^2}\,(\Box\,\varphi)^2 + \left[4\,\frac{U'\,U''}{U} - \frac{3}{2}\,\frac{(U')^3}{U^3}\right](\Box\,\varphi)\,(\varphi_{,\mu}\varphi^{,\mu}) + \left[\frac{9}{4}\,\frac{(U')^4}{U^4} - \frac{9}{2}\,\frac{(U')^2\,U''}{U^3} + 3\,\frac{(U'')^2}{U^2}\right](\varphi_{,\mu}\varphi^{,\mu})^2\Big\} \tag{5.183}$$

After having applied these transformation laws to all structures appearing in (5.176), not forgetting the remaining transformation of the overall factor $\sqrt{\hat{g}}$ and the inverse metric $\hat{g}^{\mu\nu}$ contained in the kinetic structure $\hat{g}^{\mu\nu}\,\varphi_{,\mu}\varphi_{,\nu} = U_0/U\,g^{\mu\nu}\varphi_{,\mu}\varphi_{,\nu}$, we find an expression for the effective action which still contains four redundant structures. Integration by parts (neglecting surface terms) converts these structures into contributions to other structures determined by the following transition formulas

$$\alpha\,(\Box\varphi) = -\,\alpha'\,(\varphi_{,\mu}\varphi^{,\mu}) \tag{5.184}$$

$$\alpha\,(\varphi^{;\mu\nu}\varphi_{;\mu\nu}) = \frac{1}{2}\,\alpha''\,(\varphi_{,\mu}\varphi^{,\mu})^2 + \frac{3}{2}\,\alpha'\,(\varphi_{,\mu}\varphi^{,\mu})(\Box\varphi) + \alpha\,(\Box\varphi)^2 - \alpha\,(R^{\mu\nu}\,\varphi_{,\mu}\,\varphi_{,\nu}) \tag{5.185}$$

$$\alpha\,(\varphi^{;\,\mu\nu}\varphi_{,\,\mu}\varphi_{,\,\nu}) = -\frac{1}{2}\alpha'\,(\varphi_{,\,\mu}\varphi^{,\,\mu})^2 - \frac{1}{2}\alpha\,(\varphi_{,\,\mu}\varphi^{,\,\mu})\,(\Box\phi) \tag{5.186}$$

$$\alpha\,(R^{\mu\nu}\,\varphi_{;\,\mu\nu}) = \frac{1}{2}\alpha'\,R\,(\varphi_{,\,\mu}\varphi^{,\,\mu}) - \alpha'\,(R^{\mu\nu}\,\varphi_{,\,\mu}\,\varphi_{,\,\nu}) + \frac{1}{2}\alpha\,R\,(\Box\varphi) \tag{5.187}$$

These formulas can be obtained for the single field case $N = 1$ by using the formulas given in Appendix F for the general multiplet case and they do coincide with those given in [22]. Using these formulas, we can finally bring the Einstein frame effective action (5.176) expressed in the Jordan frame parametrization into its canonical form with the minimal set of eleven independent structures

$$\begin{aligned}\Gamma^{\text{div, A}}_{1-\text{loop}} = {}& \frac{1}{32\pi^2(\omega-2)}\int d^4x\, g^{1/2}\Big\{U^{\text{A}}_{1-\text{loop}}\,R + G^{\text{A}}_{1-\text{loop}}\,(\varphi_{,\,\nu}\varphi^{,\,\nu}) + V^{\text{A}}_{1-\text{loop}} \\ & + \alpha^{\text{A}}_1\,R_{\mu\nu}R^{\mu\nu} + \alpha^{\text{A}}_2\,R^2 + \alpha^{\text{A}}_3\,R\,(\Box\,\varphi) + \alpha^{\text{A}}_4\,R\,(\varphi_{,\,\nu}\,\varphi^{,\,\nu}) + \alpha^{\text{A}}_5\,R^{\mu\nu}\,\varphi_{,\,\mu}\,\varphi_{,\,\nu} \\ & + \alpha^{\text{A}}_6\,(\Box\,\varphi)\,(\Box\,\varphi) + \alpha^{\text{A}}_7\,(\Box\,\varphi)\,(\varphi_{,\,\nu}\,\varphi^{,\,\nu}) + \alpha^{\text{A}}_8\,(\varphi_{,\,\nu}\,\varphi^{,\,\nu})^2\Big\}. \end{aligned} \tag{5.188}$$

The one-loop contributions to U, G and V are given by

$$\begin{aligned}U^{\text{A}}_{1-\text{loop}} = {}& \text{V}\left[\text{s}^2\left(-\frac{\text{G}'\text{U}'}{6\text{U}} - \frac{(\text{U}')^2\,\text{U}''}{\text{U}^2} + \frac{(\text{U}')^4}{2\text{U}^3}\right) + \text{s}\left(\frac{\text{U}''}{3\text{U}} - \frac{5\,(\text{U}')^2}{6\text{U}^2}\right) - \frac{13}{3\text{U}}\right] \\ & + \text{V}'\left[\text{s}^2\left(\frac{\text{G}'}{12} + \frac{\text{U}'\text{U}''}{2\text{U}} - \frac{(\text{U}')^3}{4\text{U}^2}\right) + \frac{7\text{s}\text{U}'}{12\text{U}}\right] - \frac{\text{V}''\,\text{s}}{6}, \end{aligned} \tag{5.189}$$

$$\begin{aligned}G^{\text{A}}_{1-\text{loop}} = {}& \text{V}''\left[\text{s}^2\left(\frac{3\text{G}'\text{U}'}{4\text{U}} + \frac{9\,(\text{U}')^2\,\text{U}''}{2\text{U}^2} - \frac{9\,(\text{U}')^4}{4\text{U}^3}\right) + \frac{2\text{s}\,(\text{U}')^2}{\text{U}^2} - \frac{2}{\text{U}}\right] \\ & + \text{V}'\left[\text{s}^3\left(-\frac{6\text{G}'\,(\text{U}')^2\,\text{U}''}{\text{U}^2} + \frac{3\text{G}'\,(\text{U}')^4}{\text{U}^3} - \frac{(\text{G}')^2\,\text{U}'}{2\text{U}} + \frac{18\,(\text{U}')^5\,\text{U}''}{\text{U}^4}\right.\right. \\ & \left. - \frac{18\,(\text{U}')^3\,(\text{U}'')^2}{\text{U}^3} - \frac{9\,(\text{U}')^7}{2\text{U}^5}\right) + \text{s}^2\left(\frac{\text{G}''\text{U}'}{4\text{U}} - \frac{19\text{G}'\,(\text{U}')^2}{8\text{U}^2}\right. \\ & \left. - \frac{18\,(\text{U}')^3\,\text{U}''}{\text{U}^3} + \frac{3\text{U}'\,(\text{U}'')^2}{2\text{U}^2} + \frac{69\,(\text{U}')^5}{8\text{U}^4} + \frac{3\text{U}'''\,(\text{U}')^2}{2\text{U}^2}\right) \\ & \left. + \text{s}\left(\frac{\text{G}'}{\text{U}} + \frac{35\text{U}'\text{U}''}{4\text{U}^2} - \frac{65\,(\text{U}')^3}{8\text{U}^3}\right) - \frac{6\text{U}'}{\text{U}^2}\right] \\ & + \text{V}\left[\text{s}^3\left(\frac{12\text{G}'\,(\text{U}')^3\,\text{U}''}{\text{U}^3} - \frac{6\text{G}'\,(\text{U}')^5}{\text{U}^4} + \frac{(\text{G}')^2\,(\text{U}')^2}{\text{U}^2} - \frac{36\,(\text{U}')^6\,\text{U}''}{\text{U}^5}\right.\right. \end{aligned}$$

$$+\frac{36\left(U'\right)^4\left(U''\right)^2}{U^4}+\frac{9\left(U'\right)^8}{U^6}\Bigg)+s^2\Bigg(-\frac{G''\left(U'\right)^2}{2U^2}-\frac{3G'U'U''}{2U^2}$$
$$+\frac{13G'\left(U'\right)^3}{4U^3}+\frac{63\left(U'\right)^4U''}{2U^4}-\frac{12\left(U'\right)^2\left(U''\right)^2}{U^3}-\frac{51\left(U'\right)^6}{4U^5}$$
$$-\frac{3U'''\left(U'\right)^3}{U^3}\Bigg)+s\Bigg(-\frac{2G'U'}{U^2}-\frac{37\left(U'\right)^2U''}{2U^3}+\frac{49\left(U'\right)^4}{4U^4}+\frac{U^{(3)}U'}{U^2}\Bigg)$$
$$+\frac{G}{U^2}+\frac{25\left(U'\right)^2}{2U^3}+\frac{4U''}{U^2}\Bigg]-V'''\frac{sU'}{2U}, \tag{5.190}$$

$$V^{A}_{1-\text{loop}}=V^2\Bigg[s^4\Bigg(\frac{6G'\left(U'\right)^3U''}{U^3}-\frac{3G'\left(U'\right)^5}{U^4}+\frac{\left(G'\right)^2\left(U'\right)^2}{2U^2}-\frac{18\left(U'\right)^6U''}{U^5}$$
$$+\frac{18\left(U'\right)^4\left(U''\right)^2}{U^4}+\frac{9\left(U'\right)^8}{2U^6}\Bigg)+s^3\Bigg(-\frac{2G'U'U''}{U^2}+\frac{5G'\left(U'\right)^3}{U^3}$$
$$+\frac{36\left(U'\right)^4U''}{U^4}-\frac{12\left(U'\right)^2\left(U''\right)^2}{U^3}-\frac{15\left(U'\right)^6}{U^5}\Bigg)+s^2\Bigg(-\frac{10\left(U'\right)^2U''}{U^3}$$
$$+\frac{25\left(U'\right)^4}{2U^4}+\frac{2\left(U''\right)^2}{U^2}\Bigg)-\frac{8s\left(U'\right)^2}{U^3}+\frac{5}{U^2}\Bigg]$$
$$+VV'\Bigg[\frac{s^4}{2U^5}\Big(-12U^3G'\left(U'\right)^2U''+6U^2G'\left(U'\right)^4$$
$$-U^4\left(G'\right)^2U'+36U\left(U'\right)^5U''-36U^2\left(U'\right)^3\left(U''\right)^2-9\left(U'\right)^7\Big)$$
$$+\frac{s^3}{2U^5}\Big(2U^4G'U''-12U^3G'\left(U'\right)^2-78U^2\left(U'\right)^3U''+12U^3U'\left(U''\right)^2$$
$$+36U\left(U'\right)^5\Big)+\frac{s^2}{2U^5}\Big(14U^3U'U''-35U^2\left(U'\right)^3\Big)+\frac{8sU'}{U^2}\Bigg]$$
$$+VV''\Bigg[\frac{s^3}{U^3}\Big(U^2G'U'+6U\left(U'\right)^2U''-3\left(U'\right)^4\Big)$$
$$+\frac{s^2}{U^3}\Big(5U\left(U'\right)^2-2U^2U''\Big)\Bigg]$$
$$+\left(V'\right)^2\Bigg[s^4\Bigg(\frac{3G'U'U''}{2U}-\frac{3G'\left(U'\right)^3}{4U^2}+\frac{\left(G'\right)^2}{8}-\frac{9\left(U'\right)^4U''}{2U^3}$$
$$+\frac{9\left(U'\right)^2\left(U''\right)^2}{2U^2}+\frac{9\left(U'\right)^6}{8U^4}\Bigg)+s^3\Bigg(\frac{7G'U'}{4U}+\frac{21\left(U'\right)^2U''}{2U^2}$$
$$-\frac{21\left(U'\right)^4}{4U^3}\Bigg)+\frac{49s^2\left(U'\right)^2}{8U^2}-\frac{2s}{U}\Bigg]$$

$$+\mathrm{V}'\mathrm{V}''\left[\mathrm{s}^3\left(-\frac{\mathrm{G}'}{2}-\frac{3\mathrm{U}'\mathrm{U}''}{\mathrm{U}}+\frac{3\left(\mathrm{U}'\right)^3}{2\mathrm{U}^2}\right)-\frac{7\mathrm{s}^2\mathrm{U}'}{2\mathrm{U}}\right]+\frac{1}{2}\left(\mathrm{V}''\right)^2\mathrm{s}^2. \quad (5.191)$$

Except for the differences that can be traced back to the missprint $5/2\,V^2$ in the Einstein frame effective action calculated in [2] (instead of the correct $5\,V^2$), the results for (5.189), (5.190) and (5.191) coincide with the one derived in [2]. The coefficients of the remaining structures are given by

$$\alpha_1^{\mathrm{A}}=\frac{43}{60},\quad \alpha_2^{\mathrm{B}}=\frac{1}{40},\quad \alpha_3^{\mathrm{B}}=-\frac{19}{12}\frac{U'}{U},\quad \alpha_4^{\mathrm{B}}=-\frac{G}{3U}-\frac{5}{24}\frac{(U')^2}{U^2}-\frac{19}{12}\frac{U''}{U},$$
$$\alpha_5^{\mathrm{B}}=0,\quad \alpha_6^{\mathrm{B}}=\frac{19}{8}\frac{(U')^2}{U^2},\quad \alpha_7^{\mathrm{B}}=\frac{G\,U'}{U^2}+\frac{5}{8}\frac{(U')^3}{U^3}+\frac{19}{4}\frac{U'\,U''}{U^2}$$
$$\alpha_8^{\mathrm{B}}=\frac{5}{4}\frac{G^2}{U^2}+7\frac{G\,(U')^2}{U^3}+\frac{331}{32}\frac{(U')^4}{U^4}+\frac{G\,U''}{U^2}+\frac{5}{8}\frac{(U')^2\,U''}{U^3}+\frac{19}{8}\frac{(U'')^2}{U^2} \quad (5.192)$$

For the comparison with "method B", we will restrict ourselves to present only the general structure of the effective action and the contribution $V^{\mathrm{B}}_{1-\mathrm{loop}}$ that can be derived from the general calculation in the Jordan frame parametrization for the scalar multiplet (5.102) in the single field limit $N=1$. The one-loop contribution to the effective potential $V^{\mathrm{B}}_{1-\mathrm{loop}}$ can serve as indicator of a possible frame dependence because it cannot receive any contributions of any other operator structure present in the effective action from the integration by parts operation. Thus, a difference in the coefficient $V^{\mathrm{A}}_{1-\mathrm{loop}}$ and $V^{\mathrm{B}}_{1-\mathrm{loop}}$ is already sufficient to show the frame dependence of quantum corrections. The one-loop divergences for the direct Jordan frame calculations in the limit $N=1$ are

$$\Gamma^{\mathrm{div,\,B}}_{1-\mathrm{loop}}=\frac{1}{32\pi^2(\omega-2)}\int d^4x\,g^{1/2}\Big\{U^{\mathrm{B}}_{1-\mathrm{loop}}\,R+G^{\mathrm{B}}_{1-\mathrm{loop}}\,(\varphi,_\nu\varphi^{,\nu})+V^{\mathrm{B}}_{1-\mathrm{loop}}$$
$$+\alpha_1^{\mathrm{B}}\,R_{\mu\nu}R^{\mu\nu}+\alpha_2^{\mathrm{B}}\,R^2+\alpha_3^{\mathrm{B}}\,R\,(\Box\varphi)+\alpha_4^{\mathrm{B}}\,R\,(\varphi,_\nu\varphi^{,\nu})+\alpha_5^{\mathrm{B}}\,R^{\mu\nu}\,\varphi,_\mu\varphi,_\nu$$
$$+\alpha_6^{\mathrm{B}}\,(\Box\varphi)\,(\Box\varphi)+\alpha_7^{\mathrm{B}}\,(\Box\varphi)\,(\varphi,_\nu\varphi^{,\nu})+\alpha_8^{\mathrm{B}}\,(\varphi,_\nu\varphi^{,\nu})^2\Big\}, \quad (5.193)$$

with the one-loop contributions to the effective potential

$$V^{\mathrm{B}}_{\text{1-loop}}=V^2\left[2\,s^2\frac{\left(U'\right)^4}{U^4}-2\,s\frac{\left(U'\right)^2}{U^3}+\frac{5}{U^2}\right]$$
$$+V\,V'\left[-8\,s^2\frac{\left(U'\right)^3}{U^3}+4\,s\frac{U'}{U^2}\right]+2\,V\,V''\,s^2\frac{\left(U'\right)^2}{U^2}$$
$$+\left(V'\right)^2\left[8\,s^2\frac{\left(U'\right)^2}{U^2}-\frac{2s}{U}\right]-4\,V'V''\,s^2\frac{U'}{U}+\frac{1}{2}\left(V''\right)^2s^2. \quad (5.194)$$

Although (5.193) and (5.188) formally have the same structure, the comparison shows that they do have different coefficients. By comparing (5.191) with (5.194), it is obvious that the off-shell effective actions calculated in the Einstein frame and in the Jordan frame do not coincide. Thus, we have shown that for this general cosmological model the procedure of calculating the one-loop quantum corrections in the Einstein and Jordan parametrizations does not yield the same result. In the next section we are going to investigate the origin of this quantum frame dependence within a more general context.

5.3.3 Geometrical Effective Action

In the last section, we have explicitly shown that the off-shell one-loop quantum corrections calculated in the cosmological Jordan and Einstein frame do not coincide for the specific model of a scalar multiplet non-minimally coupled to gravity. However, neither did we provide a *general* proof for the frame dependence of quantum corrections, nor did we locate the origin of this parametrization dependence. Although there are certain theorems regarding the reparametrization invariance of the effective action, see e.g. [9, 19, 25], all these statements only concern the *on-shell* version of the effective action. However, certain quantum phenomena require the study of the off-shell extension. An important example is the off-shell effective potential in the scenario of spontaneous symmetry breaking [11]. How can we explain the parametrization dependence of our off-shell result in view of the on-shell equivalence theorems?

By noticing that the inequivalence between the cosmological frames is just a very special aspect of the more general gauge and field parametrization dependence of the conventional effective action, we can embed our result in a more general framework, which reveals the origin of the parametrization dependence. Already in 1984, G. A. Vilkovisky [26] analysed this problem based on the work of DeWitt [13] and detected the source of the frame dependence. Details of this formalism can be found e.g. in [5, 21, 26].

At this point, we will only present the rudimentary concepts of this approach and discuss its implications with respect to the cosmological debate of Jordan frame versus Einstein frame.

In order to outline the basic idea, we will again use the condensed DeWitt notation in which the generalized field ϕ^i consists of a collection of basic scalar, vector, tensor, fields and the indices $i, j, \ldots$ including both, discrete indices $A, B, \ldots$ as well as continuous space-time points $x, x', \ldots$. In the case of our cosmological model with the action (5.68), this field is given by

$$\phi^i = \phi^A(x) = \begin{pmatrix} g_{\mu\nu}(x) \\ \\ \Phi^a(x) \end{pmatrix} \qquad (5.195)$$

and can be thought of as a coordinate describing a point $Q \in \mathcal{C}$ in the space of fields, the *configuration space* $\mathcal{C}$. Usually, physical requirements demand that the action (5.68) has to feature some symmetries and therefore should be invariant under transformations associated with these symmetries. General Relativity requires the invariance of (5.68) under arbitrary space-time coordinate transformations. A formalism which guarantees that the choice of a local coordinate system does not affect the physical results is called *covariant*. These transformations act on the continuous space-time points $x, x', \ldots$ as well as on the space-time indices $\mu, \nu, \ldots$. Our action (5.68) is clearly covariant with respect to arbitrary space-time coordinate transformations. In the case of the graviton-scalar sector of our model, the internal $O(N)$ symmetry is a rigid one. But for a more general setup, we would have to consider the invariance of (5.68) under such gauge transformations. These transformations act on the internal gauge index. But what about general *field transformations*? As we have argued above, ϕ^i just corresponds to a coordinate describing a point in configuration space $\mathcal{C}$. A field reparametrization $\phi^i \rightarrow \hat{\phi}^i$ is associated with a transformation of the field coordinate itself. In order to understand what is meant by invariance under field reparametrization, it is convenient to think of the configuration space $\mathcal{C}$ as a geometrical object, i.e. a manifold, in the same way as we think of space-time as a manifold $\mathcal{M}$. We can even make further use of this analogy and follow similar steps as in the construction of General Relativity. In General Relativity, an event corresponds to a point $P \in \mathcal{M}$ in the space-time manifold. To describe this point P we choose a coordinate system x^μ. However, since coordinates by themselves do not have a physical meaning, we might as well describe the point P in a different coordinate system x'^μ. The same reasoning should also hold for a point Q in configuration space $\mathcal{C}$. We can describe this point by a set of "coordinates" ϕ^i. Again, it should not make any difference whether we choose a different set of coordinates $\hat{\phi}^i$ in order to describe Q. The *physical* field content of the theory described by the action is determined by Q and not by ϕ^i or $\hat{\phi}^i$. Thus, we have to demand that the action (5.68) is invariant under the diffeomorphisms of $\mathcal{C}$. In other words, (5.68) has to be a scalar with respect to the configuration space, i.e all indices $i, j, \ldots$ have to be properly contracted with a suitable configuration space metric $G_{ij'}$. The *classical* action (5.68) is by construction gauge and reparametrization invariant

$$S[\phi] = \hat{S}[\hat{\phi}]. \tag{5.196}$$

In order to understand why this fails to hold for the *effective action* we have to analyse the construction of the conventional effective action in more detail.

As we have shown in Sect. 4.4, the derivation of the effective action involved the *coordinate difference* $\phi^i - \bar{\phi}^i$, see (4.47). However, only for a flat configuration space and Cartesian coordinates $\phi^i, \bar{\phi}^i$ this has a geometrical meaning as the tangent to the geodesic connecting two points in $\mathcal{C}$ described by the coordinates ϕ^i and $\bar{\phi}^i$. In a general curved configuration space, geodesics are not straight lines and this coordinate difference does not have any geometrical meaning anymore.

We can again make use of the analogy between coordinate transformations and field parametrization by introducing the configuration space "world functional"

$\sigma[\phi', \phi]$ as a generalized notation of the space-time world function $\sigma(x, x')$, defined in (D.1). The world functional $\sigma[\phi', \phi]$ is proportional to the square of the geodesic distance between ϕ' and ϕ. Its first derivative $\sigma^{i'}[\phi', \phi]$ with respect to the field point $\phi^{i'}$ describes the tangent to the geodesic in direction $\phi^i \to \phi^{i'}$. If we denote by ϕ' a background field and by $\bar{\phi}^i$ the conventional "mean field", we can obtain geometrical meaningful quantities by expressing ϕ^i in terms of $\sigma^{i'}[\phi', \phi]$ and $\bar{\phi}^i$ in terms of $\sigma^{i'}[\phi', \bar{\phi}]$. Then, we can consider the classical action $S[\phi]$ as a function $S[\phi'; \sigma^{i'}[\phi', \phi]]$ [21] and repeat the same steps as in the derivation of the effective action as in Chap. 4. In the derivation of the perturbative expansion of the effective action (4.47) we must therefore substitute the coordinate difference $(\phi^i - \bar{\phi}^i)$ by the geometrical meaningful expression $(\sigma^{i'}[\phi', \phi] - \sigma^{i'}[\phi', \bar{\phi}])$. Rather than thinking of the effective action $\Gamma[\bar{\phi}]$ as a functional of the mean field, we can regard the effective action as a functional of the pair $(\phi^{i'}, \sigma^{i'}[\phi', \bar{\phi}])$ and obtain a covariant geometrical definition. However, the details of this derivation are rather involved and will not contribute essential insights to the discussion of this section. Therefore, we will only consider a special case, which, however, includes the main feature of the geometrical framework.

In order to explicitly show how this geometrical derivation affects the "naive" definition of the *one-loop* effective action, we will simplify the discussion and restrict ourselves to a *flat* configuration space but with general "curvilinear" coordinates ϕ^i.[7] In the "naive" definition of the one-loop effective action (4.5) we have to calculate the trace of the operator $F_{ij} \propto \delta^2 S[\phi]/\delta\phi^i\delta\phi^j$. This, in turn, requires to calculate the second functional derivative of the classical action (5.68). But it is clear that the resulting object F_{ij} is not a tensor with respect to the configuration space coordinate transformations $\phi^i \to \hat{\phi}^i$. To restore covariance, we have to introduce a *covariant* functional derivative ∇_i, constructed on the basis of a suitable configuration space connection Γ^k_{ij} and replace the ordinary functional derivatives $\delta/\delta\phi^i$ by covariant ones, i.e.

$$\frac{\delta^2 S}{\delta\phi^i\delta\phi^j} \to \nabla_i\nabla_j S = \frac{\delta^2 S}{\delta\phi^i\delta\phi^j} - \Gamma^k_{ij}\frac{\delta S}{\delta\phi^k}. \tag{5.197}$$

So far, we have introduced the configuration space connection only at a formal level and its specific form is still arbitrary. However, under some reasonable physical assumptions [26] (see also [21] and [16] for a discussion) it can generally be constructed by the Christoffel symbol of the configuration space metric G_{ij} and an additional part connected with gauge transformations.[8] In a similar way as the metric $g_{\mu\nu}(x)$ on $\mathcal{M}$, the metric G_{ij} on $\mathcal{C}$ can be defined by the line element

[7] As in the case of spherical coordinates in a flat space, we can of course have a non-vanishing connection despite the fact that there is no curvature. The difference to a curved space is connected with the fact that we can transform the connection away *globally* in a flat space.

[8] In the following, we will neglect the issue of gauge transformations. In general, we have to separate the gauge transformations from the arbitrary field reparametrizations, which complicates the formalism and is of minor interest for the purpose of this section.

$$\mathrm{d}s^2 = G_{ij}\mathrm{d}\phi^i\,\mathrm{d}\phi^j. \tag{5.198}$$

Once we have made a choice[9] for G_{ij}, we can construct the Christoffel symbols $\Gamma^k_{ij}(G_{ij})$ and obtain a *covariant* definition of the one-loop effective action. Within such a covariant formalism, the effective actions calculated in different parametrizations would always yield the same result. In [26] it was therefore denoted "unique effective action". However, regardless of whether one agrees with these particular assumptions that would fix a "unique effective action", we can observe two important facts: Firstly, the origin of the frame dependence in the formulation of the conventional effective action is detected. It can be traced back to non-covariance of the conventional formalism. In the one-loop approximation of a flat configuration space it is connected with the additional factor

$$-\,\Gamma^k_{ij}\frac{\delta S}{\delta\phi^k} \tag{5.199}$$

not present in the expression for the "naive" definition of the effective action.

Secondly, this means that *independently* of the explicit expression for Γ^k_{ij}, the additional term (5.199) vanishes *on-shell* because of

$$\frac{\delta S}{\delta\phi^i} = 0. \tag{5.200}$$

This is in agreement with on-shell equivalence theorems [9, 19, 25].

What conclusions can we draw from this analysis for the cosmological debate of Jordan frame versus Einstein frame? First of all, it is clear that the cosmological debate is just a particular aspect of the more general problem of parametrization dependence of the "naive" effective action. The off-shell difference discussed in Sect. 5.3.2 has its natural origin in the shortcoming of the conventional formalism. Moreover, the question whether the Jordan frame or the Einstein frame is the correct "physical frame"[10] simply misses the point. It is meaningless in the same sense as it is meaningless to ask in General Relativity whether a particular coordinate system is more physical than another one. The problem is the incomplete non-covariant theoretical formalism rather than the task of finding physical arguments in favour of one or of the other parametrization.

The construction of the unique geometrical effective action along the lines of [26] is rather involved and requires controversial physical assumptions in order to fix the configuration space metric. Moreover, even if there would be a general agreement about these assumptions, the technical calculations that go beyond a purely formal level will become very complicated. However, we can ask if it is necessary at all

[9] The choice of the configuration space metric is by itself a subtle point. One of the criteria to fix G_{ij} proposed in [26] is that G_{ij} should be determined by the coefficients of the highest derivative terms, present in the classical action $S[\phi]$.

[10] A frequent argument in favour of the Jordan frame is that this is the frame in which real physical distances are measured in terms of the physical Jordan frame metric.

to use a unique effective action in order to obtain reparametrization independent results. Since our approach was mainly motivated by cosmology, we are interested in the question whether we can construct cosmological *observables* that are invariant under reparametrizations. Since we deal with quantum theory, this reduces to the question if we could construct a reparametrization invariant definition for physical expectation values *without* knowledge of the unique effective action. In this case, it would be sufficient to use the naive definition of the effective action in an arbitrary parametrization. The covariant definition of the expectation values would then automatically take care of the frame dependence contained in the "naive" effective action. In particular, this would simplify practical calculations, since we could chose a parametrization in which the action takes a very simple form.

5.4 Conclusions

In Sect. 5.1, we have presented a detailed calculation for a scalar field minimally coupled to gravity in order to explicitly demonstrate the individual steps of the general Schwinger–DeWitt algorithm. We have derived the corrected result of the calculations performed in [2] which served as an important cross-check for the more general results of Sect. 5.2.

In Sect. 5.2, we made use of the same technique as in Sect. 5.1 in order to calculate the one-loop corrections for the more general model of a $O(N)$-symmetric multiplet of scalar fields non-minimally coupled to gravity in the Jordan frame. We have presented the result in a closed form and arranged its presentation according to physical considerations.

In Sect. 5.3.1, we have constructed the explicit field transformation rules between Einstein frame and Jordan frame for such a model. However, due to a non-trivial integrability condition such a transition is not possible in general. Moreover, in Sect. 5.3.2, we expressed the result for the effective action of a single scalar field calculated in the Einstein frame parametrization in the Jordan frame parametrization and found by direct comparison with the effective action calculated in the Jordan frame parametrization that the divergent part of the off-shell one-loop effective action is a frame-dependent quantity.

Finally, in Sect. 5.3.3 we have explained how the origin of this frame dependence can be understood within the more general geometrical framework developed in [26]. We have discussed the implications of the parametrization dependence for the cosmological debate of Jordan frame versus Einstein frame. In particular, we have argued that many aspects of the hitherto debate become obsolete and the cosmological debate reduces to one single problem: the non-covariance of the conventional formalism.

References

1. Barvinsky, A.O., Kamenshchik, A.Yu.: Effective equations of motion and initial conditions for inflation in quantum cosmology. Nucl. Phys. B **532**, 339 (1998)
2. Barvinsky, A.O., Kamenshchik, A.Yu.: Renormalization group for nonrenormalizable theories: Einstein gravity with a scalar field. Phys. Rev. D **48**, 3677 (1993)
3. Barvinsky, A.O., Kamenshchik, A.Yu.: Asymptotic freedom in inflationary cosmology with a non-minimally coupled Higgs field. J. Cosmol. Astropart. Phys. **12**, 003 (2009)
4. Barvinsky, A.O., Kamenshchik, A.Yu.: Higgs boson, renormalization group, and naturalness in cosmology. Eur. Phys. J. C. **72**, 2219 (2012)
5. Barvinsky, A.O., Vilkovisky, G.A.: The generalized Schwinger-DeWitt technique and the unique effective action in quantum gravity. Phys. Lett. B **131**, 313 (1983)
6. Barvinsky, A.O., Vilkovisky, G.A.: The generalized Schwinger-DeWitt technique in gauge theories and quantum gravity. Phys. Rep. **119**, 1 (1985)
7. Bezrukov, F.L., Magnin, A., Shaposhnikov, M.: Standard Model Higgs boson mass from inflation. Phys. Lett. B **675**, 88 (2009)
8. Bezrukov, F.L., Shaposhnikov, M.: The Standard Model Higgs boson as the inflaton. Phys. Lett. B **659**, 703 (2007)
9. Borchers, H.J.: Über die Mannigfaltigkeit der interpolierenden Felder zu einer kausalen S-Matrix. Nuovo Cimento **15**, 784 (1960)
10. Capozziello, S., Faraoni, V.: Beyond Einstein Gravity. Springer, New York (2010)
11. Coleman, S.R., Weinberg, E.J.: Radiative corrections as the origin of spontaneous symmetry breaking. Phys. Rev. D **7**, 1888 (1973)
12. De Simone, A., Hertzberg, M.P., Wilczek, F.: Running inflation in the Standard Model. Phys. Lett. B **678**, 1 (2009)
13. DeWitt, B.S.: Dynamical theory of groups and fields. Blackie & Son, London (1965)
14. DeWitt, B.S.: Quantum theory of gravity. I. Canonical theory. Phys. Rev. **160**, 1113 (1967)
15. Dicke, R.H.: Mach's principle and Invariance under Transformation of Units. Phys. Rev. **125**, 2163 (1962)
16. Esposito, G., Kamenshchik, A.Yu., Pollifrone, G.: Euclidian Quantum Gravity on Manifolds with Boundary. Springer, New York (1997)
17. Faraoni, V., Gunzig, E., Nardone, P.: Conformal transformations in classical gravitational theories and in cosmology. Fund. Cosmic Phys. **20**, 121 (1999)
18. Kaiser, D.I.: Conformal transformations with multiple scalar fields. Phys. Rev. D **81**, 084044 (2010)
19. Kallosh, R.E., Tyutin, I.V.: The Equivalence theorem and gauge invariance in renormalizable theories. Yad. Fiz. **17**, 190 (1973)
20. Parker, L., Christensen, S.M.: MathTensor: A system for Doing Tensor Analysis by Computer. Addison-Wesley, Redwood City (1994)
21. Parker, L.E., Toms, D.J.: Quantum Field Theory in Curved Spactime. Cambridge University Press, Cambridge (2009)
22. Shapiro, I.L., Takata, H.: One-loop renormalization of the four-dimensional theory for quantum dilaton gravity. Phys. Rev. D **52**, 2162 (1995)
23. Steinwachs, C.F., Kamenshchik, A.Yu.: One-loop divergences for gravity nonminimally coupled to a multiplet of scalar fields: Calculation in the Jordan frame. I. The main results. Phys. Rev. D **84**, 024026 (2011)
24. 't Hooft, G. and Veltman, M. J. G., : One-loop divergencies in the theory of gravitation. Ann. Inst. Henri Poincaré A **20**, 69 (1974)
25. Tyutin, I.V.: Parametrization dependence of nonlinear quantum field theories (in Russian). Yad. Fiz. **35**, 222 (1982)
26. Vilkovisky, G.A.: The unique effective action in quantum field theory. Nucl. Phys. B **234**, 125 (1984)

Chapter 6
Non-minimal Higgs Inflation

In the introductory Chaps. 2–4 we have collected the theoretical knowledge required for the analysis in this chapter. In particular, in Chap. 2 we have shown that a crucial assumption for a successful description of modern cosmology is a phase of accelerated expansion—inflation. In most of the cosmological models based on this assumption, the inflationary mechanism is described in terms of a scalar field, the inflaton φ. However, despite its obvious success in fitting observational data and in resolving basic cosmological problems, some fundamental questions still remain unresolved. One of these questions is connected with the physical reality of φ. The origin and the nature of φ remains unexplained, although the ad hoc introduction of φ turns out to be very useful. Since so far no such scalar field φ has ever been measured, there is no identification with a really existing physical particle. Therefore, inflation as it stands has to be considered as a very successful, albeit still hypothetical, theoretical description.[1]

In Chap. 3, we have seen that the Standard Model of Particle Physics, whose predictions are experimentally tested to an outstanding level of precision, does also rely on the fundamental assumption about the existence of a scalar field. In this context, the scalar field is called Higgs boson and it is necessary to explain the origin of the masses of the fundamental particles, while maintaining gauge invariance. However, just as the inflaton, so far the Higgs boson has not been measured directly. Nevertheless, we can at least derive bounds for its mass from theoretical considerations and indirect measurements.

By noticing the formal similarities between the inflaton and the Higgs boson, it is natural to wonder whether there could be a deeper, more fundamental connection between these two hypothetical scalar particles. Finally, this leads to the question

---

[1] Although it should be mentioned that beside the striking evidence from the CMB, the explainable large scale structure of matter from tiny primordial quantum fluctuations and the coherent phase argument [30] all support the hypothesis of inflation. Moreover, it is expected that further experiments will measure primordial gravitational waves which ultimately are a measure of the energy scale of inflation. In consideration of all this evidence it seems to be very difficult to find a different mechanism which could explain all these features and does not contradict any observation.

C. F. Steinwachs, *Non-minimal Higgs Inflation and Frame Dependence in Cosmology*, Springer Theses, DOI: 10.1007/978-3-319-01842-3_6,
© Springer International Publishing Switzerland 2014

whether they could be just two manifestations of one and the same particle. The combination of the Higgs mechanism and inflation, from now on called Higgs inflation, would not only reduce the number of hypothetical particles, it would also offer an opportunity to connect two energy scales which are separated by at least 10^{14} GeV and allow for experimental tests.

6.1 Motivation for a Non-minimal Coupling

Before describing the details of the *non-minimal* Higgs inflation model, we will discuss the problems that do appear in the most natural attempt of a unification within the framework of a *minimally* coupled Higgs field. This will serve as a motivation to consider a modification by introducing a strong non-minimal coupling term. We resist the temptation to "baptise" this unified particle "Higgsflaton" and will instead use interchangeably the terms Higgs-inflaton, Higgs field, Higgs boson and inflaton, depending on the context.

6.1.1 Minimal Higgs Inflation

The most natural attempt to realize this unification by a *minimal* coupling of the Higgs-inflaton to gravity will lead to the following problem: The data of the CMB anisotropies contained in the power spectrum impose a constraint on the inflationary potential, cf. (2.122) [39, 42, 80]. For a moment we will only consider the graviton-scalar sector of the model and assume a *minimally* coupled inflaton field φ with quartic potential $V \simeq \lambda\varphi^4/4$. In this case, the CMB constraint translates into a constraint for the self-coupling constant λ

$$\lambda \propto \left(\frac{\Delta T}{T}\right)^2 \simeq 10^{-10}. \tag{6.1}$$

In addition, if φ constitutes the Higgs field, λ enters the definition of the Higgs mass via

$$M_{\mathrm{H}}^2 \propto \lambda\nu^2. \tag{6.2}$$

The vacuum expectation value $\nu := \langle\Phi\rangle$ of the Higgs field is determined by the symmetry breaking scale ν, marking the minimum of the effective potential. It is approximately given by $\nu \simeq 10^2$ GeV. From theoretical considerations and indirect results of particle collider experiments, the Higgs mass can be constrained (assuming the most conservative estimates) to lie within an interval given by [3]

$$115\,\text{GeV} \lesssim M_{\text{H}} \lesssim 180\,\text{GeV}. \tag{6.3}$$

This, in turn, implies that

$$\lambda \sim \frac{M_{\text{H}}^2}{v^2} \sim \frac{10^2\,\text{GeV}}{10^2\,\text{GeV}} \sim \mathcal{O}(1) \tag{6.4}$$

must be of order one (in fact rather 10^{-1} in order to justify the use of perturbation theory), which is in strong contradiction with the cosmological constraint (6.1). Hence, the two constraints (6.1) and (6.3) coming from different physical sectors have to be brought together in order to allow a unified description of the Higgs boson and the inflaton.

6.1.2 Non-minimal Higgs Inflation

While keeping the idea of a unified scalar field, one way out of the dilemma described in Sect. 6.1.1 is provided by the assumption of a *strong non-minimal coupling to gravity*. This was already recognized in [33, 36, 69, 78]. The most common choice of such a non-minimal coupling is a term proportional to $\varphi^2 R$. This means that in addition to the usual Einstein–Hilbert term $\propto R$, we have a coupling between the curvature and the scalar field of the form

$$S = \int \mathrm{d}^4x \sqrt{g} \left[\frac{1}{2}\left(M_{\text{P}}^2 + \xi\,\varphi^2\right) R + \mathcal{L}_\varphi \right]. \tag{6.5}$$

Here, M_{P} is the reduced Planck mass and is given by (restoring units for a moment) $M_{\text{P}} = m_{\text{P}}/\sqrt{8\pi} = \sqrt{\hbar c/8\pi G} \sim 2.43 \times 10^{18}\,\text{GeV}/c^2$. Let us now see how the non-minimal coupling can resolve the mismatch between the two constraints (6.1) and (6.4). In contrast to the minimally coupled case, it turns out (as we will show in Sect. 6.3.3) that the cosmological constraint (6.1) is then given by

$$\frac{\lambda}{\xi^2} \propto \left(\frac{\Delta T}{T}\right)^2 \simeq 10^{-10}. \tag{6.6}$$

This means that we can always achieve a simultaneous satisfaction of *both* constraints, (6.4) and (6.6) by adjusting the non-minimal coupling constant ξ in an appropriate way. Thus, relation (6.6) requires a strong non-minimal coupling $|\xi| \simeq 10^4$.

6.1.3 Beyond General Relativity

The price we have to pay for the combination of the two constraints is the introduction of a new parameter ξ. We should therefore try to justify its existence and in particular its magnitude of $|\xi| \sim 10^4$.

Firstly, such a term does not contradict any experiment so far. This is, of course, obvious, since today the Higgs field rests in its electroweak vacuum value $\varphi = \nu \simeq 10^2$ GeV and the contribution from the term $\xi \varphi^2 R$ in (6.5) is negligible compared to the $M_P^2 R$ contribution, even for a large value of $|\xi| \sim 10^4$. This also implies that the non-minimally coupled model is practically indistinguishable from General Relativity in the low energy regime. Since General Relativity is an excellent and very well-tested theory for our energy scale today, this can be considered as a required limit of every approach that goes beyond Einstein gravity such as $f(R)$ theories, scalar-tensor theories, models including higher scalar curvature invariants or models with modified space-time dimensions. In the absence of any experimental contradictions, maybe the simplest argument in favour of an extension of General Relativity is that there is no *a priori* argument *against* such an extension.[2]

Secondly, from various arguments[3] it seems evident that Einstein's theory of gravity has to be replaced by a more fundamental theory. Moreover, by comparison with the other types of interaction, it seems natural that such a more fundamental theory has to be a quantum theory. However, so far no satisfactory quantum gravitational theory does exist, although there are some promising candidates. For an overview and comparison of the different approaches see [51]. Unfortunately, the covariant perturbative approach to quantize gravity has the drawback of being non-renormalizable. However, we can still take the pragmatic point of view and look for an enlargement of General Relativity as an *effective* description of an underlying more fundamental quantum theory of gravity. In particular, some scalar-tensor models emerge naturally as effective theories in certain low energy limits of string theory (see e.g. [57] and references therein for a review). In these theories, the scalar dilaton fields ϕ give rise to a non-minimal coupling term of the form $\propto e^{-\alpha\phi} R$.

When going beyond General Relativity, we also have to face the problem of predictability. There are simply too many choices for an extension.[4] Furthermore, by introducing more and more new interaction structures within an effective description, we would obtain more and more free parameters. Having too many freely adjustable parameters (coupling constants), each model will ultimately fit the data and the theory looses its predictive power. However, there might still exist selection mechanisms, like particular symmetries, which reduce the arbitrariness of such a choice. Within the framework of an non-renormalizable effective field theory, the sequence of higher

---

[2] Ockham's razor does not apply if two competing theories make different testable predictions.

[3] This is suggested e.g. by the singularity theorems [43], the black hole radiation [41] and the general intention for unification.

[4] Even in the restricted scenario of scalar-tensor theories this freedom could correspond to different choices of the non-minimal coupling term parametrized by a general function $U(\varphi)$.

dimensional operators induced by quantum corrections will be suppressed by higher powers of a cut-off. This allows us to incorporate only a few new interaction terms, but at the same time also sets a boundary for the applicability of the theory.

Beside these considerations, a non-minimal coupling term as used in the non-minimal Higgs inflation scenario is naturally induced already at the one-loop order of quantum corrections to the Einstein–Hilbert action *minimally* coupled to a scalar field [18]. This can easily be seen from the results of Sect. 5.1, where we obtained a term $\propto R\,V''$ in (5.61). With the choice $V \propto \varphi^4$, this induces a non-minimal coupling term of the form $\propto R\,\varphi^2$.

Furthermore, we mention that a non-minimal coupling term fits very well in a purely relational picture of our world. Following the ideas of Leibniz and Mach, there is only *relative* motion between all material objects and our perception of time emerges as the sequence of different configurations of the relative positions of these material objects [5]. If we associate the locations of the materialized objects with points in a space-time manifold in order to establish contact with General Relativity, this implies that Newton's constant G which connects inertia and acceleration can vary from space-time point to space-time point. But a function of space and time is nothing else than a scalar field. Hence, if one is inclined to believe that this relational point of view has a fundamental meaning but wants to keep the advantages of the mathematical formalism of General Relativity,[5] the belief in a scalar-tensor theory is a very natural one, see the discussion in [20, 23, 29].

Since a non-minimal scalar field is coupled directly to the derivatives of the metric, one could still argue that a non-minimal coupling term violates such a fruitful physical concept as the equivalence principle. But finally the experiment has to decide whether the equivalence principle is realized in nature or not. And since the present violation by the non-minimal coupling (6.18) is so small, current experiments cannot decide whether the equivalence principle is rather an effective low energy statement than a fundamental principle.

When looking for an extension of General Relativity in the context of Higgs inflation, we have to study the consequences for the Standard Model at the same time. It is an open question whether there is a "big desert" between the electroweak energy scale and a GUT scale[6] (maybe even the Planck scale) or some new physics at an intermediate energy scale will appear. In the following section, we will see that the non-minimally coupled Higgs inflation model suggests that the Standard Model could be a reliable quantum field theory up to inflationary energy scales.

---

[5] For the relativistic idea, it is not mandatory to rely on a differentiable background manifold and the formalism of General Relativity. In fact, this is rather counter intuitive to a truly background-independent viewpoint of a purely relational setup. In General Relativity we indeed *do* have a background manifold. However "background independence" is established "afterwards" by considering only covariant equations.

[6] The scale which is distinguished by the approximate intersection point of the RG trajectories of the Standard Model gauge couplings.

6.2 The Standard Model Higgs Boson as the Inflaton

In order to discuss the concrete model of non-minimal Higgs inflation, it is useful to divide the action into two different sectors and first investigate these sectors separately. The graviton-Higgs sector which describes the inflationary dynamics and which gives this model its particular character, as well as the Standard Model matter sector, including the coupling between Higgs and gauge bosons via the covariant derivative of the Higgs field and the Yukawa coupling between Higgs field and fermions. It will turn out that the inflationary dynamics is essentially determined by the matter content. Let us therefore start with the matter sector.

6.2.1 *Matter Sector*

The matter sector of the model is dominated by the heaviest particles of the Standard Model, the $W^{\pm}$ and Z bosons and the Yukawa top-quark q_{t}. We will therefore neglect the contributions of the remaining lighter particles. Schematically, the interaction part of the matter Lagrangian takes the general form

$$\mathcal{L}_m = -\sum_{\chi} \frac{1}{2} \lambda_\chi \, \chi^2 \varphi^2 - \sum_{A} \frac{1}{2} g_A^2 \, A_\mu^2 \varphi^2 - \sum_{\Psi} y_\Psi \, \varphi \, \bar{\Psi} \Psi \tag{6.7}$$

for scalar fields χ, vector gauge bosons A_μ and Dirac spinors Ψ. The λ_χ, g_A and y_Ψ are the corresponding self, gauge and Yukawa couplings. The effective mass terms are generated by the Higgs mechanism (cf. the discussion in Sect. 3.2.4), leading to

$$m_{W^{\pm}}^2 = \frac{1}{4} g^2 \varphi^2, \quad m_{\mathrm{Z}}^2 = \frac{1}{4} (g^2 + g'^2) \varphi^2, \quad m_{\mathrm{t}}^2 = \frac{1}{2} y_t^2 \varphi^2. \tag{6.8}$$

The mass of the Goldstone bosons can be identified with the first derivative of the potential

$$m_G^2 := \frac{V'(\varphi)}{\varphi} = \lambda(\varphi^2 - \nu^2). \tag{6.9}$$

Their mass is vanishing in the electroweak vacuum $\varphi = \nu$, since their motion is connected with the angular movement in the degenerated vacuum of the Higgs potential. For higher values of φ (higher energies) their mass is no longer vanishing, but rather given by

$$m_G^2 \simeq \lambda \varphi^2. \tag{6.10}$$

This is of essential importance for the later inflationary analysis. The mass of the Higgs field can be related to the second derivative of V

$$m_{\mathrm{H}}^2 := V''(\varphi) = \lambda(3\varphi^2 - \nu^2)\,. \tag{6.11}$$

The Higgs component of the Higgs multiplet is, of course, non-vanishing in the electroweak vacuum $m_{\mathrm{H}}^2 = 2\lambda\nu^2$. It corresponds to the mode that oscillates around the ground state ν of the Higgs potential. Far from the minimum it takes the value

$$m_{\mathrm{H}}^2 \simeq 3\lambda\varphi^2. \tag{6.12}$$

Calculating the one-loop quantum corrections for all particles, we obtain the usual structure of the Coleman–Weinberg potential [27]

$$V_{1-\mathrm{loop}} = \sum_{\text{particles}} (\pm 1)\,\frac{m^4(\varphi)}{64\pi^2}\,\ln\frac{m^2(\varphi)}{\mu^2} = \frac{\lambda\,\mathbf{A}}{128\,\pi^2}\,\varphi^4\,\ln\frac{\varphi^2}{\mu^2}, \tag{6.13}$$

where the sum includes the statistics (± 1 according to bosons or fermions) as well as the sum over the polarizations. In the second equality of (6.13), we have inserted the definition of the particle masses in terms of the Higgs field. The denominator μ in the logarithm is the renormalization point. It has to have mass dimension one in order to render the argument of the logarithm dimensionless. The key quantity $\mathbf{A}$ is the *anomalous scaling*. Equation (6.13) can be regarded as defining equation for $\mathbf{A}$. It can be written as the following combination of coupling constants

$$\begin{aligned}\mathbf{A} &= \frac{2}{\lambda}\Big(\sum_{\chi}\lambda_\chi^2 + 3\sum_{A} g_A^4 - 4\sum_{\Psi} y_\Psi^4\Big)\\ &= \frac{3}{8\,\lambda}\Big(2\,g^4 + (g^2 + g'^2)^2 - 16\,y_t^4\Big) + 6\lambda.\end{aligned} \tag{6.14}$$

The factors 1, 3 and 4 in the first line take care of the number of degrees of freedom for real scalar fields, massive vector fields and charged Dirac spinors. The last term in the second line of (6.14) is due to contributions from Goldstone bosons. These contributions critically change the behaviour of the RG analysis during inflation. The contributions from the Higgs field itself and the graviton are absent in the expression for $\mathbf{A}$. This is due to a suppression by inverse powers of ξ. We will explain the suppression mechanism and the disentanglement of Higgs and Goldstone contributions in detail in the next sections. The name "anomalous scaling" has the following origin. Typically, the one-loop correction is logarithmic for a dimension-four operator (like φ^4) and thus the effective interaction in the one-loop approximation can be written as

$$\frac{\lambda}{4}\varphi^4\left(1 + \mathbf{A}\,\ln\frac{\varphi}{\mu} + \dots\right) \approx \frac{\lambda}{4}\varphi^4\,e^{\mathbf{A}\,\ln\frac{\varphi}{\mu}} = \frac{\lambda}{4}\varphi^4\left(\frac{\varphi}{\mu}\right)^{\mathbf{A}} = \frac{\lambda}{4}\,\mu^{-\mathbf{A}}\,\varphi^{4+\mathbf{A}}\,. \tag{6.15}$$

The original (tree-level) operator dimension of four can be seen as an approximation of the real "full quantum dimension" $4 + \mathbf{A}$. We note that we had to assume that perturbation theory in $\mathbf{A}$ must be valid. In a cosmological context [6], this anomalous dimension determines the peak in a quantum distribution function for the initial value of the inflaton in cosmological models [11]. In Sect. 7, this will be derived and discussed in detail.

6.2.2 Graviton-Scalar Sector

The graviton-scalar sector is described by the action already investigated in Chap. 5.2

$$S_{\mathrm{gh}}^{\mathrm{JF}}[g_{\mu\nu}, \Phi^a] = \int_{\mathcal{M}} \mathrm{d}^4x \sqrt{g} \left[U(\varphi) R + \frac{1}{2} G(\varphi) g^{\mu\nu} \partial_\mu \Phi^a \partial_\nu \Phi_a + V(\varphi) \right]. \tag{6.16}$$

This form of the action is still too general for our cosmological model. For the Standard Model Higgs inflation scenario, we assume a canonically normalized kinetic term and the common tree-level Higgs potential. The only $O(N)$ invariant quantity is

$$\varphi^2 = \Phi^a \Phi_a. \tag{6.17}$$

The dependence on φ guarantees that the $O(N)$ symmetry of (6.16) is not spoiled. Later, we will need the $N = 4$ case to model four real scalar fields, corresponding to the four degrees of freedom of the complex Higgs doublet. The non-minimal Higgs inflation model consists of the choices

$$U_{\mathrm{tree}}(\varphi) = \frac{1}{2} \left(M_{\mathrm{P}}^2 + \frac{1}{2} \xi \varphi^2 \right), \tag{6.18}$$

$$G_{\mathrm{tree}}(\varphi) = 1, \tag{6.19}$$

$$V_{\mathrm{tree}}(\varphi) = \frac{\lambda}{4} (\varphi^2 - \nu^2)^2, \tag{6.20}$$

with the non-minimal coupling $|\xi| \sim 10^4$ and the symmetry breaking scale $\nu = 246.22\,\mathrm{GeV}$. Using the general results presented in Chap. 5.2, we can easily derive the contributions from the graviton-scalar sector to the one-loop quantum corrections of the functional couplings $U(\varphi)$, $G(\varphi)$ and $V(\varphi)$. Inserting the specifications (6.18)–(6.20) into the general results (5.195)–(5.110) leads to drastic simplifications. As we will justify below, we can perform two additional approximations to simplify the loop corrections. Firstly, in view of the generalized gravitational coupling $\propto \sqrt{U(\varphi)}$, we can safely neglect contributions from pure graviton loops. Secondly, we can perform an expansion in $1/\xi$, which is justified by the assumption of a strong non-minimal coupling $|\xi| \simeq 10^4$. Taking these approximations into account and also collecting the contributions from the matter part to $V_{1-\text{loop}}$, we finally obtain the following expressions for the one-loop corrections:

$$U_{1-\text{loop}}(\varphi) = \frac{\varphi^2}{32\pi^2}\left(\mathbf{C}\ \ln\frac{\varphi^2}{\mu^2} + D\right), \tag{6.21}$$

$$G_{1-\text{loop}}(\varphi) = \frac{1}{32\pi^2}\left(E\ \ln\frac{\varphi^2}{\mu^2} + F\right), \tag{6.22}$$

$$V_{1-\text{loop}}(\varphi) = \frac{\lambda\,\varphi^4}{128\pi^2}\,\mathbf{A}\ \ln\frac{\varphi^2}{\mu^2}. \tag{6.23}$$

In this form, (6.21)–(6.23) were already considered in [6, 7, 12, 13] with φ being an abstract scalar in a GUT context. However, before we explicitly derive the model dependent form of the functions $\mathbf{A}$ and $\mathbf{C}$ and argue why we can neglect the contributions from D, E and F, we will investigate a specific effect caused by the non-minimal coupling.

6.2.3 S-Factor Suppression

As we have already seen from the one-loop calculations for the non-minimally coupled scalar fields discussed in Sect. 5.2.2, a specific combination of $U\ U'$ and G, which we have called s factor (5.98), appeared naturally due to the non-minimal coupling. In Sect. 5.2.2, we have also mentioned that each Higgs propagator will be weighted by one factor of s, which we are going to justify now. We only consider the graviton-Higgs sector of the model. The kinetic part of the operator F_{AB}, determined by the matrix C_{AB}, defined in (4.82) causes the propagation of the fluctuations $(h_{\mu\nu},\ \sigma)$ on the slowly varying background $(\bar{g}_{\mu\nu},\ \bar{\varphi})$. Due to the non-minimal coupling term $\propto R\,\varphi^2$, it has a non-diagonal structure. This term is responsible for the mixing of gravitational and scalar degrees of freedom. Symbolically (suppressing indices but indicating background quantities by a bar for a moment), we can write the kinetic structure structure as

$$\delta\Psi^A\ C_{AB}\ \bar{\nabla}\bar{\nabla}\ \delta\Psi^B \sim (M_{\text{P}}^2 + \xi^2\,\bar{\varphi}^2)\,h\,\bar{\nabla}\bar{\nabla}\,h + \xi\,\bar{\varphi}\,\sigma\,\bar{\nabla}\bar{\nabla}\,h + \sigma\,\bar{\Box}\,\sigma. \tag{6.24}$$

The inverse of (6.24) yields a propagator whose elements are suppressed by powers of $1/\xi$. For large $\bar{\varphi} \gg M_{\text{P}}/\sqrt{\xi}$, the suppression of pure graviton loops is, of course, obvious because the effective Planck mass squared exceeds by far the standard Planck mass squared, $M_{\text{P}}^2 + \xi\,\bar{\varphi}^2 \gg M_{\text{P}}^2$. Due to the mixing, this suppression proliferates to the full graviton-Higgs sector of the theory. In order to make this statement quantitative, we can perform the transition to the Einstein frame in which the propagator is diagonal and canonically normalized in the space of perturbations $(\hat{h}_{\mu\nu},\ \hat{\sigma})$ (in the background covariant DeWitt gauge). The diagonal propagator in the Einstein frame can be uplifted back to the Jordan frame by matrix multiplication with the Jacobian matrices $\partial(\bar{g}_{\mu\nu},\ \bar{\varphi})/\partial(\hat{\bar{g}}_{\mu\nu},\ \hat{\bar{\varphi}})$. Since the mixing disappears in the diagonal Einstein frame and the pure graviton loops are suppressed, we can study the consequences of the non-minimal coupling by investigating the Higgs-Higgs block $\sigma\Box\sigma$. It gets suppressed by the factor $s(\bar{\varphi})$

$$\left(\frac{\partial\varphi}{\partial\hat{\varphi}}\right)^2 \frac{1}{\hat{g}^{1/2}\,(\hat{\Box} - \hat{m}_{\rm H}^2)} = \frac{s(\varphi)}{g^{1/2}\,(\Box - m_{\rm H}^2)} \tag{6.25}$$

with the explicit expression

$$s(\varphi) := \frac{U}{G\,U + 3\,(U')^2} = \frac{M_{\rm P}^2 + \xi\,\varphi^2}{M_{\rm P}^2 + (6\,\xi + 1)\,\xi\,\varphi^2}. \tag{6.26}$$

We have again omitted the bars denoting a background quantity and also taken into account the scaling $\hat{g}^{1/2}(\hat{\Box} - \hat{m}_H^2) = (2U/M_{\rm P}^2)\, g^{1/2}(\Box - m_H^2)$ with the conformal factor $\Omega^2 = 2\,U/M_{\rm P}^2$ and $\hat{g}^{1/2} = \Omega^4\, g^{1/2}$ (disregarding space-time gradients of φ and bearing in mind conformal rescaling of particle masses $\hat{m} \propto \hat{\varphi}^2 \propto \Omega^{-2}\,\varphi^2$ [12]). It is useful to consider two limiting cases for the suppression function

$$s(\varphi) \to \begin{cases} \frac{1}{6\xi} & \text{for } \varphi \gg \frac{M_{\rm P}}{\sqrt{\xi}} \\ 1 & \text{for } \varphi \ll \frac{M_{\rm P}}{\xi}. \end{cases} \tag{6.27}$$

In view of $|\xi| \sim 10^4$, this clearly shows the suppression mechanism for large values of φ, whereas for lower values there is no suppression since s tends to 1. Thus, in the electroweak limit $\varphi \to v$ we recover the usual Standard Model where the influence of the non-minimal coupling is negligible. This mechanism was first understood in [69] when estimating the CMB generation in a non-minimal inflationary model. It was considered in [8] in the context of the generalized RG approach to nonrenormalizable theories. It also justifies the omission of graviton loops [7, 13] and modifies the beta functions of the Standard Model sector of this theory [28] at the high energy scale relevant for inflation. In fact, such a modification justifies the extension beyond the scale M_P/ξ interpreted in [4, 21] as a natural validity cut-off of the theory.[7]

Of course, the factor $s(\varphi)$ makes the counterterms of the theory non-polynomial in φ, and the theory as a whole becomes non-renormalizable. One might think that this completely invalidates the construction of the usual RG improvement and requires the generalization of the latter to a RG treatment with an infinite set of charges (perhaps of the functional nature like in [8, 49]). However, there is a shortcut to a simpler formulation based on the structure of the factor (6.26). In the logarithmic scale of the variable $t = \ln(\varphi/\mu)$, strongly compressing the transition domain $M_P/\xi \lesssim \varphi \lesssim M_P/\sqrt{\xi}$ for $s(\varphi)$ from 1 to $1/6\xi \simeq 0$, this factor looks very much like a step function

[7] The notion of a cut-off depends on the type of perturbation theory used. Typically, it is determined for the case when all dimensional quantities—fields and their derivatives – are considered on equal footing and treated perturbatively. However, as we will see in Sect. 6.6 the effective Planck mass requires a field dependent cut-off, which allows an extension of the model to scales above M_P/ξ.

$$s(t) \sim \Theta(t_{\text{step}} - t), \tag{6.28}$$

with $t_{\text{step}} \simeq \ln(\varphi_{\text{step}}/\mu) = \ln(M_P/\sqrt{6}\xi\mu)$. So, it is nearly constant for $\varphi < \varphi_{\text{step}}$ and $\varphi > \varphi_{\text{step}}$. Therefore, in both phases of such an approximation the theory is renormalizable, but has different running of coupling constants.[8] This qualitatively justifies the use of beta functions modified by s-factors smoothly interpolating between these two phases. In Sect. 6.5.1 we will see that the numerical results for the smooth s-factors and the step function nearly coincide.

There is an important subtlety with the modification of beta functions, which has been disregarded in [28] and the first version of [9]. The mnemonic rule of associating the factor $s(\varphi)$ with every propagator of the Higgs *multiplet*, as suggested in [28], turns out to be incorrect. Goldstone modes, in contrast to the Higgs particle, are not coupled to curvature and they do not have a kinetic term mixing with gravitons [15]. Therefore, their contribution is not suppressed by the s-factor of the above type. Separation of Goldstone contributions from the Higgs contribution leads to a modification of the one-loop beta functions presented in Sect. 6.4.2 below. We will show this disentanglement of Higgs and Goldstone contributions by explicit calculations in Sect. 6.2.4 and by a diagrammatic technique in Sect. 6.4.3.

6.2.4 Origin of Suppression and Goldstone Contributions

Beside the considerations of Sect. 6.2.3, the origin of the suppression factor (6.26) due to the non-minimal coupling has been shown in [8] and Appendix A of [9]. Now we will re-derive these results from the general results [82] of Sect. 5.2. The one-loop corrections for a $O(N)$ multiplet of real scalar fields non-minimally coupled to gravity—calculated in the Jordan frame – have the structure (5.109) and (5.195). Substituting the values (6.18)–(6.20) for U, G and V and further neglecting graviton loops,[9] which is justified by the considerations of the previous section, we can in addition perform an expansion in $1/\xi$ in order to obtain the dominant one-loop contributions for the Higgs inflation model. Under the additional assumption of a

[8] This is analogous to the approach of [15] where the model was approximated by matching the usual low-energy Standard Model phase with the chiral phase of the Standard Model at the inflation energy scale. This picture, however, takes place in the Einstein frame of the theory and is aggravated by the problem of transition between two different parametrizations of one quantum theory—the Cartesian coordinates in the space of the Higgs multiplet in the low-energy phase versus spherical coordinates in the chiral phase of the Standard Model. This transition is physically non-trivial.

[9] At the level of the Schwinger–DeWitt algorithm this corresponds to focusing only on the Higgs-Higgs block P_a^b of the potential matrix (4.88). On the level of the general results (5.195)–(5.110), this means neglecting terms associated with inverse powers of M_P.

slowly varying background,[10] we recover[11] the result already derived in the Appendix of [9]

$$V_{1-\text{loop}} = \frac{1}{32\,\pi^2}\left[m_{\text{G}}^4\,(N-1) + s^2\,m_{\text{H}}^4\right] + O(\xi^{-1})$$
$$= \frac{\lambda}{4}\varphi^4\left[\frac{\lambda}{32\,\pi^2}\,(6+18\,s^2)\right] + O(\xi^{-1}), \tag{6.29}$$
$$U_{1-\text{loop}} = \frac{1}{32\,\pi^2}\,R\left[m_{\text{G}}^2\,(N-1) + s^2\,m_{\text{H}}^2\right]\xi + O(\xi^0)$$
$$= \frac{1}{2}\xi\,R\,\varphi^2\left[\frac{3\,\lambda}{16\,\pi^2}(1+s^2)\right] + O(\xi^0), \tag{6.30}$$

where we have identified the Goldstone and Higgs mass according to (6.11) and (6.9) and set $N = 4$. At this point, we want to demonstrate that the disentanglement of Goldstone and Higgs contributions leads (in view of the different weighting with (6.27)) to a drastic modification of the one-loop corrections (6.29) and (6.30) for high energies. The Goldstone bosons play an essential role in the inflationary dynamics, since the contributions to (6.29) and (6.30) from the graviton-scalar sector of the model come exclusively from the Goldstone-bosons. In addition, there are no further contributions from the matter sector of the Standard Model to the correction of the non-minimal coupling (6.21). By comparison of (6.21) and (6.30), we can identify the coefficient **C** in (6.21) for high energies with

$$\mathbf{C} = 3\,\xi\lambda\,(1+s^2) + O(\xi^0) \simeq 3\,\xi\lambda + O(\xi^0). \tag{6.31}$$

We emphasize again that it is determined only by the Goldstone contributions. In contrast, the one-loop potential (6.23) also receives significant contributions from the heavy particles of the Standard Model which are summarized together with the Goldstone contributions (6.29) in one quantity, the anomalous scaling (6.14). The Goldstone contribution (6.29) to **A** for high energies is given by the last term $\lambda\,(6+18\,s^2) \simeq 6\,\lambda$ in (6.14). As already mentioned in Sect. 6.2.1, where **A** was introduced, the Higgs contributions to **A** are absent due to their suppression in (6.29) by a small factor of $s^2 \sim 1/\xi^2$.

[10] This is certainly justified for an inflationary scenario with $R \sim H^2 \sim V/6U \sim$ const. and $\varphi \sim$ const., cf. (6.36). At the level of the Schwinger–DeWitt formalism, the vanishing of gradients $\nabla\varphi = 0$ leads to $\hat{\mathcal{R}}_{\mu\nu} = 0$ and therefore to a simplified expression for the $\hat{a}_2$ coefficient (4.112).

[11] The results for (6.21) in the Appendix of [9] and (5.109) of [82] differ by a global sign which is associated with R due to the use of a different sign convention.

6.3 Cosmological Aspects of the Model

6.3.1 Slow-roll Dynamics in the Jordan Frame

Considering the formulas for the cosmological quantities within the slow-roll approximation derived in Sect. 2.69, it is obvious that the whole inflationary dynamics is determined completely by the effective potential in the Einstein frame. So far, all our expressions have been derived in the original Jordan frame, e.g. the one-loop contributions of Sect. 5.2. We have discussed the issue of transforming between the frames at the quantum level in Sect. 5.3.1. Thus, we expect to obtain different results depending on whether we calculate the effective action in the Jordan or in the Einstein frame respectively. We have calculated the effective action in the Jordan frame, whereas in [16] the Einstein frame based quantities have been used. Bearing in mind the discussion of Sect. 5.3.3 regarding the quantization in different conformal frames, we can nevertheless transform our Jordan frame based result to the Einstein frame in order to use the standard slow-roll machinery. Before investigating the effective potential in the Einstein frame, we will briefly consider the effective equations of motion in the general Jordan frame case within the cosmological slow-roll setup. For the sake of completeness, here we will briefly repeat part of the analysis of [7].

In order to derive the cosmological parameters of inflation, it is necessary to derive the equations of motion in the Jordan frame within the slow-roll approximation. Starting from the graviton-inflaton sector with the general tree-level action (6.16) for one $N = 1$ homogeneous inflaton φ, we can use (2.21)–(2.26) to specify the model to a closed Friedmann universe. Setting in addition $G(\varphi) = 1$ and varying with respect to the scale factor a, the lapse function N and the scalar field φ, we obtain the equations of motion for the homogeneous background as in [7]

$$-6\,U\,\frac{\dot{a}^2}{a^2} - \frac{6\,U}{a^2} - 6\,U'\,\dot{\varphi}\,\frac{\dot{a}}{a} + \frac{\dot{\varphi}^2}{2} + V = 0, \tag{6.32}$$

$$12\,U\,\frac{\ddot{a}}{a} + 6\,U\,\frac{\dot{a}^2}{a^2} + 12\,U'\,\dot{\varphi}\,\frac{\dot{a}}{a} + 6\,U''\,\dot{\varphi}^2 + 6\,U'\ddot{\varphi} + \frac{6\,U}{a^2} + \frac{3}{2}\,\dot{\varphi}^2 - 3\,V = 0, \tag{6.33}$$

$$\ddot{\varphi} + 3\,\frac{\dot{a}}{a}\,\dot{\varphi} - 6\,U'\,\frac{\dot{a}^2}{a^2} - 6\,U'\,\frac{\ddot{a}}{a} - \frac{6\,U'}{a^3} + V' = 0. \tag{6.34}$$

The first two equations are the first and second Friedmann equations and the last one is the Klein–Gordon equation for the scalar field. We can solve the first equation with respect to $\dot{a}^2/a^2$ and the second with respect to $\ddot{a}/a$ and insert the results into the Klein–Gordon equation. The result in terms of the Hubble parameter $H = \dot{a}/a$ is

$$\ddot{\varphi} + 3\,H\,\dot{\varphi} - \frac{\left(2\,V\,U' - U\,V' - \frac{1}{2}\,U'\,\dot{\varphi}^2 - 3\,U'\,U''\,\dot{\varphi}^2\right)}{U + 3\,U'^2} = 0. \tag{6.35}$$

In the inflationary scenario we can neglect the terms $\propto 1/a^2$, since they are exponentially suppressed for $a(t) \propto \exp(H\,t)$. Moreover, within the slow-roll approximation, the terms which contain second derivatives of the field or are quadratic in the first derivatives can be dropped. Then, the first Friedmann equation and the Klein–Gordon equation reduce to the relations

$$H^2 = \frac{V}{6\,U} - \frac{U'\,F}{3\,U} \tag{6.36}$$

$$\dot{\varphi} = \frac{F}{3\,H}. \tag{6.37}$$

We have defined the classical rolling force[12] (not to be confused with F in (6.22))

$$F(\varphi) := \frac{2\,V\,U' - V'\,U}{G\,U + 3\,U'^2} = -\frac{U^3}{U + 3\,U'^2}\frac{\mathrm{d}}{\mathrm{d}\varphi}\left(\frac{V}{U^2}\right) \tag{6.38}$$

and used relation (6.37) to substitute the $\dot{\varphi}$ in the Friedmann equation. We can make a quick check of the result by setting $U = M_{\mathrm{P}}^2/2$ and observe that the first term in F vanishes so that we obtain the standard inflationary relation for a scalar field minimally coupled to gravity. Moreover, in order to keep the discussion as general as possible, we have re-inserted the functional charge G. The correct insertion follows from dimensional and scaling reasons.

We can now upgrade all expressions to the quantum level [7] and obtain effective equations by the simple replacement of $U \to U_{\mathrm{eff}}$, $V \to V_{\mathrm{eff}}$, and $G \to G_{\mathrm{eff}}$, with the last quantities defined by

$$U_{\mathrm{eff}}(\varphi) := U_{\mathrm{tree}} + U_{1-\mathrm{loop}} = \frac{1}{2}(M_{\mathrm{P}}^2 + \xi\,\varphi^2) + \frac{\varphi^2}{32\,\pi^2}\left(\mathbf{C}\ln\frac{\varphi^2}{\mu^2} + D\right), \tag{6.39}$$

$$G_{\mathrm{eff}}(\varphi) := G_{\mathrm{tree}} + G_{1-\mathrm{loop}} = 1 + \frac{1}{32\,\pi^2}\left(F\,\ln\frac{\varphi^2}{\mu^2} + E\right), \tag{6.40}$$

$$V_{\mathrm{eff}}(\varphi) := V_{\mathrm{tree}} + V_{1-\mathrm{loop}} = \frac{\lambda}{4}(\varphi^2 - \nu^2)^2 + \frac{\lambda\,\varphi^4}{128\,\pi^2}\,\mathbf{A}\,\ln\frac{\varphi^2}{\mu^2}, \tag{6.41}$$

using (6.18)–(6.20) and (6.21)–(6.23). The justification of this simple substitution follows from the analysis of Chap. 5.2, where we have calculated the divergent part of the effective action (5.102) for the general model (5.68). At the one-loop level, structures originally present in the tree-level action were reproduced and received quantum corrections which are summarized in (6.39)–(6.41). It was shown, cf. (5.106) that additional higher-dimensional structures are suppressed [7]. Therefore, we can just replace the tree-level expression for U, G and V by their corrected quantum versions

[12] The name rolling force stems from the fact that the Klein–Gordon equation (re-introducing the second derivative term $\ddot{\varphi}$) corresponds to the equation for a damped harmonic oscillator with a damping term $\propto H\dot{\varphi}$ and a driving force $\propto F$. The inflationary picture suggests that this force drives the field to roll down the potential V.

(6.39)–(6.41). The parts in (6.21)–(6.23), summarized in the coefficients D, E and F have been calculated in [7], where it has been shown that they are suppressed by inverse powers of ξ.[13] We can therefore safely neglect them in our Higgs inflation model and obtain the truncated effective action

$$S^{\text{eff}}_{\text{trunc}}[g_{\mu\nu},\,\Phi^a] = \int_{\mathcal{M}} \mathrm{d}^4x\,\sqrt{g}\,\left(U_{\text{eff}}\,R + \frac{1}{2}\,G_{\text{eff}}\,g^{\mu\nu}\,\partial_\mu\Phi^a\partial_\nu\Phi_a + V_{\text{eff}} + \ldots\right). \tag{6.42}$$

and the "effective quantum rolling force" by

$$\begin{aligned} F_{\text{eff}}(\varphi) &= \frac{2\,V_{\text{eff}}\,U'_{\text{eff}} - V'_{\text{eff}}\,U_{\text{eff}}}{G_{\text{eff}}\,U_{\text{eff}} + 3\,U'^2_{\text{eff}}} \simeq -\frac{M_{\text{P}}^2}{6}\frac{\lambda}{\xi^2}\,\varphi\left(1 + \frac{\mathbf{A_I}}{64\,\pi^2}\frac{\xi\,\varphi^2}{M_{\text{P}}^2}\right) \\ &= -\frac{M_{\text{P}}^2}{6}\frac{\lambda}{\xi^2}\,\varphi\left(1 + \frac{\varphi^2}{\varphi_I^2}\right), \end{aligned} \tag{6.43}$$

where we have used $\varphi \gg \frac{M_{\text{P}}}{\xi} \gg \nu$ in the second equality. The quantity $\mathbf{A_I} := \mathbf{A} - 12\,\lambda$ and its origin will be derived below in detail. Furthermore, we have introduced a specific scale

$$\varphi_I^2 := \frac{64\,\pi^2\,M_{\text{P}}^2}{\xi\,\mathbf{A_I}}, \tag{6.44}$$

in the last equality of (6.43) which was derived in [7] as the maximum of a quantum probability distribution resulting from the cosmological tunnelling scenario.[14] We will come back to this point in Chap. 7, where we will explicitly derive this probability distribution. We want to emphasize that for large (inflationary) field values φ, the classical terms (largest in φ) cancel out identically and F_{eff} gets dominated by the quantity $\mathbf{A_I}$, which is purely of quantum origin.

6.3.2 *Effective Potential and Slow-Roll in the Einstein Frame*

In order to derive the cosmological parameters (2.69)–(2.69) for our model, we need to transform the effective potential (6.41) to the Einstein frame. Since we are only interested in the effective potential of the Higgs-inflaton φ and since the influence of the Goldstone bosons is already effectively contained in (6.41), we can use the

---

13 In [7] only a single scalar field has been considered. Therefore, the Goldstone contributions to **A** and **C** did not appear.

14 The whole formalism presented here has been derived in [7] for a single scalar field. Thus, there are neither Goldstone boson contributions to V_{eff} (no terms $\propto\, 6\,\lambda$ in (6.14)) nor to U_{eff} ($\mathbf{C} \ll 1$ and thus $\mathbf{A} = \mathbf{A_I}$).

transition formulas for the single field

$$\hat{g}_{\mu\nu} = \frac{2\,U(\varphi)}{M_{\rm P}^2}\, g_{\mu\nu}, \quad \left(\frac{{\rm d}\hat{\varphi}}{{\rm d}\varphi}\right)^2 = \frac{M_{\rm P}^2}{2}\, \frac{GU + 3U'^2}{U^2}, \tag{6.45}$$

$$\hat{V}(\hat{\varphi}) = \left(\frac{M_{\rm P}^2}{2}\right)^2 \frac{V(\varphi)}{U^2(\varphi)}\bigg|_{\varphi=\varphi(\hat{\varphi})}. \tag{6.46}$$

instead of the general formulas derived in Sect. 5.3.1. In the inflationary regime $\varphi > M_{\rm P}/\sqrt{\xi} \gg v$, we obtain for the effective potential in the Einstein frame

$$\begin{aligned}\hat{V}_{\rm eff} &= \left(\frac{M_{\rm P}^2}{2}\right)^2 \frac{V_{\rm eff}}{U_{\rm eff}^2} \simeq \frac{M_{\rm P}^4\,\lambda}{4\,\xi^2}\, \frac{(1+\frac{\mathbf{A}}{16\,\pi^2}\ln\frac{\varphi}{\mu})}{(1+\frac{2\,M_{\rm P}^2}{\xi\,\varphi^2}+\frac{\mathbf{C}}{4\,\pi^2\,\xi}\ln\frac{\varphi}{\mu})}\\ &\simeq \frac{M_{\rm P}^4\,\lambda}{4\,\xi^2}\left(1 - \frac{2\,M_{\rm P}^2}{\xi\,\varphi^2} + \frac{(\mathbf{A}-\frac{4\,\mathbf{C}}{\xi})}{16\,\pi^2}\ln\frac{\varphi}{\mu}\right)\\ &= \frac{M_{\rm P}^4\,\lambda}{4\,\xi^2}\left(1 - \frac{2\,M_{\rm P}^2}{\xi\,\varphi^2} + \frac{\mathbf{A_I}}{16\,\pi^2}\ln\frac{\varphi}{\mu}\right),\end{aligned} \tag{6.47}$$

with $\varphi = \varphi(\hat{\varphi})$. Equation (6.47) *defines* the inflationary key quantity

$$\mathbf{A_I} := \mathbf{A} - \frac{4\,\mathbf{C}}{\xi} = \frac{3}{8\,\lambda}\left(2\,g^4 + (g^2+g'^2)^2 - 16\,y_{\rm t}^4\right) - 6\lambda. \tag{6.48}$$

We will see that this quantity determines the fate of the model entirely, since the predictions of the Higgs mass crucially depend on the inflationary anomalous scaling $\mathbf{A_I}$. Due to the transition to the Einstein frame, the effective potential receives the Goldstone quantum corrections $\mathbf{C}$ contained in $U_{1-{\rm loop}}$. These are responsible for the difference between $\mathbf{A}$ and $\mathbf{A_I}$ summarized in the factor $4\,\mathbf{C}/\xi = -12\,\lambda$. Since $\mathbf{A_I}$ is a one-loop correction, its magnitude reflects the strength of quantum corrections. The usual slow-roll analysis with this effective potential has been performed in [12].[15]

Let us first repeat the derivation of the most important cosmological quantities in the formalism presented in [12] on the basis of the slow-roll formalism presented in Sect. 2.69. In consequence of (2.136)–(2.144), the central quantity of all cosmological parameters is the effective potential in the Einstein frame (6.47). Since the cosmological parameters (2.136)–(2.144) are formulated in terms of the slow-roll parameters (2.76) and (2.79), we can easily calculate these parameters for the potential (6.47) of our model. The effective Einstein frame slow-roll parameters can be expressed as

[15] In [12], the contributions from Goldstone bosons have not been taken into account and all parameters have been calculated at the electroweak scale $v = 246$ GeV only. However, in view of (6.9) and $\lambda(0) \ll 1$ this does not make not make a difference at this scale.

$$\hat{\varepsilon}_{\text{eff}} := \frac{M_P^2}{2}\left(\frac{1}{\hat{V}_{\text{eff}}}\frac{\mathrm{d}\hat{V}_{\text{eff}}}{\mathrm{d}\hat{\varphi}}\right)^2 \simeq \frac{4}{3}\left(\frac{M_P^2}{\xi\,\varphi^2} + \frac{\mathbf{A_I}}{64\,\pi^2}\right)^2, \tag{6.49}$$

$$\hat{\eta}_{\text{eff}} := \frac{M_P^2}{\hat{V}_{\text{eff}}}\frac{\mathrm{d}^2\hat{V}_{\text{eff}}}{\mathrm{d}\hat{\varphi}^2} \simeq -\frac{4}{3}\frac{M_P^2}{\xi\,\varphi^2}. \tag{6.50}$$

The inflationary phase $\varphi > \varphi_{\text{end}}$ ends when $\varphi = \varphi_{\text{end}}$. In turn, this moment is determined by the condition $\hat{\varepsilon}_{\text{end}} \simeq O(1)$ and it is usually fixed by defining $\hat{\varepsilon}_{\text{end}} := 1$. However, in favour of the elegance of the formalism, we define the end of inflation by the relation

$$\hat{\varepsilon}_{\text{end}} := \frac{3}{4}. \tag{6.51}$$

Then, under the natural assumption that the perturbative loop expansion is applicable, i.e. $\mathbf{A_I}/64\,\pi^2 \ll 1$ (which will also be justified numerically below), Eq. (6.49) implies that the value of the inflaton field at the end of inflation equals

$$\varphi_{\text{end}} \simeq 2\,\frac{M_P}{\sqrt{3\,\xi}}. \tag{6.52}$$

The number of e-folds is a measure of the duration of inflation. The name "*e*-folds" stems from the fact that under an exponential expansion of the universe $a \propto \exp(H\,t)$, $H = \text{const.}$, the amount of expansion between two moments of time t and t_{end} can be measured by comparing the scale factor at the beginning and at the end of the exponential expansion $N := \ln a_{\text{end}}/a = H\,(t_{\text{end}} - t)$. The general formula with a varying $H(t)$ is given by the integral over the inflationary stage, beginning with t and ending with t_{end}

$$N = \int_t^{t_{\text{end}}} \mathrm{d}t\,\frac{\dot{a}}{a} = \int_t^{t_{\text{end}}} \mathrm{d}t\,H(t) = \int_\varphi^{\varphi_{\text{end}}} \mathrm{d}\varphi\,\frac{H(\varphi)}{\dot{\varphi}} \tag{6.53}$$

$$\simeq \int_{\varphi_{\text{in}}}^{\varphi_{\text{end}}} \mathrm{d}\varphi\,\frac{3\,H^2(\varphi)}{F(\varphi)}, \tag{6.54}$$

where we have changed the variables of integration from t to the Jordan frame field φ and used the slow-roll relations (6.37) and (6.36). We can upgrade this expression again by taking into account the replacement $H \to H_{\text{eff}}$ and $F \to F_{\text{eff}}$ which leads to

$$N_{\text{eff}} \simeq \int_{\varphi_{\text{in}}}^{\varphi_{\text{end}}} \mathrm{d}\varphi\,\frac{3\,H_{\text{eff}}^2(\varphi)}{F_{\text{eff}}(\varphi)} \simeq \frac{48\,\pi^2}{\mathbf{A_I}}\ln\left(1 + \frac{\varphi^2}{\varphi_I^2}\right). \tag{6.55}$$

It is now convenient to introduce a new parameter x defined by

$$\frac{\varphi^2}{\varphi_I^2} =: e^x - 1. \tag{6.56}$$

The parameter x can then be read off by comparing (6.56) with (6.55) to be

$$x = \frac{N_{\text{eff}}\, \mathbf{A_I}}{48\, \pi^2} \tag{6.57}$$

The advantage of this notation is that it absorbs all numerical factors and the number of e-folds (later taken to be $N_{\text{eff}} \simeq 60$) into one quantity x which is proportional to the inflationary anomalous scaling $\mathbf{A_I}$. With this new definition, we can easily calculate the correction terms to the cosmological parameters like the spectral index. Moreover the quantity x is a direct measure of the strength of quantum corrections, since $\mathbf{A_I}$ is entirely of quantum nature.

6.3.3 Cosmological Parameters in Non-minimal Higgs Inflation

Using the definitions of the last sections, we can express the slow-roll parameters in the Einstein frame (6.49)–(6.50) in terms of x. From now on, we will drop the subscript "eff", bearing in mind that all cosmological parameters are to be derived by the effective quantities. Then the first slow-roll parameter becomes

$$\begin{aligned}\hat{\varepsilon}_{\text{v}} &\simeq \frac{4}{3}\left(\frac{M_{\text{P}}^2}{\xi\,\varphi^2} + \frac{\mathbf{A_I}}{64\,\pi^2}\right)^2 = \frac{4}{3}\left(\frac{M_{\text{P}}^2}{\xi\,\varphi^2} + \frac{M_{\text{P}}^2}{\xi\,\varphi_I^2}\right)^2 = \frac{4}{3}\frac{M_{\text{P}}^4}{\xi^2\,\varphi^4}\left(1 + \frac{\varphi^2}{\varphi_I^2}\right)^2 \\ &= \frac{4}{3}\frac{\mathbf{A_I}^2}{(64\,\pi^2)^2}\frac{\varphi_I^4}{\varphi^4}\,e^{2x} = \frac{3}{4\,N^2}\left(\frac{x\,e^x}{e^x - 1}\right)^2,\end{aligned} \tag{6.58}$$

and the second slow-roll parameter can be written as

$$\hat{\delta}_{\text{v}} \simeq -\frac{4\,M_{\text{P}}^2}{3\,\xi\,\varphi^2} = -\frac{4}{3}\frac{\mathbf{A_I}}{64\,\pi^2}\frac{\varphi_I^2}{\varphi^2} = -\frac{1}{N}\frac{x}{e^x - 1}. \tag{6.59}$$

Now we can calculate the quantities (2.135)–(2.145) by simply substituting the expressions (6.58) and (6.59). Thus, the amplitude of the scalar curvature perturbation reads

$$\mathcal{P}_\zeta := \zeta^2(k) \simeq \frac{\hat{V}}{24\,\pi^2\,M_{\text{P}}^4\,\hat{\varepsilon}_{\text{v}}} \simeq \frac{\lambda}{96\,\pi^2\,\xi^2\,\hat{\varepsilon}_{\text{v}}} = \frac{N^2}{72\,\pi^2}\frac{\lambda}{\xi^2}\left(\frac{e^x - 1}{x\,e^x}\right)^2. \tag{6.60}$$

Using the observational constraint of the combined WMAP + BAO + N data with the normalization $\zeta^2 \simeq 2.5 \times 10^{-9}$ at the pivot point $k_0 = 0.002\ \text{Mpc}^{-1}$ [45, 53] which we choose to correspond to $N \simeq 60$, the result of the quantum corrected cosmological constraint for the ration of coupling constants obtained form (6.60) can be expressed as

$$\frac{\lambda}{\xi^2} \simeq 0.5 \times 10^{-9} \left(\frac{x\, e^x}{e^x - 1} \right)^2 . \tag{6.61}$$

The multiplicative correction factor is entirely of quantum nature with the limits

$$\left(\frac{x\, e^x}{e^x - 1} \right)^2 \sim \begin{cases} 1 & \text{for} \quad x \ll 1 \\ x^2 & \text{for} \quad x \gg 1 \end{cases} . \tag{6.62}$$

Thus, for fixed $N \simeq 60$ and small quantum corrections $\mathbf{A_I} \ll 1$ we recover the "classical" constraint (6.6) in the limit $x \to 0$. Similarly, we can express the scalar spectral index

$$\begin{aligned} n_s := & \ 1 + \frac{\mathrm{d} \ln \zeta^2(k)}{\mathrm{d} \ln k} \simeq 1 - 6\,\hat{\varepsilon}_v + 2\,\hat{\delta}_v \\ \simeq & \ 1 - \frac{2}{N} \frac{x}{e^x - 1}, \end{aligned} \tag{6.63}$$

the power spectrum of the primordial tensor perturbations (2.141)

$$h_g^2(k) \simeq \frac{\lambda}{6\,\pi^2 \xi^2}\,, \tag{6.64}$$

the tensor to scalar ratio (2.144)

$$r = \frac{h_g^2}{\zeta^2(k)} \simeq 16\,\hat{\varepsilon}_v = \frac{12}{N^2} \left(\frac{x\, e^x}{e^x - 1} \right)^2 \tag{6.65}$$

in terms of multiplicative quantum corrections terms which are functions of x.

6.4 Higgs Inflation and Renormalization Group

6.4.1 Higgs Inflation at the Electroweak Scale

For our purpose, the most important quantity is the spectral index of the primordial scalar perturbations n_s. It measures the variation of the perturbation amplitude contained in the power-spectrum with a given wave number or mode k. It is given by the combination of slow-roll parameters (6.63) to first order in the slow-roll approximation. For small quantum contributions $x \ll 1$ and $N = 60$ it attains the "classical" $n_s = 1 - 2/N \simeq 0.967$.

On the basis of this quantum corrected cosmological formalism, the authors of [12] obtained from the WMAP + BAO + SN constraint of [53] at the 2σ confidence level

$$0.934 < n_s(k_0) < 0.988, \tag{6.66}$$

the *cosmological* restriction:

$$-1.57 < x < 1.79 \quad \Leftrightarrow \quad -12.4 < \mathbf{A} < 14.1. \tag{6.67}$$

However, the authors of [12] also obtained a constraint from the Standard Model by using the known values for m_W, m_Z, m_t at the electroweak scale $\varphi \simeq \nu$. Substituting these masses in the definition of $\mathbf{A}$ yields

$$\mathbf{A} \simeq \frac{12}{m_H^2 \nu^2} (m_Z^4 + 2\, m_W^4 - 4\, m_t^4). \tag{6.68}$$

For the Standard Model bounds $115\,\text{GeV} < m_H < 180\,\text{GeV}$ and with $m_Z = 91\,\text{GeV}$, $m_W = 80\,\text{GeV}$, $m_t = 171\,\text{GeV}$ and $\nu = 247\,\text{GeV}$, in [12] it has been found:

$$-48 < \mathbf{A} < -20. \tag{6.69}$$

Certainly, the two ranges for $\mathbf{A}$ of (6.67) and (6.69) do not overlap, so that we cannot simultaneously satisfy both—the cosmological constraint as well as the Standard Model constraint. However, this does not mean that we have to abandon the Higgs inflation model. Up to this point, we have only considered the one-loop quantum corrections at the electroweak scale, while in the Higgs inflation scenario, we intend to connect two energy scales separated by many orders of magnitude. Therefore we have to connect them by the RG flow from the electroweak vacuum up to inflationary energy scales [9, 16, 28]. In the following, we will show in detail that we even can simply adopt the cosmological formalism developed in [12], but with $\mathbf{A}$ replaced by $\mathbf{A_I}$ and all quantities evaluated at the inflationary instead of the electroweak scale.

6.4.2 Renormalization Group Improvement

Since we intend to connect two phenomena at two different energy scales separated by many orders of magnitude, it is not sufficient to calculate the relevant quantum corrections. In view of the electroweak vacuum of the Standard Model at $\varphi = \nu \simeq 10^2\,\text{GeV}$ and the inflationary phase in the early universe at $\varphi \simeq M_P/\sqrt{\xi} \simeq 10^{17}\,\text{GeV}$, we should take into account the dependence of the coupling constants on the energy scale. This means that we have to consider the *running* couplings

$$g_i \to g_i(t), \quad g_i = \{\lambda,\ \xi,\ y_t,\ g,\ g'\ g_s\} \tag{6.70}$$

with the logarithmic parameter

$$t = \ln\frac{\varphi}{\mu}, \tag{6.71}$$

determining the RG flow. The running couplings are the solutions of the coupled system of six ordinary differential equations which are determined by the beta-functions

$$\frac{\mathrm{d}\,g_i}{\mathrm{d}t} = \beta_i(g,\ g'\ g_s,\ y_t,\ \lambda,\ \xi). \tag{6.72}$$

The beta functions are explicitly given by

$$\beta_\lambda = \frac{\lambda}{16\,\pi^2}\,(18\,s^2\,\lambda + \mathbf{A}) - 4\,\gamma\,\lambda, \tag{6.73}$$

$$\beta_\xi = \frac{6\,\xi}{16\,\pi^2}\,(1+s^2)\,\lambda - 2\,\gamma\,\xi, \tag{6.74}$$

$$\beta_{y_t} = \frac{y_t}{16\,\pi^2}\left[-\frac{2}{3}\,g'^2 - 8\,g_s^2 + (1+\frac{s}{2})\,y_t^2\right] - \gamma\,y_t, \tag{6.75}$$

$$\beta_g = -\frac{g^3}{16\,\pi^2}\,\frac{(39-s)}{12}, \tag{6.76}$$

$$\beta_{g'} = \frac{g'^3}{16\,\pi^2}\,\frac{(81+s)}{12}, \tag{6.77}$$

$$\beta_{g_s} = -\frac{7\,g_s^3}{16\,\pi^2}. \tag{6.78}$$

The beta functions and in particular the correct s factor insertion were obtained by a combined method of detailed calculations, cf. Sect. 5.2, and diagrammatic techniques, cf. Sect. 6.4.3. For the general connection between the one-loop corrections and the beta functions see Sect. 4.3.3. The anomalous dimension of the Higgs field γ is given by the standard expression in the Landau gauge

$$\gamma = \frac{1}{16\pi^2}\left(\frac{9}{4}\,g^2 + \frac{3}{4}\,g'^2 - 3\,y_t^2\right) \tag{6.79}$$

and determines the field renormalization factor $Z(t)$, which is the solution of the RG equation

$$\frac{\mathrm{d}\,Z(t)}{\mathrm{d}t} = \gamma(t)\,Z(t). \tag{6.80}$$

Once we know a solution for the couplings $g_i(t)$, we can express $\gamma(t)$ in terms of these solutions and calculate $Z(t)$. Therefore, the RG equation for $Z(t)$ can be considered separately from the RG system (6.73)–(6.78). We also note that the beta functions are usually defined as [27].

$$\frac{\mathrm{d}\, g_i}{\mathrm{d}t} = \frac{\beta_i}{(1-\gamma)} \tag{6.81}$$

The reason for the absence of the factor in the one-loop approximation is the following:

$$\frac{\beta_i}{(1-\gamma)} \approx \beta_i\,(1+\gamma) = \beta_i + \beta_i\,\gamma. \tag{6.82}$$

The last term is a product of two one-loop beta functions $\propto 1/16\,\pi^2$ and therefore effectively a two-loop order quantity $\propto 1/(16\,\pi^2)^2$ and can be neglected at one-loop order. According to the Coleman–Weinberg technique [27], the one-loop RG improved effective action has the form (6.16) with the RG improved functional couplings

$$V_{\mathrm{RG}}(\varphi) := \frac{\lambda(t)}{4}\, Z^4(t)\,\varphi^4, \tag{6.83}$$

$$U_{\mathrm{RG}}(\varphi) := \frac{1}{2}\left(M_{\mathrm{P}}^2 + \xi(t)\, Z^2(t)\,\varphi^2\right), \tag{6.84}$$

$$G_{\mathrm{RG}}(\varphi) := Z^2(t). \tag{6.85}$$

The beta-functions can be derived in several ways. The most straightforward would be to explicitly calculate the divergent part of the one-loop corrections for the whole model, including the Standard Model matter part. However, this would result in an enormously complicated calculation. This type of calculation has only been done for the graviton-Higgs sector and general couplings U, V and G in [82]. For the Higgs inflation model this was discussed in Sect. 6.2.4. The calculations of (6.30) and (6.29) have explicitly shown the separation of Higgs and Goldstone contributions and the suppression of Higgs modes due to the non-minimal coupling. However, once we have analysed the suppression mechanism and have shown that it behaves like a step function and only affects the Higgs component of the multiplet, we can follow another strategy to calculate the coefficients of the beta functions. They can be calculated using diagrammatic techniques and counting symmetry factors of the corresponding diagrams where each Higgs loop is accompanied by a suppression factor s. Since the results of both methods coincide, the pragmatic method of [28] can be used. We simply modify the well-known Standard Model one-loop beta-functions for the couplings λ, g, g', g_s, g_{Y_t} taken from [32] and for ξ taken from [21] by weighting each Higgs propagator with a factor of s in the corresponding diagrams. In contrast to [28], we separate the Higgs and Goldstone contributions and weight them differently

[9]. Our modified beta functions [9] coincide with those used in [25], but are different from those used in [16, 28].

6.4.3 Diagrammatic Insertion Technique of s-Factors

In Chap. 5.2, we have explicitly calculated the one-loop counterterms (5.109) and (5.195) for the graviton-Higgs sector of the model (6.16). These results have been obtained from the "off-shell" effective action and directly enter the definition of the beta functions β_λ and β_ξ for $\lambda(t)$ and $\xi(t)$. They are essential for the model because they determine the running of the RG improved effective potential in the Einstein frame from which all cosmological parameters are derived. It is important to stress again that our counterterms (and thus the beta functions) originate from the quantization in the Jordan frame. In order to demonstrate that the diagrammatic technique leads to the same results, we will now re-derive the result (6.29) by means of the λ^2 counterterm only from diagrammatic techniques. This relation between calculational and diagrammatic techniques will be of great importance, as based on this coincidence we can use the conventional results for the Standard Model beta functions and simply modify them by the proper s factor insertion in order to obtain the beta functions (6.73)–(6.78).

6.4.3.1 Feynman Diagrams for O(N) Theory

We consider again the $O(N)$ symmetric multiplet of scalar fields Φ^a with

$$\varphi^2 := \Phi^a\,\Phi^b\,\delta_{ab} \quad \text{and} \quad \mathcal{L}_{\text{int}} = \frac{\lambda}{4}\,\varphi^4. \tag{6.86}$$

In order to quantize the theory and derive Feynman rules we should start with the diagrammatic representation of the tree-level interaction (Fig. 6.1).

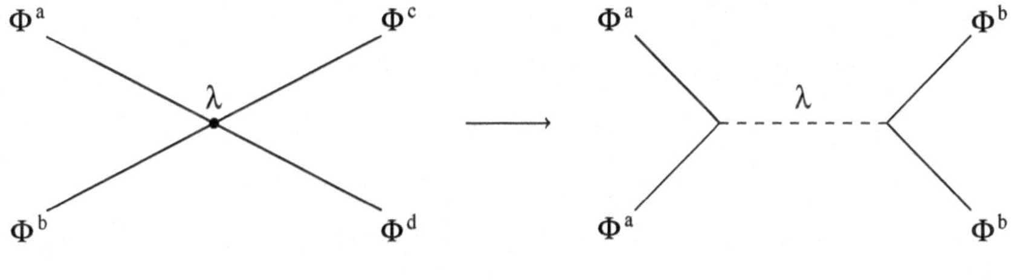

Fig. 6.1 Representation of the $O(N)$ self interaction vertex

We can express the $O(N)$ symmetry in the diagrams by stretching the point corresponding to an interaction vertex to a line. This graphical representation takes care of the $O(N)$ symmetry in the following way: Two external legs which are attached to the dashed "interaction line" at the same side, each representing a component of the field Φ^a, have to be identical (only allowing interactions that are monomials in φ^2).

In the following, we are interested in the one-loop effective potential and therefore in the corresponding one-loop four-point function $\Gamma^{(4)}$. In the scalar $O(N)$ theory, there is only one one-loop 1PI diagram. In the new representation, this diagram decomposes into a sum of three topologically distinct diagrams as shown in Fig. 6.2 below.

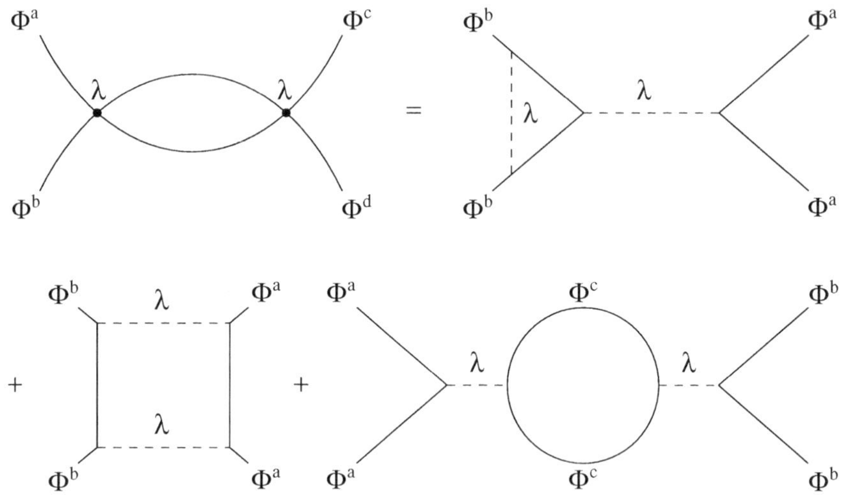

Fig. 6.2 The only one-loop 1PI diagram decomposes into a sum of three topologically different diagrams

The important point is that only in the third diagram all N fields are running in the loop. Thus, for large N this one will give the dominant contribution to $\Gamma^{(4)}$. We have to calculate the individual symmetry factors of these three diagrams. For the combinatorics, it makes a difference which external legs (particles associated with components of Φ) participate in the interaction. We will only consider the case where all external legs are equal, since we are ultimately interested in the case $N = 4$, where only the radial Higgs-component appears as a physical particle. Thus, we set $\Phi^a \equiv \varphi$ for all external legs in all diagrams.

6.4.3.2 Symmetry Factors

In order to figure out the individual contributions of these diagrams we have to calculate the corresponding symmetry factors. This can easily be done with the following formula (a modified version of the formula presented in [52])

$$\mathcal{S} := \frac{N_{\rm E} \cdot N_{\rm I}}{N_{\rm V} \cdot \prod_i N_{\rm L}^i}. \tag{6.87}$$

The single pieces of this formula can be calculated as follows: We start with a diagram and decompose it into a "pre-diagram" with the vertices and the external lines isolated. We want to figure out how many possibilities we have to construct the diagram from the "pre-diagram", while keeping the shape of the diagram intact. We start with attaching the external lines to the vertices and then combine the remaining internal lines:

- $N_{\rm E}$ = Number of possibilities to attach successively all external lines to the lines emanating from the vertices.
- $N_{\rm I}$ = Number of possibilities to combine the remaining internal lines.
- N_V = Number of vertices (equivalent to the power of the coupling constant or order in perturbation theory).
- $N_{\rm L}^i$ = Number of (equal) emanating lines of vertex i ($i = 1, \ldots, N_{\rm V}$).

For the first diagram we find

$$\begin{aligned} N_{\rm E} &= 8 \times 2 \times 4 \times 2, \quad N_{\rm I} = 2 \times 1, \quad N_{\rm V} = 2, \quad N_{\rm L}^i = 4, \\ \Rightarrow \mathcal{S}_1 &= \frac{8 \times 4 \times 2 \times 2 \times 2}{2 \times 4 \times 4} = 8 \end{aligned} \tag{6.88}$$

and the same symmetry factor $\mathcal{S}_2 = 8$ for the second diagram. For the third diagram, we obtain

$$\mathcal{S}_3 = \frac{8 \times 1 \times 4 \times 1 \times 2N \times 1}{2 \times 4 \times 4} = 2N. \tag{6.89}$$

The contributions to the four-point vertex function $\Gamma^{(4)}$ are given by

$$\lambda^2 \mathcal{S} = \lambda^2(\mathcal{S}_1 + \mathcal{S}_2 + \mathcal{S}_3) = 2\,(N+8)\,\lambda^2. \tag{6.90}$$

In the case of the Higgs multiplet $N = 4$, we therefore obtain $\Gamma^{(4)} = 24\,\lambda^2$.

Now we come to the crucial point. Since we know from the results of our direct calculations [9, 82] (cf. (6.25) and (6.29)) that each Higgs propagator—and only the Higgs propagator—gets suppressed by a factor of s, we have to associate each solid internal Higgs line in the diagram with a factor of s. If we do this for all three distinct diagrams separately, we have to modify the corresponding symmetry factors $\mathcal{S} \to \mathcal{S}^{\rm sup}$. For the first diagram, we find $\mathcal{S}_1^{\rm sup} = 8\,s^2$. For the second diagram, we find a similar factor $\mathcal{S}_2^{\rm sup} = 8\,s^2$. However, in the third diagram, only one of the N particles running in the loop corresponds to the Higgs field and therefore we obtain $\mathcal{S}_3^{\rm sup} = 2(N - 1 + s^2)$. Due to the different s-factor weighting, we finally obtain the one-loop correction of the four-point vertex

$$\Gamma^{(4)}_{\rm sup} = \mathcal{S}^{\rm sup}\lambda^2 = \lambda^2(\mathcal{S}_1^{\rm sup} + \mathcal{S}_2^{\rm sup} + \mathcal{S}_3^{\rm sup}) = 2\,(N - 1 + 9s^2) \tag{6.91}$$

and for the Higgs multiplet with $N = 4$, we find

$$\Gamma^{(4)}_{\text{sup}} = (6 + 18\, s^2)\, \lambda^2\,, \tag{6.92}$$

which is in perfect agreement with the formula (6.29) for $V_{1-\text{loop}}$ derived by the rigorous calculations in [82] and the Appendix of [9].

A crucial difference between our results [9] and similar approaches [16, 28] consists in the treatment of the correct s factor insertion in the beta functions (6.73)–(6.78). In [28], *each* propagator was weighted with the suppression factor *for all components of* Φ^a and thereby the authors did not distinguish between Goldstone and Higgs modes. In this case, the result reads

$$\Gamma^{(4)}_{\text{fullsupp}} = 24\, s^2\, \lambda^2. \tag{6.93}$$

During inflation, this leads to a sudden stop of the growth of $\lambda(t)$, since from $t = t_{\text{step}}$ on, all important λ^2 contributions in (6.73) are suppressed by $s^2 \sim 1/\xi^2 \sim 0$. In contrast, the contributions $\propto 6\, \lambda^2$ from the Goldstone bosons running in the loop are present in our model (6.92). Therefore, (6.92) lies between the conventional Standard Model result without non-minimal coupling (corresponding to $s = 1$) [32]

$$\Gamma^{(4)}_{\text{SM}} = 24\lambda^2, \tag{6.94}$$

where no suppression at all is present and the result (6.93). There Higgs and Goldstone modes are *both* weighted with factors of s with the effect that all components of Φ^a are suppressed equally. The ongoing growth of $\lambda(t)$ in the inflationary regime due to the Goldstone contributions (6.92) in (6.73) sensitively affects the estimation of M_{H}. It is exactly this crucial suppression mechanism, which has its origin in the non-minimal coupling, that makes the Higgs inflation model work. The suppression of the Higgs contributions protects the beta functions (6.73)–(6.78) up to inflationary energy scales from running into a Landau pole[16] and leads to a phenomenon similar to *asymptotic freedom*. Of course, this does not really correspond to the scenario of asymptotic freedom, since that would require that $g_i(t) \to 0$ in the limit $t \to \infty$. However, it can be thought of as a "window of asymptotic freedom" which ensures at least the finiteness of the couplings g_i up to inflationary energy scales. In addition, the different weighting justifies the separation of the matter sector, *including* the Goldstone bosons and the graviton-Higgs sector and shows that by a strong minimal coupling of the Higgs field to gravity, the fate of the model is mainly dominated by quantum corrections from the *matter sector*, an effect already discussed in [7].

The fact that only the radial Higgs component is influenced by the effect of the non-minimal coupling has its origin in the $O(N)$ symmetry of the multiplet. In the calculations of Sect. 5.2.2, this can be traced back to the different structures in the

[16] At a Landau pole, the running coupling constants $g_i(t)$ become infinite for finite values of the RG parameter t.

effective action arising due to the projectors $(\delta_{ab} - n_a n_b)$. However, there is a "fast and easy way" to understand this, even without these explicit calculations. In the context of the transition between Jordan frame and Einstein frame, see Sect. 5.3.1, we have defined the symmetric matrix

$$M_{ab} = \frac{M_{\mathrm{P}}^2}{2\,U^2}\left(3\,U'^2\,n_a n_b + G\,U\delta_{ab}\right), \tag{6.95}$$

which arose due to the non-minimal coupling. The eigenvalues of this matrix can easily be calculated and yield

$$\lambda_N = \frac{M_{\mathrm{P}}^2}{2\,U}\left(\frac{G\,U + 3\,U'^2}{U}\right) = s\,\frac{M_{\mathrm{P}}^2}{2\,U}, \quad \lambda_1 = \lambda_2 = \ldots = \lambda_{N-1} = \frac{G\,M_{\mathrm{P}}^2}{2\,U}. \tag{6.96}$$

This clearly shows that only the radial component gets suppressed by the s-factor and all other $(N-1)$ angular components remain unsuppressed.

The coincidence of calculational and diagrammatic methods justifies the modification of the known beta functions for the gauge and Yukawa couplings [32] by the correct diagrammatic s-factor insertion technique.

6.4.4 Running Potential at the Inflationary Scale

In order to establish the contact between the RG improved potential during inflation with the effective potential at the electroweak scale, it is useful to find an expression for the RG improved Jordan frame action so that it takes the form of the tree-level Jordan frame action, but with the constant couplings replaced by running couplings. Then we perform the transition to the Einstein frame. Finally, we even obtain an analytical expression for the running Einstein frame potential, when we exploit the fact that the duration of inflation is very short in units of the logarithmic scaling parameter t, compared to the post-inflationary stage.

We are going to calculate the *exact* RG flow of the effective potential in the next section by solving the RG equations *numerically*. However, in order to show the relation of $\hat{V}_{\mathrm{RG}}$ to the effective potential (6.47), we can rewrite the graviton-Higgs sector of the RG improved effective action (which is simply (6.16) with the running couplings (6.70) substituted) in terms of a new field

$$\phi := Z(t)\,\varphi. \tag{6.97}$$

The kinetic part in the graviton-inflaton sector of the RG improved effective action in terms of the new field can be calculated by using

$$\partial_\mu \phi = \partial_\mu \left[Z(t)\, \varphi \right] = \varphi\, \partial_\mu Z(t) + Z(t)\, \partial_\mu \varphi = \varphi \, \frac{\mathrm{d}Z(t)}{\mathrm{d}t}\, \partial_\mu t + Z(t)\, \partial_\mu \varphi \tag{6.98}$$

$$= \left[\frac{\mathrm{d}Z(t)}{\mathrm{d}t} + Z(t) \right] \partial_\mu \varphi = Z(t) \left[1 + \gamma(t) \right] \partial_\mu \varphi \simeq Z(t)\, \partial_\mu \varphi, \tag{6.99}$$

which is justified by the slow running of $Z(t)$. As a consequence, the RG improved effective action in terms of the new field ϕ has a canonical normalized kinetic term.

$$S_{\mathrm{RG}} = \int_{\mathcal{M}} \mathrm{d}^4 x\, \sqrt{g} \left(U_{\mathrm{RG}}(t)\, R + \frac{1}{2}\, g^{\mu\nu}\, \partial_\mu \phi \partial_\nu \phi + V_{\mathrm{RG}}(t) \right). \tag{6.100}$$

Comparing (6.100) with the tree-level action (6.16), we note that they both give rise to exactly the same structure, but with the constant couplings g_i replaced by the running couplings $g_i(t)$ and the Higgs-inflaton field φ replaced by the new field ϕ. We are ultimately interested in the running potential $\hat{V}_{\mathrm{RG}} \propto V_{\mathrm{RG}} / U_{\mathrm{RG}}^2$ in the Einstein frame because it contains all the information needed to calculate the cosmological parameters. Thus, it is in particular the running of the ratio $\lambda(t)/\xi^2(t)$ which is important. Now, we can further simplify the expressions for the running couplings $\lambda(t)$ and $\xi(t)$ during inflation by the following observation: The inflationary era lasts very shortly in units of Higgs field e-foldings

$$\Delta t := t - t_{\mathrm{end}} = \ln \phi / \phi_{\mathrm{end}} \simeq \ln \varphi / \varphi_{\mathrm{end}} \ll 1. \tag{6.101}$$

The term $\ln Z/Z_{\mathrm{end}}$ in (6.101) can be neglected again due to the slow running of $Z(t)$. Performing an approximation of the couplings up to terms linear in Δt, leads to

$$g_i(t) = g_i(t_{\mathrm{end}}) + \left. \frac{\mathrm{d}\, g_i(t)}{\mathrm{d}t} \right|_{t = t_{\mathrm{end}}} \Delta t + O((\Delta t)^2) \approx g_i(t_{\mathrm{end}}) + \beta_i(t_{\mathrm{end}})\, \Delta t, \tag{6.102}$$

where we have chosen the end of inflation t_{end} as expansion point. With the expressions (6.73) and (6.74), the solution for the couplings λ and ξ in the inflationary era ($s \simeq 0$) can be written approximately as

$$\lambda(t) \approx \lambda_{\mathrm{end}} \left(1 - 4\, \gamma_{\mathrm{end}}\, \Delta t + \frac{\mathbf{A}_{\mathrm{end}}}{16\, \pi^2}\, \Delta t \right), \tag{6.103}$$

$$\xi(t) \approx \xi_{\mathrm{end}} \left(1 - 2\, \gamma_{\mathrm{end}}\, \Delta t + \frac{6\, \lambda_{\mathrm{end}}}{16\, \pi^2}\, \Delta t \right), \tag{6.104}$$

where the subscript "end" means "evaluated at t_{end}". Transforming to the Einstein frame, we obtain the RG improved potential in the Einstein frame during inflation $\phi \gg M_{\mathrm{P}}/\sqrt{\xi}$

$$\hat{V}_{\rm RG}(t) = \left(\frac{M_{\rm P}^2}{2}\right)^2 \frac{V_{\rm RG}(t)}{[U_{\rm RG}(t)]^2} = M_{\rm P}^4 \frac{\lambda(t)}{4\,\xi^2(t)} \left(1 + \frac{M_{\rm P}^2}{\xi(t)\,\phi^2}\right)^{-2}$$
$$\simeq M_{\rm P}^4 \frac{\lambda_{\rm end}}{4\,\xi_{\rm end}^2} \left(1 - 2\,\frac{M_{\rm P}^2}{\xi_{\rm end}\,\phi^2} + \frac{\mathbf{A}_{\mathbf{I}\rm end}}{16\,\pi^2} \ln \frac{\phi}{\phi_{\rm end}}\right). \tag{6.105}$$

In the last step, we have used the approximation (6.103) and (6.104), linear in $\Delta\,t$ and it is understood that $\phi = \phi(\hat{\phi})$. Comparison with (6.47) shows that the RG improved effective potential during inflation corresponds to the effective potential (6.47), but for the field ϕ instead of φ and with the couplings at the electroweak scale $t_{\rm EW} \simeq 0$ replaced by the values at the inflationary scale $t_{\rm end}$ and a specific choice of the renormalization point $\mu = \phi_{\rm end}$. This proves, that we can simply adopt the formalism already developed in [12] and presented in Sect. 6.3.3 to calculate the parameters of the CMB, but with $\mathbf{A}$ replaced by $\mathbf{A_I}$ and all quantities evaluated at the energy scale of inflation $t_{\rm end}$ rather than at the electroweak scale. Therefore, we have to find the RG flow $\mathbf{A_I}(t_{\rm EW}) \to \mathbf{A_I}(t_{\rm end})$ interpolating between $t_{\rm EW} = 0$ and $t_{\rm end}$. This opens the possibility to resolve the mismatch of the cosmological and Standard Model constraints mentioned in Sect. 6.4.1, since $\mathbf{A_I}(0) \neq \mathbf{A_I}(t_{\rm end})$. In contrast to the short inflationary phase Δt, the post-inflationary running $t_{\rm end} - t_{\rm EW}$ is very long and must be calculated numerically.

However, before analysing the RG flow numerically in the next section, we would like to stress what the important differences between our approach [9] and the approaches followed in [16] and [28] already outlined in Sect. 6.2.4 and Sect. 6.4.3, imply for the running potential $\hat{V}_{\rm RG}(t)$. The RG improved effective potential in the Einstein frame during inflation

$$\hat{V}_{\rm RG}(t) \propto \frac{V_{\rm RG}(t)}{U_{\rm RG}^2(t)} \propto \frac{\lambda(t)}{\xi^2(t)} \tag{6.106}$$

has a RG running

$$\frac{{\rm d}\,\hat{V}_{\rm RG}(t)}{{\rm d}t} \propto \frac{\rm d}{{\rm d}t}\left(\frac{\lambda(t)}{\xi^2(t)}\right) \propto \frac{\xi(t)\,\beta_\lambda - 2\,\lambda(t)\,\beta_\xi}{\xi^3(t)}. \tag{6.107}$$

As we have already mentioned in Sect. 6.2.3, the separation between Higgs and Goldstone modes [9] is an essential feature of the Higgs inflation scenario. We can now retrieve its consequences when considering the running potential. By substituting the beta-functions (6.73) and (6.74) in (6.107), we find

$$\frac{{\rm d}\,\hat{V}_{\rm RG}(t)}{{\rm d}t} \propto \frac{1}{16\,\pi^2}\,\frac{\lambda}{\xi^2}\left[(6\,s^2 - 12)\,\lambda + \mathbf{A}\right] \propto \frac{\mathbf{A_I}(t)}{16\,\pi^2}\,\hat{V}_{\rm RG}(t). \tag{6.108}$$

We see again the origin of the quantity $\mathbf{A_I}$ for inflationary energies $t > t_{\rm end}$ with $s \simeq 0$. The solution with initial condition $\hat{V}_{\rm RG,\,0} = \hat{V}_{\rm RG}(t_{\rm end})$ is therefore simply

$$\hat{V}_{\mathrm{RG}}(t) \propto \hat{V}_{\mathrm{RG}}(t_{\mathrm{end}})\, e^{\frac{\mathbf{A_I}(t)}{16\pi^2}} \approx (1 + \mathbf{A_I}(t_{\mathrm{end}})\, \Delta t + O(\Delta t^2))\, \hat{V}_{\mathrm{RG}}(t_{\mathrm{end}}), \tag{6.109}$$

which recovers (6.105) in the approximation linear in Δt. Now we are going to confront this result with the running of $\hat{V}_{\mathrm{RG}}(t)$ in [16] and [28]. In order to do so, we have to insert their beta functions in (6.107).

In [28], the authors used the following beta functions for λ and ξ, with all components of the multiplet equally suppressed by s:

$$\beta_\lambda \propto (24\,\lambda^2 s^2 + ...), \quad \beta_\xi \propto (12\,\lambda\,\xi\, s^2 + ...). \tag{6.110}$$

We have focused on the differences in the important terms $\propto \lambda^2$ and "+..." denotes the terms $\propto \gamma$ as in (6.73) and (6.74), which anyway cancel out identically as in (6.108). Thus, in this case the RG running of the potential yields

$$\frac{\mathrm{d}\,\hat{V}_{\mathrm{RG}}(t)}{\mathrm{d}t} \propto \frac{(24\,\lambda^2\,\xi\, s^2 - 24\,\lambda^2 \xi\, s^2)}{\xi^3} = 0\,. \tag{6.111}$$

Consequently, $\hat{V}_{\mathrm{RG}}(t) = \text{const.}$ *independently* of the value for s.

In [16], the authors used the beta functions without any suppression ($s = 1$)

$$\beta_\lambda \propto (24\,\lambda^2 + ...), \quad \beta_\xi \propto (\xi + \frac{1}{6})(12\,\lambda + ...) \simeq (12\,\lambda\,\xi + ...), \tag{6.112}$$

and therefore also obtain

$$\frac{\mathrm{d}\,\hat{V}_{\mathrm{RG}}(t)}{\mathrm{d}t} \propto \frac{(24\,\lambda^2\,\xi - 24\,\lambda^2 \xi)}{\xi^3} = 0\,, \tag{6.113}$$

with the effect that the potential $\hat{V}_{\mathrm{RG}}(t)$ is constant during inflation. This clearly shows that the difference for $\hat{V}_{\mathrm{RG}}(t)$ in [9, 16] and [28] is due to the *different weighting* of Higgs and Goldstone contributions with s. Only if Higgs and Goldstone contributions are treated differently, the cancellation of the important terms in (6.111) and (6.113) is prevented and the correct dependence of $\hat{V}_{\mathrm{RG}}(t)$ on $\mathbf{A_I}$ as in (6.108) is revealed.

6.5 Numerical Analysis

The system of six coupled ordinary differential Eq. (6.72) is too complicated to obtain exact analytic solutions. But once the arbitrary renormalization point μ is fixed and the initial conditions for the couplings $g_{j,0} := g_j(t_0)$ at some initial t_0 are known, we can solve this system numerically. However, there are several subtleties in this approach, which we will discuss in the following.

6.5.1 Exact Numerical Solutions of the RG Improved Model

First of all, we need to fix the renormalization point μ. The choice of μ is of course arbitrary, but for a better comparison with the results of [16, 28] and in order to adopt the values of these couplings at the electroweak scale $t_{\rm EW}$ from [3], we have fixed the $t_0 \simeq t_{\rm EW} = 0$ initial conditions for the couplings at the top-quark scale[17]

$$\mu \equiv M_{\rm t} = 171 \text{ GeV}. \tag{6.114}$$

Later, we have varied the top-quark mass within the allowed error bars to investigate the sensitivity of the RG system on this input. The initial values for the couplings of the weak and strong interactions at the top-quark scale are given by [3]

$$g^2(0) = 0.4202, \quad g'^2(0) = 0.1291, \quad g_{\rm s}^2(0) = 1.3460 \tag{6.115}$$

where $g^2(0)$ and $g'^2(0)$ were obtained by a simple one-loop RG flow from the conventional values of $\alpha(M_{\rm Z}) \equiv g^2(M_{\rm Z})/4\,\pi = 0.0338$ and $\alpha'(M_{\rm Z}) \equiv g'^2(M_{\rm Z})/4\,\pi = 0.0102$ at the $M_{\rm Z}$ scale.[18] The initial value for the strong interaction at the $M_{\rm t}$ scale was generated by the numerical program of [3].

The initial values for the Higgs coupling λ and the Yukawa top-quark coupling $y_{\rm t}$ are determined by the "pole mass matching" scheme developed in [77, 84]

$$\lambda(0) = \frac{M_{\rm H}^2}{2\,v^2}\,[1 + 2\,\Delta_{\rm H}(M_{\rm H})]\,, \tag{6.116}$$

$$y_{\rm t}(0) = \frac{\sqrt{2}\,M_{\rm t}}{v}\,[1 + \Delta_{\rm t}(M_{\rm H})]\,. \tag{6.117}$$

The correction terms $\Delta_{\rm H}(M_{\rm H})$ and $\Delta_{\rm t}(M_{\rm H})$ comprise relevant mass operator corrections in the effective Higgs and top-quark propagators (cf. relations (6.12) and (6.8) for the masses without these corrections), which depend on $M_{\rm H}$ as well as on $M_{\rm t}$, $M_{\rm Z}$ and the Weinberg angle $\theta_{\rm W}$. Their explicit expressions can be found in the Appendix of [32] but for the sake of completeness, we also present them up to one-loop order in Appendix E. The numerical values for the Fermi constant $G_{\rm F}$, the top-quark mass $M_{\rm t}$, the mass of the gauge boson $M_{\rm Z}$ and the Weinberg angle $\theta_{\rm W}$ can also be obtained from [3].

In order to solve the coupled system of differential Eq. (6.72), we still need the initial condition for $\xi(t)$, which is not directly known. However, it can be determined from the CMB normalization condition for the amplitude of the power spectrum (6.61). As the quantum perturbations of φ and the new field ϕ are obviously related

---

[17] In the inflationary analysis this choice might seem unnatural. A more natural choice would be $\mu = M_{\rm P}/\sqrt{\xi}$ in order to keep the one-loop logarithms small.

[18] The analytical algorithm of transition between different scales for g_s^2 was presented in [64, 75, 76].

by $\zeta_\varphi = \zeta_\phi / Z(t)$, Eq. (6.61) immediately yields the following estimate for the ratio of the two coupling constants[19]

$$\frac{1}{Z_{\rm in}^2}\,\frac{\lambda_{\rm in}}{\xi_{\rm in}^2} \simeq 0.5 \times 10^{-9} \left(\frac{x_{\rm in}\,\exp(x_{\rm in})}{\exp(x_{\rm in}) - 1}\right)^2 \tag{6.118}$$

at the moment of horizon crossing with the number of e-folds $N = 60$. We call this moment the "beginning of inflation" and label it by $t_{\rm in}$. Thus, (6.118) yields rather a *final* condition $\xi_{\rm in}$ than an initial condition for the non-minimal coupling. The moment $t_{\rm in}$, in turn, can be determined by combining (6.44), (6.56) and (6.57). With the relation

$$\varphi_{\rm in} = M_{\rm t}\,\exp t_{\rm in} \tag{6.119}$$

this leads to

$$M_{\rm t}^2\,\exp(2\,t_{\rm in}) = \frac{4\,N}{3\,\xi_{\rm in}}\,M_{\rm P}^2\,\frac{\exp(x_{\rm in}) - 1}{x_{\rm in}} \tag{6.120}$$

or

$$t_{\rm in} = \ln\frac{M_{\rm P}}{M_{\rm t}} + \frac{1}{2}\ln\frac{4\,N}{3\,\xi_{\rm in}} + \frac{1}{2}\,\frac{\exp(x_{\rm in}) - 1}{x_{\rm in}}. \tag{6.121}$$

The system of the RG equations (6.72) for the six couplings $(g,\ g'\,g_s,\ y_t,\ \lambda,\ \xi)$ with five initial conditions (6.115)–(6.116) and one final condition (6.118) at $t_{\rm in}$, determined by the bootstrap equation (6.121), uniquely define the RG flow for given values of the Higgs mass $M_{\rm H}$ and top-quark mass $M_{\rm t}$. It is clear that the boundary conditions for this flow are rather involved and go beyond the usual Cauchy problem, since even the definition of the final moment of time (6.121) is non-local in t, i.e., it contains the whole history from $t = 0$ to $t = t_{\rm in}$. The RG flow parameter t can be interpreted as time, but opposite to the physical direction of time.

Nevertheless, the program `Mathematica` enables us to solve this problem numerically in the following way. We take the system of the six couplings with the initial conditions (6.115)–(6.117) for the five couplings λ, $y_{\rm t}$, g, g', $g_{\rm s}$. We note that the initial conditions of λ and $y_{\rm t}$ depend parametrically on the Higgs mass $M_{\rm H}$, whereas $M_{\rm t}$ is fixed by (6.114). In order to obtain an initial condition ξ_0 for ξ, we can choose a specific but arbitrary ξ_0.[20] Then, we choose a specific value for $M_{\rm H} = M_{\rm H}^*$ and solve the complete system (6.72) with these initial conditions, we calculate the moment of time $t_{\rm in}$ by solving the bootstrap equation (6.121) and insert the solutions $\lambda(t)$ and $\xi(t)$ at that moment of time $t_{\rm in}$ in the final condition (6.118).

---

[19] No modification of the spectral index due to Z occurs in the one-loop RG running because the difference $n_{\rm s}^{\varphi} - n_{\rm s}^{\phi} = 2\gamma dt/dN = (\gamma\,\mathbf{A}_{\rm I}/48\pi^2)e^x/(e^x - 1)$ belongs to the two-loop order.

[20] It is clear that $|\xi_0|$ should roughly be of order 10^4 to satisfy (6.118), but for the algorithm we describe here it does not matter in principle, albeit choosing a very different ξ_0 will increase the number of iterations as we will see in a moment.

Now, we can check how well (6.118) is satisfied for the pair of values (ξ_0, $M_{\rm H}^*$) and repeat this procedure by holding $M_{\rm H}^*$ fixed and choosing a different ξ_0. In that way, we can scan over the full range of ξ_0 for one specific $M_{\rm H^*}$, finding exactly the value $\xi_0 = \xi_0^{\rm opt}$ so that the pair of values ($\xi_0^{\rm opt}$, $M_{\rm H}^*$) satisfies the final condition (6.118) best. We save this pair of values ($\xi_0^{\rm opt}$, $M_{\rm H}^*$) and repeat the whole scanning procedure again, but this time for a different initial value $M_{\rm H}^{'*}$, in order to obtain another optimal pair of values ($\xi_0^{'\rm opt}$, $M_{\rm H}^{'*}$). We proceed in the same way and scan the whole range of 115 GeV $< M_{\rm H} <$ 180 GeV. This finally results in a table of optimal data points to which we can fit the function $\xi_0(M_{\rm H})$. This function gives the missing initial condition for the RG system. Once determined, this function $\xi(M_{\rm H})$ allows the RG system (6.72) to be solved as usual by six initial conditions (g, g' g_s, $y_t(M_{\rm H})$, $\lambda(M_{\rm H})$, $\xi_0(M_{\rm H})$), with the solutions $g_i(t; M_{\rm H})$ depending parametrically on $M_{\rm H}$ via the initial conditions.

As mentioned before, we determine the end of inflation by the condition $\hat{\varepsilon}_{\rm end} = 3/4$. Comparing with (6.49), we obtain

$$\varphi_{\rm end} \simeq \frac{2\,M_{\rm P}}{\sqrt{3\,\xi}} \tag{6.122}$$

under the assumption

$$\frac{\mathbf{A_I}(t_{\rm end})}{64\,\pi^2} \ll 1, \tag{6.123}$$

which will be justified below. Using again the relation $t = \ln\,\varphi/M_{\rm t}$, we can determine the moment of time $t_{\rm end}$ when inflation ends

$$t_{\rm end} = \ln\,\frac{M_{\rm P}}{M_{\rm t}} + \frac{1}{2}\,\ln\,\frac{4}{3\,\xi_{\rm end}}. \tag{6.124}$$

Thus, we see that the end of inflation in units of inflaton field e-folding $t_{\rm in} - t_{\rm end} = \ln\,(\varphi_{\rm in}/\varphi_{\rm end})$ is very short

$$t_{\rm in} - t_{\rm end} = \frac{1}{2}\,\ln\,N + \frac{1}{2}\,\ln\,\frac{\xi_{\rm in}}{\xi_{\rm end}} + \frac{1}{2}\,\ln\,\frac{e^{x_{\rm in}} - 1}{x_{\rm in}} \simeq \frac{1}{2}\,\ln\,N \simeq 2, \tag{6.125}$$

compared to the post-inflationary evolution $t_{\rm end} - t_{\rm EW} \simeq 35$. In (6.125) we have taken into account that $\xi_{\rm in} \simeq \xi_{\rm end}$ and $x = \mathcal{O}(1)$. This is, of course, typical for large-field inflationary models with $N \sim \varphi^2/M_{\rm P}^2$. The difference (6.125) also restricts the range of the approximation (6.101) linear in the logarithms. The linearisation (6.103) and (6.104) imply the bound $|\mathbf{A_I}(t_{\rm end})|\,\Delta t/16\,\pi^2 \ll 1$, with $\Delta t < t_{\rm in} - t_{\rm end} \simeq (\ln\,N)/2$. Thus, this approximation holds for

$$\frac{|\mathbf{A_I}(t_{\rm end})|\,\Delta t}{16\,\pi^2} \ll \frac{2}{\ln\,N} \simeq 0.5. \tag{6.126}$$

6.5.2 Discussion of Numerical Results

The running of $\mathbf{A_I}(t)$ strongly depends on the behaviour of $\lambda(t)$. It is well known that for small Higgs masses the usual RG flow in the Standard Model leads to an instability of the electroweak vacuum caused by negative values of $\lambda(t)$ in a certain range of t, see e.g. [32, 74] and references therein. The same happens to the modified RG for the non-minimally coupled Higgs field considered here.

The numerical solution for $\lambda(t)$ in the setting of the previous section is shown in Fig. 6.3 for five values of the Higgs mass and the value of top-quark mass $M_t = 171$ GeV. The lowest one corresponds to the critical value

$$M_H^{inst} \simeq 134.27 \text{ GeV}. \tag{6.127}$$

This is the boundary of the instability window where $\lambda(t)$ bounces back to positive values after vanishing at $t_{inst} \sim 41.6$ or $\varphi_{inst} \sim 80 M_P$.[21] It turns out that the corresponding $\xi(t) \sim 5000$ is nearly constant (see below), so that the factor (6.26) at t_{inst} is very small, $s \simeq 1/6\xi \sim 0.00005$. Thus, the situation is different from the usual Standard Model with $s = 1$, and numerically the critical value turns out to be higher than the known Standard Model stability bound ~ 125 GeV [32].

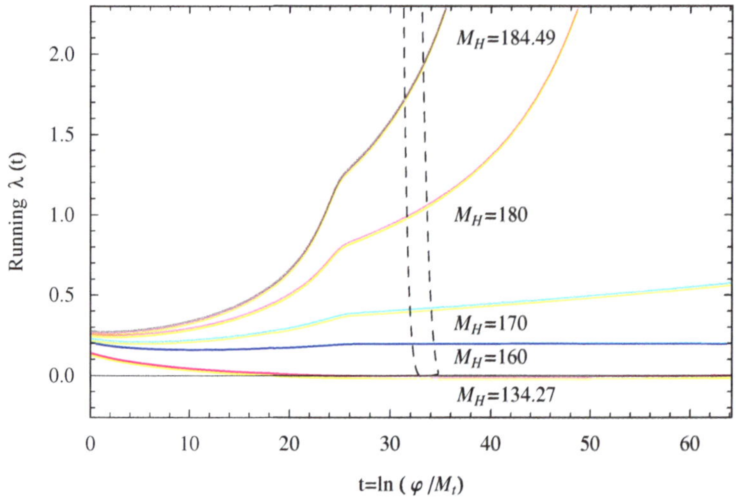

Fig. 6.3 Running $\lambda(t)$ for five values of the Higgs mass above the instability threshold. *Dashed curves* mark the boundaries of the inflation domain $t_{end} \le t \le t_{in}$

Figure 6.3 shows that near the instability threshold $M_H = M_H^{inst}$, the running coupling $\lambda(t)$ stays very small for all scales t relevant to the observable CMB,

---

[21] An over-Planckian scale of φ does not signify a breakdown of the semi-classical expansion, because the energy density $\sim 10^{-10} M_P^4$ stays far below the Planckian value.

(6.121)–(6.124). This follows from the fact that the positive running of $\lambda(t)$ caused by the term $(18s^2+6)\lambda^2$ in β_λ, (6.73), is much slower for $s \ll 1$ than that of the usual Standard Model driven by the term $24\lambda^2$. For larger $M_{\rm H}$, this suppression in the term $18s^2\lambda^2$ is responsible for the origin of the bump on the plots at $t_{\rm step} \sim 26$ where the value of s drops from 1 to 0 and $\lambda(t)$ continues growing but by a slower rate. The smallness of λ of course improves the efficiency of perturbation theory. However, at the critical value of the Higgs mass, we encounter a singularity of both anomalous scalings $\mathbf{A}(t)$ and $\mathbf{A_I}(t)$ at the instability point $t_{\rm inst}$ where $\lambda(t_{\rm inst}) = 0$.

This is depicted in Fig. 6.4 for $\mathbf{A_I}(t)$ by the red curve corresponding to the slightly overcritical value 134.271 GeV (vertical red line and quasi-hyperbolic curves to the left and to the right of it). Other overcritical curves with small $M_{\rm H} > M_{\rm H}^{\rm inst}$ run through zero and change the sign of $\mathbf{A_I}(t)$ from negative to positive. For larger $M_{\rm H}$, when λ also becomes larger, the term -6λ in (6.48) for $\mathbf{A_I}$ does not let it become positive at any t, and $\mathbf{A_I}$ starts decreasing after reaching some maximal negative value (like for the green curve of $M_{\rm H} = 180\,$GeV).

An important observation is that for all Higgs masses in the range $M_{\rm H}^{\rm inst} = 134.27\,{\rm GeV} < M_{\rm H} < 185\,{\rm GeV}$, the inflation range $t_{\rm end} < t < t_{\rm in}$ is always below the instability value $t_{\rm inst} = 41.6$ (numerics gives that $t_{\rm in} < 34.8$), so that $\mathbf{A_I}(t)$ is always *negative* during inflation. Its running is depicted in Fig. 6.4.

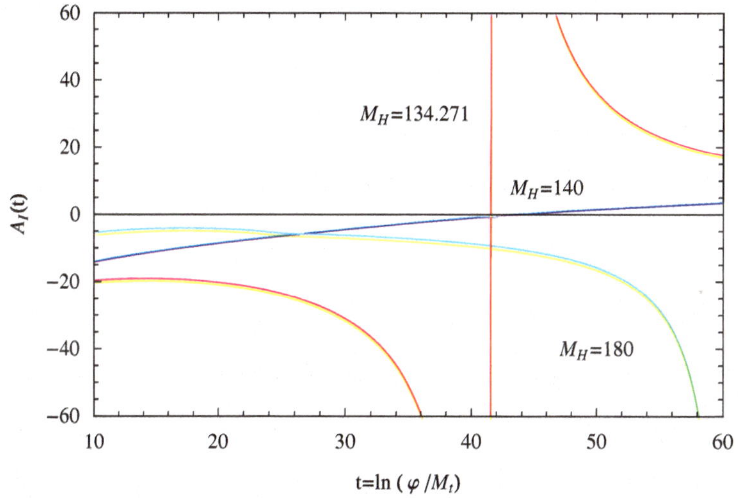

Fig. 6.4 Running anomalous scaling for the critical Higgs mass (the *red curve* with a *vertical segment* at the singularity with $t_{\rm inst} \sim 41.6$) and for two masses in the stability domain (*blue and green curves*)

$\mathbf{A_I}(t)$ runs from big negative values $\mathbf{A_I}(0) < -20$ at the electroweak scale to small but also negative values at the inflationary scale below $t_{\rm inst}$. This makes the CMB data compatible with the generally accepted Higgs mass range. Indeed, the knowledge of the RG flow immediately allows us to obtain $\mathbf{A_I}(t_{\rm end})$ and $x_{\rm end}$ and thus find the parameters of the CMB power spectrum (6.63)–(6.65) as functions of M_H. The parameter of primary interest—the spectral index—is given by Eq. (6.63) with $x = x_\mathrm{end} := N\,\mathbf{A_I}\,(t_\mathrm{end})/48\pi^2$ and depicted in Fig. 6.5.

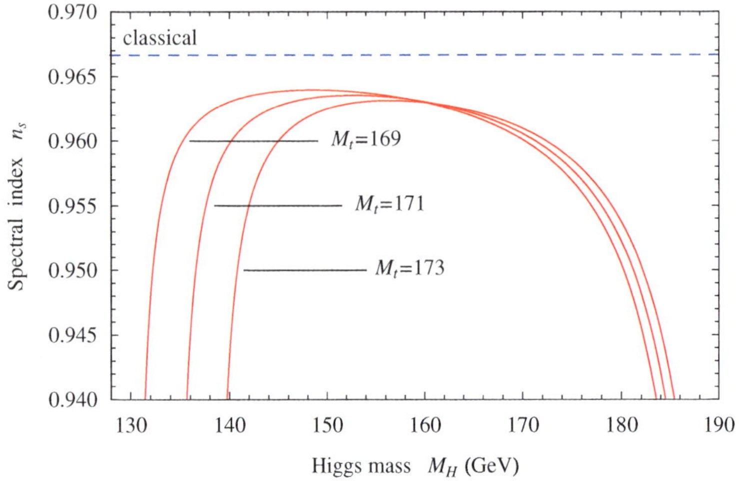

Fig. 6.5 The spectral index n_s as a function of the Higgs mass M_H for three values of the top-quark mass

Even for low values of the Higgs mass above the stability bound, n_s falls into the range admissible by the CMB constraint, now existing at the 2σ confidence level (based on the combined WMAP + BAO + SN data [45]),

$$0.94 < n_\mathrm{s}(k_0) < 0.99. \tag{6.128}$$

The spectral index becomes too small (that is, dropping below 0.94) only for large x_end or large negative $\mathbf{A_I}(t_\mathrm{end})$, which only happens when M_H either approaches the instability bound or exceeds 180 GeV at the decreasing branch of the n_s graph.

Thus, we get lower and upper bounds on the Higgs mass, which both follow from a comparison of the *lower* bound of the spectral index in (6.128) with (6.63). It is surprising that the sharpest bounds on M_H, both lower and upper bounds, are determined by the CMB only.

In [17, 28], similar bounds were obtained by the instability criterion

$$\lambda(t;\ M_\mathrm{H})|_{M_\mathrm{H}=M_\mathrm{H}^\mathrm{min}} = 0 \tag{6.129}$$

for the lower bound $M_\mathrm{H}^\mathrm{min}$ on the Higgs mass and by the criterion that perturbation theory remains applicable

$$\left.\frac{\lambda(t;\ M_{\rm H})}{16\,\pi^2}\right|_{M_{\rm H}=M_{\rm H}^{\rm max}} \ll 1 \tag{6.130}$$

for the upper bound $M_{\rm H}^{\rm max}$ of the Higgs mass. The determination of *both* bounds by the CMB has the advantage to yield *exact* bounds, whereas especially the criterion (6.130) is numerically not precisely fixed. The numerical analysis for the corresponding $x_{\rm end} \simeq -1.4$ gives for $M_{\rm t} = 171\,{\rm GeV}$ the following range for a CMB-compatible Higgs mass:

$$135.62\ {\rm GeV} \lesssim M_{\rm H} \lesssim 184.49\ {\rm GeV}. \tag{6.131}$$

Both bounds belong to the non-linear domain of Eq. (6.63) because they correspond to the value $x_{\rm end} = -1.4 < -1$. However, their calculation is still within the domain of our linear-in-logs approximation because the quantity $\mathbf{A_I}(t_{\rm end})/16\pi^2 \simeq -0.07$ satisfies the restriction (6.126) for $x_{\rm end} = -1.4$. For a smaller $|\mathbf{A_I}(t_{\rm end})|$, the expression (6.63) for $n_{\rm s}$ can be linearised in x and takes a particularly simple form,

$$n_{\rm s} = 1 - \frac{2}{N} + \frac{\mathbf{A_I}(t_{\rm end})}{48\pi^2}, \quad \mathbf{A_I}(t_{\rm end}) \ll \frac{48\pi^2}{N} \sim 8. \tag{6.132}$$

It is applicable for a wide range of the Higgs mass within the domain of (6.131).

As we can see, the upper bound on $n_{\rm s}$ in (6.128) does not generate any restrictions on $M_{\rm H}$, and it will not affect the $M_{\rm H}$-range unless it is lowered by future CMB observations to about 0.964—the top of the $n_{\rm s}$-graphs in Fig. 6.5. The lower CMB bound in (6.131) is slightly higher than the instability bound $M_{\rm H}^{\rm inst} = 134.27\,{\rm GeV}$. This bound depends on the initial data for weak and strong couplings and, even more important, on the top-quark mass $M_{\rm t}$ that is known with less precision. The bound $M_{\rm H}^{\rm inst}$ given above was obtained for $M_{\rm t} = 171\,{\rm GeV}$. In the following, we consider this value of the top-quark mass. Results for the neighbouring values $M_{\rm t} = 171 \pm 2\,{\rm GeV}$ are presented in Fig. 6.5 in order to show how strongly the plot gets shifted along the $M_{\rm H}$ axis. The general pattern of this shift follows from the dependence of $M_{\rm H}^{\rm inst}$ on $M_{\rm t}$—the instability bound is larger for a larger top-quark mass, which can be explained by a negative contribution $-y_t^4 \sim -M_{\rm t}^4$ to $\mathbf{A}$, and via $\mathbf{A}$ to β_λ, (6.73).

Viability of the inflation scenario also implies that the system safely evolves from the end of inflation to the electroweak vacuum through the periods of thermalization (reheating), radiation and matter domination. This mainly requires the possibility to roll down from $\varphi_{\rm end} = M_{\rm t}\exp(t_{\rm end})$ to $\varphi = v$ or a permanent positive slope of the effective potential [17] (we disregard here the tunnelling scenario). The shape of the graph of this potential depicted in Fig. 6.6 for the instability threshold confirms the danger of having a negative slope for $M_{\rm H}$ close to $M_{\rm H}^{\rm inst}$—the formation of a false vacuum at the instability scale.

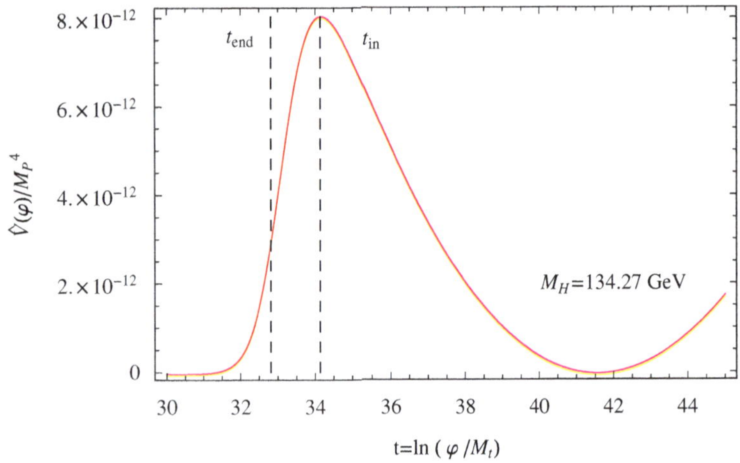

Fig. 6.6 The Einstein frame effective potential for the instability threshold $M_{\rm H}^{\rm inst} = 134.27\,\text{GeV}$. A false vacuum occurs at the instability scale $t_{\rm inst} \simeq 41.6$, $\varphi \sim 80\ M_{\rm P}$, which is much higher than the inflation scale $\varphi < \varphi_{\rm in} \simeq 0.04\ M_{\rm P}$. The *dashed lines* indicate a hypothetical inflation domain (ruled out by the lower $n_{\rm s}$ CMB bound and the requirement of a positive slope of $\hat{V}$ at $t \le t_{\rm in}$)

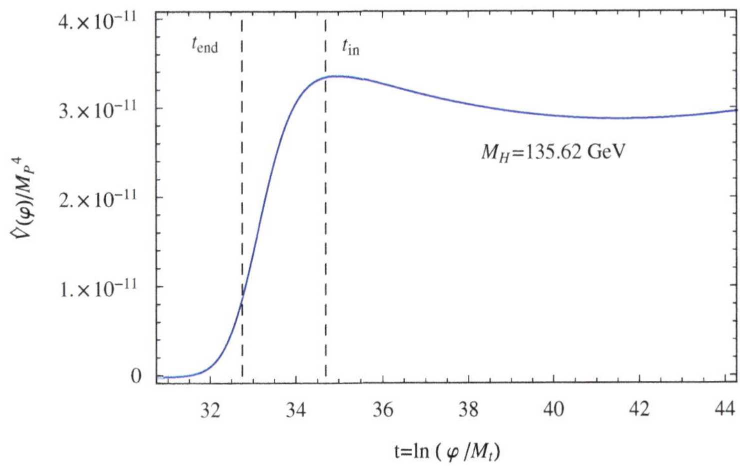

Fig. 6.7 Inflaton potential at the lower CMB compatible value of $M_{\rm H}$. A metastable vacuum exists at $t \simeq 42$, which is much higher than $\hat{V}$ during inflation

At higher Higgs masses until about 160 GeV we first get a family of metastable vacua at scales $\gtrsim t_{\rm inst}$. An example is the plot for the lower CMB bound $M_{\rm H} = 135.62\,\text{GeV}$ depicted in Fig. 6.7.

For even larger $M_{\rm H}$ these metastable vacua get replaced by a negative slope of the potential which interminably decreases to zero at large t (at least within the perturbation theory range of the model), see Fig. 6.8.

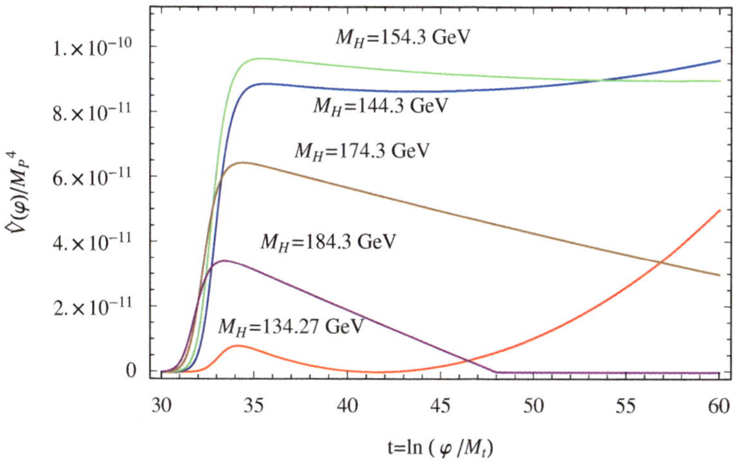

Fig. 6.8 The succession of effective potential graphs for $M_{\mathrm{H}}^{\mathrm{inst}} < M_{\mathrm{H}} < 184.3\,\mathrm{GeV}$

The graphs of the running effective potential depicted in Fig. 6.8 first show the occurrence of a metastable vacuum for increasing M_{H} and then the formation of a negative slope branch for even higher M_{H}. Local peaks of $\hat{V}$ situated at $t = 34 \div 35$ grow with M_{H} for $M_{\mathrm{H}} \lesssim 160\,\mathrm{GeV}$ and start decreasing for larger M_{H}.

However, these metastable vacua and the negative slope of the potential are not dangerous for the inflationary scenario whose scales $t_{\mathrm{in}} < 34.8$ are much lower than the metastability scale and belong to the positive slope of $\hat{V}$ to the left of its peak at $t = 34 \div 35$. Thus, even if the metastable vacuum occurs for low M_{H}, it exists before the inflationary stage probed by current CMB observations.[22]

Let us finally focus on the running of $\xi(t)$ depicted for five values of the Higgs mass in Fig. 6.9, starting with the lower bound of the range (6.157) below. The running is very slow for low values of the Higgs mass near the instability threshold, which of course follows from the smallness of the running $\lambda(t)$ in this domain. Another property of the ξ-behaviour is that the normalization of the power spectrum (6.118) leads to a value $\xi \sim 5000$ for small Higgs masses, which is smaller than the old estimate $\sim 10^4$ [6, 7, 9, 13, 17, 69, 78]. This is caused by a decrease of $\lambda(t)$ which at t_{in} becomes much smaller than $\lambda(0)$—an observation made in [28]. This relaxes the situation with the problem of "unnaturalness" in this model and brings ξ closer to a very subjective borderline between "natural" and "unnatural" values of coupling constants.

For large M_{H} close to the upper bound of the CMB range, $\xi(t)$ grows like $\lambda(t)$ to the Landau pole and eventually leaves the perturbation theory domain. However, it grows much faster than $\lambda(t)$, so that $\hat{V}(\varphi) \sim \lambda(t)/\xi^2(t)$ tends to zero, as shown by the negative slope branches of the potential in Fig. 6.8.

[22] The existence of this vacuum can perhaps be probed by wavelengths longer than that of a pivotal $N \simeq 60$, but this requires a deeper analysis.

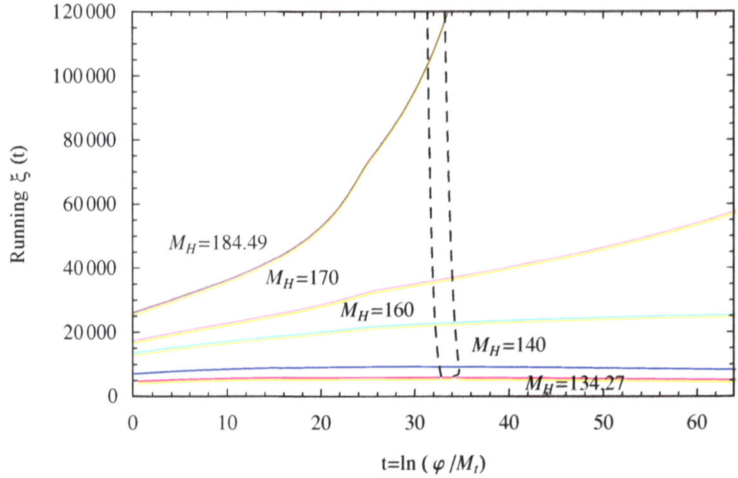

Fig. 6.9 Plots of running $\xi(t)$

A final comment of this section concerns the change of numerical results for a "two-phase renormalizable" model of the system, when the $s(t)$-function is replaced by the step function $\Theta(t_{\rm step} - t)$, discussed in Sect. 6.2.3, in order to justify the modified RG scheme. This replacement leads to a negligible change in numerics – the instability threshold remains nearly the same. The lower CMB bound on $M_{\rm H}$ grows by one in the second digit after the decimal point, while the upper bound decreases by 1 GeV.

6.6 Validity of the Model: Naturalness and Unitarity

6.6.1 Running Cut-off

The expression (6.42) is a truncation of the curvature and derivative expansion of the full effective action. It was repeatedly claimed that with large ξ the weak field version of this expansion on a flat (and empty) space background has a cut-off $4\pi M_{\rm P}/\xi$ [4, 21]. This scale is essentially lower than the Higgs field during inflation $\varphi \sim M_{\rm P}/\sqrt{\xi}$ and therefore seems to invalidate predictions based on (6.42) without unnatural suppression of higher-dimensional operators. The attempt to improve the situation by transition to the Einstein frame [56] was claimed to fail [22, 44, 46] in view of a multiplet nature of the Higgs field involving Nambu–Goldstone modes.[23]

[23] We have shown in Sect. 5.3.1 that such a transition between Jordan frame and Einstein frame is always possible for a $O(N)$ symmetric multiplet of scalar fields non-minimally coupled to gravity, in contrast to the claim of [46].

Here we show that these objections against naturalness are not conclusive. First, as already mentioned, a large value of φ during inflation does not really indicate a large *physical* scale of the problem. In contrast to curvature and energy density, the inflaton itself is not a physical observable, but rather a *configuration space coordinate* of the model, cf. the discussion in Sect. 5.3.3. Secondly, we are about to show now that the inflation scale actually lies below the gradient expansion cut-off and this justifies the naturalness of the obtained results. No transition to another conformal frame is needed for that, but rather the resummation accounting for the transition to a large φ background.

Indeed, the main peculiarity of the model (6.16) is that in the background field method with small derivatives, the effective Planck mass is determined by $\sqrt{M_{\rm P}^2+\xi\varphi^2}$. The power-counting method of [21] underlying the derivation of the cut-off $4\pi M_{\rm P}/\xi$ also applies here, but with the Planck mass $M_{\rm P}$ replaced by the effective one

$$M_{\rm P} \to \sqrt{M_{\rm P}^2+\xi\varphi^2} > \sqrt{\xi}\varphi. \tag{6.133}$$

The resulting cut-off is thus bounded from below by

$$\Lambda(\varphi) = \frac{4\pi\,\varphi}{\sqrt{\xi}}, \tag{6.134}$$

and this bound can be used as a *running* cut-off of the gradient and curvature expansion. The origin of this cut-off can be demonstrated in the one-loop approximation. When calculated in the *Jordan* frame, the one-loop divergences quadratic in the curvature R have a contribution strongest in ξ

$$\xi^2\frac{R^2}{16\pi^2}. \tag{6.135}$$

This can be seen from the Appendix of [9] or the coefficient (5.113) of the general result derived in Sect. 5.2.2. As compared to the tree-level part linear in the curvature $\sim (M_{\rm P}^2+\xi\varphi^2)R$, the one-loop R^2-term turns out to be suppressed by the above cut-off factor $16\pi^2(M_{\rm P}^2+\xi\varphi^2)/\xi^2 \simeq \Lambda^2$.

The on-shell curvature estimate at the inflation stage reads $R \sim V/U \sim \lambda\varphi^2/\xi$ in the Jordan frame, so that the resulting curvature expansion runs in powers of

$$\frac{R}{\Lambda^2} \sim \frac{\lambda}{16\pi^2} \tag{6.136}$$

and remains efficient in the usual perturbation theory range of the Standard Model, $\lambda/16\pi^2 \ll 1$. This works perfectly well in our Higgs inflation model because during the full CMB compatible range of the Higgs mass $\lambda < 2$ (see Fig. 6.3).

From the viewpoint of the gradient expansion for φ, this cut-off is even more efficient. Indeed, the inflaton field gradient can be expressed in terms of the inflaton potential $\hat{V}$ and the inflation smallness parameter $\hat{\varepsilon}$ taken in the Einstein frame, $\dot{\varphi} \simeq (\varphi^2/M_P^2)(\xi\hat{\varepsilon}_V\hat{V}/18)^{1/2}$. With $\hat{V} \simeq \lambda M_P^4/4\xi^2$ this immediately yields the gradient expansion in powers of

$$\frac{\partial}{\Lambda} \sim \frac{1}{\Lambda}\frac{\dot{\varphi}}{\varphi} \simeq \frac{\sqrt{\lambda}}{48\pi}\sqrt{2\,\hat{\varepsilon}_V}, \tag{6.137}$$

which is even better than (6.136) by the factor ranging from $1/N$ at the beginning of inflation to $O(1)$ at the end of it.

Equations (6.136) and (6.137) justify the effective action truncation in (6.42) in the inflationary domain. Thus, only multi-loop corrections to the coefficient functions $V(\varphi)$, $U(\varphi)$ and $G(\varphi)$ might stay beyond control in the form of higher-dimensional operators $(\varphi/\Lambda)^n$ and violate the flatness of the effective potential necessary for inflation. However, in view of the form of the running cut-off (6.134), they might be large, but do not affect the shape of these coefficient functions because of the field independence of the ratio φ/Λ. In the next section, we will argue that those contributions are controlled by an approximate shift symmetry of the system. Only the logarithmic running of the couplings in (6.83)–(6.85) determined by the RG dominates the quantum input in the inflationary dynamics and its CMB spectra.[24]

A similar approach was undertaken in [14]. However, the authors obtain an even higher cut-off in the Jordan frame $\Lambda_J \propto \sqrt{\xi}\,\varphi$, which differs from (6.134) by a factor of ξ. The origin of this difference can be traced back to the fact that the results in [9] and [14] were obtained by calculating the effective actions in different frames.

6.6.2 Scale-Invariance and Shift-Symmetry

We consider a rigid scale transformation (dilatation) of the coordinates

$$\alpha: \quad x^\mu \xrightarrow{\alpha} \alpha\, x^\mu. \tag{6.138}$$

If the theory is invariant under diffeomorphism, we can write (6.138) as an transformation, acting linearly on the fields [37]

$$\alpha: \quad \phi^A \xrightarrow{\alpha} \alpha^{d_A}\,\phi^A. \tag{6.139}$$

The transformation of each component of the generalized field ϕ^A is weighted by its canonical mass dimension $d_A = [\phi^A]_M$. In particular, we are interested in the graviton scalar sector of the model, i.e those components of ϕ^A that correspond to the fields φ and $g_{\mu\nu}$. They transform like:

---

[24] Like the logarithmic term of (6.47) which dominates over the nearly flat classical part of the inflaton potential and qualitatively modifies tree-level predictions of the theory [9].

$$\varphi \xrightarrow{\alpha} \alpha\,\varphi, \quad g^{\mu\nu} \xrightarrow{\alpha} \alpha^2\, g^{\mu\nu}, \quad g_{\mu\nu} \xrightarrow{\alpha} \alpha^{-2}\, g_{\mu\nu}. \tag{6.140}$$

Furthermore, we obtain the scaling behaviour of the curvature and the determinant

$$R \xrightarrow{\alpha} \alpha^2\, R, \quad g \xrightarrow{\alpha} \alpha^{-8}\, g, \quad \sqrt{g} \xrightarrow{\alpha} \alpha^{-4}\, \sqrt{g}. \tag{6.141}$$

Let us consider the tree-level action in the Jordan frame $S^{\text{JF}}_{\text{tree}}$ (6.16) for high inflationary energies $\varphi \gg M_{\text{P}}/\sqrt{\xi} \gg \nu$. In this limit

$$U_{\text{tree}} = \frac{1}{2}\,(M_{\text{P}}^2 + \xi\,\varphi^2) \simeq \frac{1}{2}\,\xi\,\varphi^2, \tag{6.142}$$

$$V_{\text{tree}} = \frac{\lambda}{4}\,(\varphi^2 - \nu^2)^2 \simeq \frac{\lambda}{4}\,\varphi^4. \tag{6.143}$$

This corresponds to the limit of *induced gravity* $(M_{\text{P}},\, \nu) \longrightarrow 0$ [1, 68, 88]. The important feature of this limit in (6.142) and (6.143) is that the tree-level action becomes independent of any mass scale

$$S^{\text{ind}}_{\text{tree}} := \lim_{\substack{\nu\to 0\\ M_{\text{P}}\to 0}} S^{\text{JF}}_{\text{tree}} = \int \mathrm{d}^4x\,\sqrt{g}\,\left(\frac{1}{2}\,\xi\,\varphi^2\,R - \frac{1}{2}\,g^{\mu\nu}\,\partial_\mu\varphi\,\partial_\nu\varphi - \frac{\lambda}{4}\,\varphi^4\right). \tag{6.144}$$

As a consequence, it is easy to see that $S^{\text{ind}}_{\text{tree}}$ is invariant under the scaling operation (6.138). This means that the tree-level action in the Jordan frame possesses a nearly exact scale-invariance $S^{\text{JF}}_{\text{tree}} \xrightarrow{\alpha} S^{\text{JF}\,()}_{\text{tree}} \simeq S^{\text{JF}}_{\text{tree}}$ for inflationary energy scales which becomes *exact* in the induced gravity limit.

What does this correspond to in the Einstein frame? In order to answer this question, let us investigate how the field φ in the Jordan frame parametrization changes by a transition to the Einstein frame parametrization $\hat{\varphi}$. Both fields are related by the differential Eq. (6.45). In the limit $\varphi \gg M_{\text{P}}/\sqrt{\xi}$ it takes the form

$$\left(\frac{\mathrm{d}\hat{\varphi}}{\mathrm{d}\varphi}\right)^2 \simeq 6\,\left(\frac{M_{\text{P}}}{\varphi}\right)^2. \tag{6.145}$$

This equation can easily be solved by the solution

$$\hat{\varphi} \simeq \sqrt{6}\,M_{\text{P}}\,\ln\,\frac{\varphi}{\varphi_0}. \tag{6.146}$$

The initial value φ_0 is determined by the inflationary scale $\varphi_0 = M_{\text{P}}/\sqrt{\xi}$. Therefore, we obtain the following relation between the Jordan frame and the Einstein frame parametrizations of the inflaton

$$\hat{\varphi} = \sqrt{6}\,M_{\text{P}}\,\ln\,\frac{\sqrt{\xi}\,\varphi}{M_{\text{P}}} \quad\longleftrightarrow\quad \varphi = \frac{M_{\text{P}}}{\sqrt{\xi}}\,e^{\frac{\hat{\varphi}}{\sqrt{6}\,M_{\text{P}}}}. \tag{6.147}$$

By subjecting the Jordan frame field φ to a scale transformation $\varphi \stackrel{\alpha}{\longrightarrow} \alpha\,\varphi$, we find for the Einstein frame field

$$\hat{\varphi} \stackrel{\alpha}{\rightarrow} \sqrt{6}\,M_{\rm P}\,\ln\frac{\sqrt{\xi}\,\alpha\,\varphi}{M_{\rm P}} = \hat{\varphi} + \ln\,\alpha. \tag{6.148}$$

We notice that the scaling operation of the inflaton field with α in the Jordan frame corresponds to a shift operation with a constant $\ln\alpha$ in the Einstein frame. Thus, for the inflaton field there is a duality between a scale invariance in the Jordan frame and a shift symmetry in the Einstein frame

$$\varphi \stackrel{\alpha}{\longrightarrow} \alpha\,\varphi \quad \longleftrightarrow \quad \hat{\varphi} \stackrel{\alpha}{\longrightarrow} \hat{\varphi} + \ln\,\alpha. \tag{6.149}$$

The appearance of a shift symmetry in the context of inflation is a key feature in string theory inspired models, see e.g. [47]. The duality between scale invariance and shift symmetry in the different conformal frames was first presented in [14]. The shift symmetry requires a constant potential $\hat{V}$ in the Einstein frame, as $\hat{V} \stackrel{\alpha}{\longrightarrow} \hat{V}^{(\alpha)} = \hat{V}$. If this symmetry was exact even on the full quantum level, it would provide a mechanism to protect the inflaton potential from dangerous quantum corrections up to the UV limit. However, it is clear that the logarithmic corrections in the Jordan frame $\propto \ln\varphi/\mu$ introduce a mass scale via μ. Thus, the scale invariance is not exact anymore, but it is only *weakly* broken by the logarithms.[25] Thus, in contrast to the hypothesis made in [14] that the scale invariance in the Jordan frame and the corresponding shift symmetry in the Einstein frame should be *exact* with *all* quantum corrections included. In our scenario, this is an *approximate* symmetry which only becomes exact at the tree-level, but is weakly broken by the loop corrections and will be totally broken at a very large curvature scale, possibly at $R \sim M_{\rm P}^2$. In this scenario, the approximate shift symmetry arises rather naturally from the model itself, i.e. by the duality with the scale invariance of the Jordan frame in the limit $\varphi \gg M_{\rm P}/\sqrt{\xi}$.

By comparison with [14], we thus find ourselves in a situation where these symmetries are interpreted in a complementary way. In our hypothesis, the *exact classical* scale invariance is destroyed the more quantum corrections we include and the higher scales we move towards $R \longrightarrow M_{\rm P}^2$, whereas in [14] the opposite conclusion is drawn and the symmetry becomes exact with *all* quantum corrections included in the UV.

This discrepancy between the hypothesis of [14] and the scenario presented here, i.e. the different limits for which the scale invariance in the Jordan frame becomes exact, can also be traced back to the conformal anomaly by quantizing in different frames. We have calculated the effective action in the Jordan frame and then transformed the effective potential to the Einstein frame in order to derive all cosmological

---

[25] It is clear that even in the absence of quantum corrections the scale invariance in the Jordan frame is broken by the mass scale $M_{\rm P}$, when inflation comes to an end (remember that this scale was determined by the condition $\hat{\varepsilon} = 3/4$). This is of course necessary since an exact scale invariance would imply a strictly constant (for all times) potential in the Einstein frame and would therefore not permit a phase of reheating.

parameters. In contrast, the authors of [14] transformed the action to the Einstein frame already *before* quantization. The two conformal frames are related by the conformal factor

$$\Omega^2(\varphi) := \frac{2\,U(\varphi)}{M_{\mathrm{P}}^2} = 1 + \frac{\xi\,\varphi^2}{M_{\mathrm{P}}^2}\,. \tag{6.150}$$

In the Jordan frame, the particle masses are generated via the Higgs mechanism

$$m_{\mathrm{par}}(\varphi) \sim \varphi. \tag{6.151}$$

Under a conformal transformation, a scalar field behaves as

$$\varphi \to \hat{\varphi} = \Omega^{(2-d)/2}\,\varphi = \frac{\varphi}{\Omega}, \tag{6.152}$$

in $d = 4$ space-time dimensions. Thus, under a conformal transformation to the Einstein frame the particle masses get rescaled as

$$m_{\mathrm{par}}(\varphi) \to \hat{m}_{\mathrm{par}}(\varphi) = \frac{m_{\mathrm{par}}(\varphi)}{\Omega(\varphi)}. \tag{6.153}$$

For large values $\varphi \gg M_{\mathrm{P}}/\sqrt{\xi}$, the conformal factor (6.150) becomes $\Omega \sim \sqrt{\xi}\varphi/M_{\mathrm{P}}$, so that the particle masses in the Einstein frame become field independent

$$\hat{m}_{\mathrm{par}}(\varphi) \simeq \frac{M_{\mathrm{P}}\,m_{\mathrm{par}}(\varphi)}{\sqrt{\xi}\,\varphi} \propto \frac{M_{\mathrm{P}}}{\sqrt{\xi}} = \text{const.} \tag{6.154}$$

The one-loop quantum corrections in the effective action (in particular, the Coleman–Weinberg potential (6.13)) have the logarithmic structure

$$g^{1/2}\,m_{\mathrm{par}}^4(\varphi)\,\ln\frac{m_{\mathrm{par}}^2(\varphi)}{\mu^2}. \tag{6.155}$$

Using $g^{1/2} \to \hat{g}^{1/2} = \Omega^4(\varphi)\,g^{1/2}$, the transition to the Einstein frame yields

$$\hat{g}^{1/2}\,\hat{m}_{\mathrm{par}}^4(\varphi)\,\ln\frac{\hat{m}_{\mathrm{par}}^2(\varphi)}{\mu^2} = g^{1/2}\,m_{\mathrm{par}}^4(\varphi)\,\ln\frac{m_{\mathrm{par}}^2(\varphi)}{\Omega^2(\varphi)\,\mu^2} \simeq g^{1/2}\,m_{\mathrm{par}}^4(\varphi) \times \text{const.} \tag{6.156}$$

As already pointed out in [12], due to the cancellation of conformal factors, the pre-logarithmic coefficients do not get suppressed, while the field dependence in the argument of the logarithm cancels. Therefore, the quantum corrections for large φ calculated in the Einstein frame differ from those of the Jordan frame by replacing the logarithms by a constant. In other words, the conformal anomaly displays itself as a weak logarithmic frame dependence. Thus, within the procedure of [14], there are no large logarithmic quantum corrections. The procedure presented here does not insist on a *fundamental scale-invariance* at the full quantum level but rather identifies the scale-invariance as an *approximate* classical symmetry. As long as the field-dependent cut-off (6.134) is above the energy scale of inflation and the approximate symmetry is not completely broken at this scale, the Higgs inflation model is a consistent and reliable theory.

6.7 Variants of Higgs Inflation

Since the idea of non-minimal Higgs inflation has become an active field of research, we would like to provide a short collection of references regarding the variety in which this idea has been embedded. For a comparison of *numerical* predictions of M_{H} in different approaches, see [70]. In [25, 55] an additional scalar field has been considered in order to describe dark matter. In a similar model [63], the scalar WIMP abundance was used to predict a tensor-to-scalar ratio measurable by further cosmological experiments. The Higgs inflation model is also applied in the context of the asymptotic safety approach to gravity within the mathematical formalism of the functional RG [71], a very active and interesting field initiated by Weinberg in [87] and promoted by [26, 54, 65, 66] and others. In [62], the authors propose to consider Higgs inflation within a minimal extended Standard Model with seesaw mechanism. In [83] non-Gaussianities in the Higgs inflationary scenario were studied. Higgs inflation has also attracted some interest in the SUSY/SUGRA community [2, 31, 34, 35]. Recently it was also investigated within a $f(R)$ theory [86]. In [58], the authors proposed to leave the parameter space determined by (ξ, ν) unspecified at first and to investigate certain limits which yield a surprisingly successful description in agreement with cosmological data. There are also some variants which do not use a non-minimal coupling of the form $\varphi^2 R$, but rather different ideas. Among them, models including a non-minimal coupling of the kinetic term to the Einstein tensor [38] or a non-linear derivative interaction, denoted Higgs G-inflation, were proposed [48]. In a very interesting series of papers [19, 37, 72, 73] the authors considered in addition to the Higgs-inflaton a dilaton field. In these models the idea of scale invariance at a fundamental level plays a dominant role and moreover due to the additional scalar dilaton these models try not only to connect the Standard Model with the inflationary phase of the early universe but also with the present acceleration, i.e. the dark energy problem. Summing up, we can say that the non-minimal Higgs inflation scenario with all its variants is an interesting field of research. All these models share the promising feature of being testable in the near future.

6.8 Conclusions

We have found that our model is remarkably consistent with CMB observations in the Higgs-mass range

$$135.6\,\text{GeV} \lesssim M_{\text{H}} \lesssim 184.5\,\text{GeV}, \tag{6.157}$$

which is very close to the widely accepted range dictated by electroweak vacuum stability and perturbation theory bounds.

Both bounds follow from the lower WMAP bound for the CMB spectral index $n_{\text{s}}(k_0) > 0.94$ (the combined WMAP+BAO+SN data at the pivot point $k_0 = 0.002$ Mpc^{-1} corresponding to $N \simeq 60$ e-folds [45, 53]). The lower bound on M_{H} is very close to the instability threshold which in this model equals $M_{\text{H}}^{\text{inst}} \simeq 134.27\,\text{GeV}$. This is higher than the conventional Standard Model estimate $\sim 125\,\text{GeV}$, which is caused by a strong non-minimal back reaction of gravity and/or the simplicity of our approximation disregarding higher loop orders.[26] During the process of writing up this thesis there has been an official CERN press release, with the announcements that the LHC could have found signatures of the Higgs boson. It was claimed that strong evidence has been found that the Higgs mass is concentrated around the expected value of 125 GeV [24]. However, even if this first evidence turns out to be true, this does not mean that the non-minimal Higgs inflation model has to be abandoned. The general qualitative features of the model, such as the suppression mechanism which leads to finite values of the running couplings during inflationary energy scales, seem to be robust. An obvious improvement with respect to the numerical predictions can be made if more exact values of the top-quark mass become available and by including two-loop corrections. Numerical test runs and the results of the two-loop analysis in [28] and [15] clearly show that including two-loop contributions generically leads to a shift of the Higgs mass interval to lower values, very close to the recently announced value of 125 GeV [24].

Remarkably, the upper bound in (6.157) does not follow from a demand of applicability of perturbation theory, but is rather imposed by the CMB data. This CMB mechanism works in a trustable perturbation regime with $\lambda(t_{\text{in}}) < 2$. The current upper CMB bound $n_{\text{s}} < 0.99$ does not impose restrictions on the Higgs mass, but it will generate bounds if the observed n_{s} essentially drops below the "classical" value $n_{\text{s}} = 1 - 2/N \simeq 0.967$. Vice versa, this model will be falsified if the lower observational bound on n_{s} exceeds 0.964—the top of the n_{s}-graphs in Fig. 6.5.

Our approach [9] represents the RG improvement of the analytical results in [12]. In fact, we have completely recovered the analytic formalism of [12] for all inflation parameters, which only gets modified by the RG mapping between the coupling constants at the electroweak scale and those at the scale of inflation. A peculiarity of this formalism is that for large $\xi \gg 1$ the effect of the Standard Model

[26] It is well known that despite the perturbative range of coupling constants, the two-loop RG improvement essentially lowers the electroweak instability threshold compared to the one-loop RG running [32].

phenomenology on inflation is universally encoded in one quantity—the anomalous scaling $\mathbf{A_I}$. It has been suggested earlier in [6] for a generic gauge theory. In the Standard Model it is dominated by the contributions of heavy particles—$(W^{\pm}, Z)$-bosons, top-quark and Goldstone modes. This quantity is forced to run in view of the RG resummation of leading logarithms, and this running raises a large negative electroweak value of $\mathbf{A_I}$ to a *small negative* value at the inflation scale. Ultimately, this leads to the admissible range of Higgs masses (6.157) very close to the conventional Standard Model range.

In fact, this mechanism resembles the phenomenon of asymptotic freedom because $\mathbf{A_I}/64\pi^2$ determines the strength of quantum corrections in the inflationary dynamics [7, 12]. Usually, asymptotic freedom is associated with the asymptotic decrease of $\lambda(t)$ to zero. Here, this phenomenon is more subtle because it occurs within the range (6.157) and fails near its lower and upper boundaries. For small Higgs masses in the range (6.157), as can be seen in Figs. 6.3 and 6.4, the running $\lambda(t)$ becomes very small when approaching the inflationary region $t \sim 32 \div 35$ and generates a large $\mathbf{A_I}(t)$ because of its $1/\lambda$ part induced by vector boson and top-quark particles. For large masses, the coupling λ grows towards the Landau pole (though much slower than in the non-gravitating Standard Model because of the suppression of the major part of the λ^2-term in β_λ). Therefore, $\mathbf{A_I}(t)$ becomes large and negative again, this time due to the negative contribution -6λ of the Goldstone modes, and finally falls out of the CMB range at $M_{\rm H} \simeq 185$ GeV.

Quantum effects are small only in the intermediate part of (6.157) with a moderately small λ where $n_{\rm s}$ is close to the "classical" limit $1 - 2/N \simeq 0.967$ for $x := N\,\mathbf{A_I}/48\pi^2 \ll 1$. Here, the original claim of [16] on smallness of quantum corrections is correct, but this smallness, wherever it takes place, is achieved via a RG summation of large leading logarithms.

Qualitatively, our main conclusions are close to those of [28] and [15], although the RG treatment in these works is very different from ours. In contrast, however, we do not try to suggest error bars for the Higgs mass range (6.157) and specify corrections due to the uncertainty in the top-quark mass $M_{\rm t}$. We only present our results for the intermediate value of the top mass $M_{\rm t} = 171$ GeV and demonstrate the sensitivity of the results to this choice by two plots of $n_{\rm s}$ with the two neighbouring values $M_{\rm t} = 169$ GeV and $M_{\rm t} = 173$ GeV (see Fig. 6.5).

Furthermore, the unresolved issue with the general problem of parametrization and gauge dependence of the off-shell extended effective action remains. This has already been discussed in Sect. 5.3.3. Unfortunately, we cannot formulate the calculation of CMB parameters as on-shell amplitudes or expectation values of *physical observables uniquely defined at the quantum level*. The power spectrum represents the quantum correlation function calculated at a special field-dependent moment of time—the horizon crossing. By using the Heisenberg equations of motion (both for linear quantum perturbations and the non-linear background) this function expresses as $\zeta^2 = \hat{V}/24\pi^2 M_{\rm P}^4 \hat{\varepsilon}_{\rm V}$ entirely in terms of the inflaton potential $\hat{V}(\varphi)$ and its gradients [61, 79, 81], cf. Eq. (6.58) for $\hat{\varepsilon}_{\rm V}$. So the main inflation observable is the inflaton potential itself, but its quantum version $\hat{V}^{\rm eff}$ and its off-shell extension with $(\hat{V}^{\text{eff}}(\varphi))' \neq 0$—the effective potential—is gauge-dependent.[27] Moreover, the off-shell extension of the effective potential (and more generally of the effective action) also depends on the parametrization of quantum fields as discussed in detail in Chap. 5.3.3. A particular example of this dependence is given by different results for the CMB characteristics obtained in different conformal frames—the original Jordan frame [9, 10] versus the Einstein frame [15, 16]. Another important example is the Standard Model renormalization with different parametrizations of the Higgs multiplet. In the conventional quantization in Cartesian coordinates for the complex $SU(2)$ doublet or real $O(4)$ multiplet the theory is renormalizable and has one set of quantum corrections. In the spherical coordinate system (used, for example, for the description of the chiral phase of the Standard Model [15, 60]) the theory is non-renormalizable and has another set of quantum corrections [50]. The S-matrices in both parametrizations coincide, but their off-shell effective actions are different.

Both off-shell aspects of the above mentioned type turn out to be important. The Einstein frame differs from the Jordan one by a field-dependent conformal rescaling of the metric and a very non-linear change of the Higgs field—the radial variable of the Higgs multiplet. Quantization in these new variables has two major effects which are absent in the original Jordan frame. One effect is the change of the argument of the logarithms in quantum corrections, caused by conformal rescaling of the mass parameters [12, 16, 72]. This leads to nearly constant logarithms not contributing to quantum corrections. Another effect is the change of the coefficients of these logarithms [8, 82]. Altogether, this leads to a big difference between the results in the Jordan frame and the Einstein frame—there are no large quantum corrections at high values of M_{H} when imposing the upper CMB bound on the Higgs range—a situation less sensitive to the CMB data, as claimed in [15].

Here, we are forced to reconcile with the gauge and parametrization dependence of the obtained results. Fortunately, this dependence which enters through the renormalization function Z_{in} in (6.118) is rather weak. The value of Z_{in} in the Feynman gauge differs from that of the Landau gauge (used throughout here) by about 6 %, which ultimately affects the value of spectral index in the third digit after the decimal point.

We have also shown the naturalness of the gradient and curvature expansion in this model, which is guaranteed within the conventional perturbation theory range of the Standard Model, $\lambda/16\pi^2 \ll 1$, and holds for the whole range of the CMB-compatible Higgs mass (6.157)—the latter property being a consequence of the asymptotic freedom in the sense described above. This result is achieved by the background field resummation of weak field perturbation theory leading to the replacement of the fundamental Planck mass in the known cut-off $4\pi M_P/\xi$ [4, 21] by the effective one. In addition, we have shown how the inflaton potential in the Einstein frame is protected against higher quantum corrections by an approximate shift symmetry related to an approximate scale invariance in the Jordan frame. All these aspects refute the objections of [4, 21] based on the analysis of scattering amplitudes in the

[27] In fact, this is a problem of consistent transition from $\langle O[V(\varphi)]\rangle$ to $O[V^{\text{eff}}(\langle\varphi\rangle)]$ which brings the problem of gauge dependence in the formalism of the mean field $\langle\varphi\rangle$ and its effective action.

electroweak vacuum background. Smallness of the cut-off in this background does not contradict physical bounds on the Higgs mass originating from CMB data for the following reasons: The determination of $M_{\rm H}$ takes place at the TeV scale much below the non-minimal Higgs cut-off $4\pi M_P/\xi$, whereas inflationary dynamics and CMB formation occur for $\lambda/16\pi^2 \ll 1$ below the *running* cut-off $\Lambda(\varphi) = 4\pi\varphi/\sqrt{\xi}$. It is the phenomenon of inflation which connects these two scales due to the exponentially large stretching and allows us to probe the physics of underlying Standard Model by CMB observations at the 500 Mpc wavelength scale.

Finally, in addition to a good agreement of the spectrum of cosmological perturbations with the CMB data, our model also describes the mechanism of generating the cosmological *background* itself upon which these perturbations exist. This mechanism will be discussed in the last part of this thesis, Chap. 7, and is based on the no-boundary or tunnelling cosmological wave function [40, 59, 67, 85, 89] possibly prescribing initial conditions for inflation.

References

1. Adler, S.L.: Einstein gravity as a symmetry-breaking effect in quantum field theory. Rev. Mod. Phys. **54**, 729 (1982)
2. Alvarez-Gaume, L., Gomez, C., Jimenez, R.: A minimal inflation scenario. J. Cosmol. Astropart. Phys. **03**, 027 (2011)
3. Amsler, C. et al. (Particle Data Group): Review of particle physics. Phys. Lett. B **667**, 1 (2008)
4. Barbon, J.L.F., Espinosa, J.R.: On the naturalness of Higgs inflation. Phys. Rev. D **79**, 081302 (2009)
5. Barbour, J., Bertotti, B.: Mach's principle and the structure of dynamical theories. Proc. R. Soc. Lond. A **382**, 295 (1982)
6. Barvinsky, A.O., Kamenshchik, A.Yu.: Quantum scale of inflation and particle physics of the early universe. Phys. Lett. B **332**, 270 (1994)
7. Barvinsky, A.O., Kamenshchik, A.Yu.: Effective equations of motion and initial conditions for inflation in quantum cosmology. Nucl. Phys. B **532**, 339 (1998)
8. Barvinsky, A.O., Kamenshchik, A.Yu.: Renormalization group for nonrenormalizable theories: Einstein gravity with a scalar field. Phys. Rev. D **48**, 3677 (1993)
9. Barvinsky, A.O., Kamenshchik, A.Yu., Kiefer, C., Starobinsky, A.A., Steinwachs, C.: Asymptotic freedom in inflationary cosmology with a non-minimally coupled Higgs field. J. Cosmol. Astropart. Phys. **12**, 003 (2009)
10. Barvinsky, A.O., Kamenshchik, A.Yu.: Higgs Boson, renormalization group, and naturalness in cosmology. Eur. Phys. J. C. **72**, 2219 (2012)
11. Barvinsky, A.O., Kamenshchik, A.Yu.: Tunneling cosmological state revisited: origin of inflation with a nonminimally coupled standard model Higgs inflaton. Phys. Rev. D **81**, 043530 (2010)
12. Barvinsky, A.O., Kamenshchik, A.Yu.: Inflation scenario via the Standard model Higgs Boson and LHC. J. Cosmol. Astropart. Phys. **11**, 021 (2008)
13. Barvinsky, A.O., Nesterov, D.V.: Effective equations in quantum cosmology. Nucl. Phys. B **608**, 333 (2001)
14. Bezrukov, F., Magnin, A., Shaposhnikov, M., Sibiryakov, S.: Higgs inflation: consistency and generalisations. J. High Energy Phys. **01**, 016 (2011)
15. Bezrukov, F., Shaposhnikov, M.: Standard model Higgs boson mass from inflation: two loop analysis. J. High Energy Phys. **07**, 089 (2009)

16. Bezrukov, F.L., Magnin, A., Shaposhnikov, M.: Standard model Higgs boson mass from inflation. Phys. Lett. B **675**, 88 (2009)
17. Bezrukov, F.L., Shaposhnikov, M.: The standard model Higgs boson as the inflaton. Phys. Lett. B **659**, 703 (2007)
18. Birrell, N.D., Davies, P.C.W.: Quantum fields in curved space. Cambridge University Press, Cambridge (1982)
19. Blas, D., Shaposhnikov, M., Zenhausern, D.: Scale-invariant alternatives to general relativity. Phys. Rev. D **84**, 044001 (2011)
20. Brans, C., Dicke, R.H.: Mach's principle and a relativistic theory of gravitation. Phys. Rev. **124**, 925 (1961)
21. Burgess, C.P., Lee, H.M., Trott, M.: Power-counting and the validity of the classical approximation during inflation. J. High Energy Phys. **09**, 103 (2009)
22. Burgess, C.P., Lee, H.M., Trott, M.: On Higgs inflation and naturalness. J. High Energy Phys. **07**, 007 (2010)
23. Capozziello, S., Faraoni, V.: Beyond Einstein Gravity. Springer (2010)
24. CERN Press Release (2011). ATLAS and CMS experiments present Higgs search status. URL (cited on January 30, 2012): http://press.web.cern.ch/press/PressReleases/Releases2011/PR25.11E.html
25. Clark, T.E., Liu, B., Love, S.T., ter Veldhuis, T.: Standard model Higgs boson-inflaton and dark matter. Phys. Rev. D **80**, 075019 (2009)
26. Codello, A., Percacci, R., Rahmede, C.: Investigating the ultraviolet properties of gravity with a Wilsonian renormalization group equation. Ann. Phys. (NY) **324**, 414 (2009)
27. Coleman, S.R., Weinberg, E.J.: Radiative corrections as the origin of spontaneous symmetry breaking. Phys. Rev. D **7**, 1888 (1973)
28. De Simone, A., Hertzberg, M.P., Wilczek, F.: Running inflation in the standard model. Phys. Lett. B **678**, 1 (2009)
29. Dicke, R.H.: Mach's principle and Invariance under transformation of units. Phys. Rev. **125**, 2163 (1962)
30. Dodelson, S.: Coherent phase argument for inflation. AIP Conf. Proc. **689**, 184 (2003)
31. Einhorn, M.B., Jones, D.R.T.: Inflation with non-minimal gravitational couplings in supergravity. J. High Energy Phys. **03**, 026 (2010)
32. Espinosa, J.R., Giudice, G.F., Riotto, A.: Cosmological implications of the Higgs mass measurement. J. Cosmol. Astropart. Phys. **05**, 002 (2008)
33. Fakir, R., Unruh, W.G.: Improvement on cosmological chaotic inflation through nonminimal coupling. Phys. Rev. D **41**, 1783 (1990)
34. Ferrara, S., Kallosh, R., Linde, A., Marrani, A., Van Proeyen, A.: Jordan frame supergravity and inflation in the NMSSM. Phys. Rev. D **82**, 045003 (2010)
35. Ferrara, S., Kallosh, R., Linde, A., Marrani, A., Van Proeyen, A.: Superconformal symmetry, NMSSM, and inflation. Phys. Rev. D **83**, 025008 (2011)
36. Futamase, T., Maeda, K.: Chaotic inflationary scenario of the Universe with a nonminimally coupled "inflaton" field. Phys. Rev. D **39**, 399 (1989)
37. Garcia-Bellido, J., Rubio, J., Shaposhnikov, M. and Zenhausern, D. Higgs-dilaton cosmology: from the early to the late Universe. Phys. Rev. D 84, 123504 (2011)
38. Germani, C., Kehagias, A.: New model of inflation with nonminimal derivative coupling of standard model Higgs Boson to gravity. Phys. Rev. Lett. **105**, 011302 (2010)
39. Guth, A.H., Pi, S.Y.: Fluctuations in the new inflationary universe. Phys. Rev. Lett. **49**, 1110 (1982)
40. Hartle, J.B., Hawking, S.W.: Wave function of the universe. Phys. Rev. D **28**, 2960 (1983)
41. Hawking, S.W.: Particle creation by Black Holes. Commun. Math. Phys. **43**, 199 (1975)
42. Hawking, S.W.: The development of irregularities in a single bubble inflationary universe. Phys. Lett. B **115**, 295 (1982)
43. Hawking, S.W., Penrose, R.: The singularities of gravitational collapse and cosmology. Proc. R. Soc. Lond. A **314**, 529 (1970)
44. Hertzberg, M.P.: On inflation with non-minimal coupling. J. High Energy Phys. **11**, 023 (2010)

45. Hinshaw, G., et al.: Five-Year Wilkinson Microwave Anisotropy Probe (WMAP) Observations: data processing, sky maps, and basic results. Astrophys. J. Suppl. Ser. **180**, 225 (2009)
46. Kaiser, D.I.: Conformal transformations with multiple scalar fields. Phys. Rev. D **81**, 084044 (2010)
47. Kaloper, N., Sorbo, L.: A natural framework for chaotic inflation. Phys. Rev. Lett. **102**, 121301 (2009)
48. Kamada, K., Kobayashi, T., Yamaguchi, M., Yokoyama, J.: Higgs G inflation. Phys. Rev. D **83**, 083515 (2011)
49. Kazakov, D.I.: A generalization of the renormalization-group equations for quantum-field theories of arbitrary form. Theor. Math. Phys. **75**, 440 (1988)
50. Kelly, P.F., Kobes, R., Kunstatter, G.: Parametrization invariance and the resolution of the unitary gauge puzzle. Phys. Rev. D **50**, 7592 (1994)
51. Kiefer, C.: Quantum Gravity, 2nd edn. Oxford University Press, Oxford (2007)
52. Kleinert, H., Schulte-Frohlinde, V.: Critical properties of ϕ^4-theories. World Scientific, River Edge (2001)
53. Komatsu, E., et al.: Five-Year Wilkinson Microwave Anisotropy Probe (WMAP) Observations: cosmological interpretation. Astrophys. J. Suppl. Ser. **180**, 330 (2009)
54. Lauscher, O., Reuter, M.: Ultraviolet fixed point and generalized flow equation of quantum gravity. Phys. Rev. D **65**, 025013 (2002)
55. Lerner, R.N., McDonald, J.: Gauge singlet scalar as inflaton and thermal relic dark matter. Phys. Rev. D **80**, 123507 (2009)
56. Lerner, R.N., McDonald, J.: Higgs inflation and naturalness. J. Cosmol. Astropart. Phys. **04**, 015 (2010)
57. Lidsey, J.E., Wands, D., Copeland, E.J.: Superstring cosmology. Phys. Rep. **337**, 343 (2000)
58. Linde, A., Noorbala, M., Westphal, A.: Observational consequences of chaotic inflation with nonminimal coupling to gravity. J. Cosmol. Astropart. Phys. **03**, 013 (2011)
59. Linde, A.D.: Quantum creation of the inflationary universe. Lett. Nuovo Cimento **39**, 401 (1984)
60. Longhitano, A.C.: Heavy Higgs bosons in the Weinberg-Salam model. Phys. Rev. D **22**, 1166 (1980)
61. Mukhanov, V.F., Chibisov, G.V.: Quantum fluctuation and a nonsingular universe. JETP Lett. **33**, 532 (1981). (In Russian)
62. Okada, N., Rehman, M. U. and Shafi, Q.: Running standard model inflation and type I seesaw, p. 13. http://arxiv.org/abs/0911.5073 (2009). Accessed 19 Dec 2011
63. Okada, N., Shafi, Q.: WIMP dark matter inflation with observable gravity waves. Phys. Rev. D **84**, 043533 (2011)
64. Pasechnik, R.S., Shirkov, D.V., Teryaev, O.V.: Bjorken sum rule and perturbative QCD frontier on the move. Phys. Rev. D **78**, 071902 (2008)
65. Reuter, M., Saueressig, F.: Renormalization group flow of quantum gravity in the Einstein–Hilbert truncation. Phys. Rev. D **65**, 065016 (2002)
66. Reuter, M., Wetterich, C.: Effective average action for gauge theories and exact evolution equations. Nucl. Phys. B **417**, 181 (1994)
67. Rubakov, V.A.: Particle creation in a tunneling universe. JETP Lett. **39**, 107 (1984)
68. Sakharov, A.D.: Vacuum quantum fluctuations in curved space and the theory of gravitation. Sov. Phys. Dokl. **12**, 1040 (1968)
69. Salopek, D.S., Bond, J.R., Bardeen, J.M.: Designing density fluctuation spectra in inflation. Phys. Rev. D **40**, 1753 (1989)
70. Schucker, T.: Higgs-mass predictions, p. 31. http://arxiv.org/pdf/0708.3344v8.pdf (2007). Accessed 30 Jan 2012)
71. Shaposhnikov, M., Wetterich, C.: Asymptotic safety of gravity and the Higgs boson mass. Phys. Lett. B **683**, 196 (2010)
72. Shaposhnikov, M., Zenhausern, D.: Quantum scale invariance, cosmological constant and hierarchy problem. Phys. Lett. B **671**, 162 (2009)

73. Shaposhnikov, M., Zenhausern, D.: Scale invariance, unimodular gravity and dark energy. Phys. Lett. B **671**, 187 (2009)
74. Sher, M.: Electroweak Higgs potentials and vacuum stability. Phys. Rep. **179**, 273 (1989)
75. Shirkov, D.V.: Perturbative analysis of general renorm-group solutions in a massive case. Nucl. Phys. B **371**, 467 (1992)
76. Shirkov, D.V., Mikhailov, S.V.: Mass dependent α_s evolution and the light gluino existence. Z. Phys. C **63**, 463 (1994)
77. Sirlin, A., Zucchini, R.: Dependence of the Higgs coupling $\bar{h}_{\overline{\mathrm{MS}}}(M)$ on $m(\mathrm{H})$ and the possible onset of new physics. Nucl. Phys. B **266**, 389 (1986)
78. Spokoiny, B.L.: Inflation and generation of perturbations in broken-symmetric theory of gravity. Phys. Lett. B **147**, 39 (1984)
79. Starobinsky, A.A.: A new type of isotropic cosmological models without singularity. Phys. Lett. B **91**, 99 (1980)
80. Starobinsky, A.A.: Dynamics of phase transition in the new inflationary universe scenario and generation of perturbations. Phys. Lett. B **117**, 175 (1982)
81. Starobinsky, A.A.: The perturbation spectrum evolving from a nonsingular initially de Sitter cosmology and the microwave background anisotropy. Sov. Astron. Lett. **9**, 302 (1983)
82. Steinwachs, C.F., Kamenshchik, A.Yu.: One-loop divergences for gravity nonminimally coupled to a multiplet of scalar fields: Calculation in the Jordan frame. I. The main results. Phys. Rev. D **84**, 024026 (2011)
83. Sugiyama, N., Futamase, T.: Non-Gaussianity generated in the inflationary scenario with nonminimally coupled inflaton field. Phys. Rev. D **81**, 023504 (2010)
84. Tarrach, R.: The pole mass in perturbative QCD. Nucl. Phys. B **183**, 384 (1981)
85. Vilenkin, A.: Quantum creation of universes. Phys. Rev. D **30**, 509 (1984)
86. Watanabe, Y.: Rate of gravitational inflaton decay via gauge trace anomaly. Phys. Rev. D **83**, 043511 (2011)
87. Weinberg, S.: Ultraviolet divergences in quantum theories of gravitation. In: Hawking S. W., Israel W. (eds.) General relativity: an Einstein centenary survey, pp. 790–831 (1979). (Chapter 16)
88. Zee, A.: Broken-symmetric theory of gravity. Phys. Rev. Lett. **42**, 417 (1979)
89. Zeldovich, Y.B., Starobinsky, A.A.: Quantum creation of a universe in a nontrivial topology. Sov. Astron. Lett. **10**, 135 (1984)

Chapter 7
Quantum Cosmology

In this chapter we will investigate the cosmos as a whole. First, we will try to approach the fundamental questions about the cosmos by resorting to various ideas and conceptional questions in order to understand the environment we live in. Then, we will investigate the tunnelling scenario in the framework of quantum cosmology. Finally, we will apply the results of the previous chapter to derive initial conditions for inflation.

7.1 Conceptual Questions

In order to avoid semantic misunderstandings in the context of "multiverses", we will from now on use "cosmos" synonymously with "universe as a whole" and use "universe" only for a particular patch of the cosmos.

7.1.1 Observations and Speculations

Making scientific statements about the cosmos as a whole seems to be a desperate endeavour at first glance. Our experimentally accessible environment is not only limited by our insufficient technical capabilities (e.g. too weak astronomical telescopes and finite energy resources for colliders), but also by *principle* bounds. The Theory of General Relativity predicts that there are regions in the cosmos, we will never be able to observe, not even in principle because these regions stir out from us faster than light. Thus, we will never be able to exchange information with these causally disconnected regions beyond our cosmological horizon. Moreover, for the region where we can perform experiments, General Relativity is a very well confirmed theory. Thus, we have to take this as a serious limitation for any theory that tries to go beyond

C. F. Steinwachs, *Non-minimal Higgs Inflation and Frame Dependence in Cosmology*, Springer Theses, DOI: 10.1007/978-3-319-01842-3_7,
© Springer International Publishing Switzerland 2014

General Relativity. How can we ever think about making scientific statements[1] about regions that we will never be able to observe, not even in principle? Even if we had a theory of the cosmos that would yield predictions, what scientific value does it have if we are unable to empirically test these predictions by physical experiments? However, we will argue that even in such a situation, we could still draw some useful information from theories about the cosmos as a whole. But this comes at a price. We have to make the fundamental *assumption* that our physical laws can be extrapolated to the unobservable regions of the cosmos. By physical laws, we mean fixed relations between physical quantities.[2] Our confidence in the established theories is so strong that we dare to push their scope of applicability beyond the frontiers of observability and extend them to the cosmos or at least beyond our cosmological horizon. Mathematical self-consistency then becomes the most important guideline. However, we want to emphasize that it is nevertheless possible and logically consistent to make testable statements about those regions which are even in principle unobservable. In general, in such a situation, one has to start with some assumptions which have to be justified a posteriori. On the basis of these assumptions, we can use our mathematical theory to make statements about such unobservable regions and it is possible that we can deduce predictions from this theory, which can be tested in *our observable* patch. Since a priori it is not clear whether these theoretical predictions coincide with our empirical observations, we have at least the option of falsification. In that sense, it is legitimate to carefully extend the frontiers beyond our observable part of the cosmos.

Independently of whether we consider a theory of the cosmos or limit ourselves to our observable patch, it is assumed that we need a more fundamental theory. At least, there are convincing arguments for the need of a *quantized* theory of gravity [42]. Although there are promising candidates for a quantum theory of gravity (see [43] for an overview), none of these approaches has ultimately succeeded so far.[3] Therefore, the most pragmatic approach seems to be to apply these ideas to the symmetry reduced scenario of cosmology, but to try to stay as close as possible to the concepts and mathematical framework of the "full" candidate theory. Even if there could be artefacts due to the drastic symmetry reduction,[4] the mathematical problems become at least more feasible. Within such a setup, we can focus on the conceptual questions and even hope for some experimental input from effective or indirect astrophysical observations. Independently of the specific candidate theory

---

[1] According to Karl Popper [51], a scientific theory has to make predictions that can be falsifiable.

[2] This does not mean that the coupling constants in our patch of the cosmos must have the same values in different regions. Some authors even question if there is a fundamental meaning of physical laws at all or if they can also vary in different regions of the cosmos [61], but we will not follow these ideas.

[3] There are approaches, such as string theory, which are even more ambitious and try to go beyond a quantization of gravity to find a "theory of everything". See e.g. [49] for an overview and a summary of literature about string theory.

[4] In cosmological minisuperspace models, the infinitely many degrees of freedom contained in the metric field $g_{\mu\nu}(x)$ are usually reduced to a single degree of freedom, the homogeneous scale factor $a(t)$.

of quantum gravity we choose to work with when considering questions regarding the cosmos, the acceptance of the fundamental *quantum character* of the cosmos entails immediate *conceptual* implications about the nature of the cosmos. We will discuss these issues in more detail in Sects. 7.1.2 and 7.1.3 because the further semi-classical cosmological treatment in Sects. 7.2.2–7.3 has to be justified by preceding considerations at a more fundamental quantum level.

7.1.2 Decoherence and the Everett Many-Worlds Interpretation

By definition, the cosmos is a closed physical system. In fact, it is the only closed physical system in a strict sense. All other systems have to be considered as open subsystems. This unique property of the cosmos has drastic consequences. In particular, in the context of quantum cosmology, the role of the observer cannot be considered any longer as observing the system from outside the system. He himself is rather part of the system. Thus, there is no clear separation between observer and system. Accepting this, we are inevitably drawn towards the conclusion that there can only exist one full quantum state of the cosmos, containing all possible information, including the observer. However, this is in contradiction with an essential assumption of the Copenhagen interpretation of quantum mechanics, formulated to resolve the measurement problem. This interpretation needs a clear separation between a quantum and a classical realm in order to define a classical observer [22] or even a "classical consciousness" [71]. In this interpretation, it is assumed that the unitary evolution of the linear Schrödinger equation is followed by a non-unitary instantaneous reduction of the state vector into a "classical" (ensemble) mixture of possible outcomes induced by the measurement. Following [74], we can illustrate the dilemma by considering an isolated two spin system with two independent solutions $|\Uparrow\rangle$ and $|\Downarrow\rangle$ of the general Schrödinger equation

$$i\frac{\partial}{\partial t}|\Psi\rangle = \hat{H}\,|\Psi\rangle \quad\Rightarrow\quad |\Psi\rangle_F = \hat{U}|\Psi\rangle_I, \quad \hat{U} = \mathcal{T}\,e^{\int \mathrm{d}t\,\hat{H}(t)}, \tag{7.1}$$

with $\mathcal{T}$ being the time ordering operator. From the linearity it follows that each superposition $\alpha\,|\Uparrow\rangle + \beta\,|\Downarrow\rangle;\ \ \alpha,\beta\in\mathbb{C},\ |\alpha|^2+|\beta|^2 = 1$ is again a solution of the Schrödinger equation, so that both solutions span the two dimensional Hilbert space $\mathcal{H}_{\uparrow\downarrow}$. With the "Schrödingers cat" Gedankenexperiment [56] in mind, we can think of a detector that points $|D_\Uparrow\rangle$ "alive" if it measures the system in $|\Uparrow\rangle$ and $|D_\Downarrow\rangle$ "dead" if it measures the system in $|\Downarrow\rangle$. The process of measurement can be described dynamically by the unitary evolution of the detector D, initially in the "neutral state" $|D_0\rangle$

$$\hat{U}\,|\Uparrow\rangle|D_0\rangle := |\Uparrow\rangle|D_\Uparrow\rangle, \quad \hat{U}\,|\Downarrow\rangle|D_0\rangle := |\Downarrow\rangle|D_\Downarrow\rangle. \tag{7.2}$$

In the case of a superposition of the system states, the entanglement with the detector leads after the unitary evolution to a "macroscopic" superposition of the pointer of the measurement device

$$\hat{U}|\Psi\rangle_I := \hat{U}\,(\alpha\,|\Uparrow\rangle + \beta\,|\Downarrow\rangle)|D_0\rangle = \alpha\,|\Uparrow\rangle|D_\uparrow\rangle + \beta\,|\Downarrow\rangle|D_\downarrow\rangle =: |\Psi\rangle_F. \tag{7.3}$$

This is certainly not what we observe in our everyday life. The collapse postulate solves this problem by fixing a basis of specific macroscopic outcomes, i.e. dead *or* alive, and thereby *postulating* that all *quantum correlations* do vanish by the measurement process. In order to quantify the part that has been neglected by this procedure, it is convenient to describe the system in terms of a density matrix $\rho_I := |\Psi\rangle_I\langle\Psi|_I$

$$\begin{aligned}\hat{U}\rho_I\hat{U}^\dagger &= |\alpha|^2\,|\Uparrow\rangle\langle\Uparrow|\,|D_\uparrow\rangle\langle D_\uparrow| + \alpha\,\beta^*\,|\Uparrow\rangle\langle\Downarrow|\,|D_\uparrow\rangle\langle D_\downarrow| \\ &\quad + \alpha^*\,\beta\,|\Downarrow\rangle\langle\Uparrow|\,|D_\downarrow\rangle\langle D_\uparrow| + |\beta|^2\,|\Downarrow\rangle\langle\Downarrow|\,|D_\downarrow\rangle\langle D_\downarrow| =: \rho_F \end{aligned}\tag{7.4}$$

We have used the integrated version of the von Neumann evolution equation $\dot{\rho} = i/\hbar\,[\rho, \hat{H}]$ in order to obtain the dynamical evolution of ρ_I. The Copenhagen postulate demands that the quantum correlations (the non-diagonal terms in (7.4)) are absent after a measurement and only the reduced density matrix remains:

$$\rho_R := |\alpha|^2\,|\Uparrow\rangle\langle\Uparrow|\,|D_\uparrow\rangle\langle D_\uparrow| + |\beta|^2\,|\Downarrow\rangle\langle\Downarrow|\,|D_\downarrow\rangle\langle D_\downarrow|. \tag{7.5}$$

From an operational point of view, this has the advantage that we only have to deal with the "*classical*" probability that the final pointer state is *either* $|D_\uparrow\rangle$ *or* $|D_\downarrow\rangle$. However, rephrasing the postulate in other words implies that we are already aware of the fact that there are only two possible outcomes, even *before* the measurement is actually performed. By accepting this, we can ascribe each of these orthogonal possibilities an *exclusive* "classical ensemble" probability $|\alpha|^2$ or $|\beta|^2$, which is exactly the Born rule of quantum mechanics. Even if this Copenhagen algorithm is sufficient for most practical situations, there are several serious drawbacks. First of all, an "instantaneous collapse" obviously contradicts the relativity principle. Moreover, it is certainly not satisfactory to be obliged to introduce some additional ad hoc postulates rather than to *derive* them from the theory itself. This becomes especially dubious if this postulate contradicts some of the most important features of the theory, the superposition principle and the unitary evolution.

In 1957, Hugh Everett proposed a solution to this dilemma [31]. His idea was to take the fundamental *mathematical* principles of linearity and superposition seriously and to demand that these properties also have to have a real *physical* meaning. By this step, Everett exposed the whole "classical" construction as an unfortunate historical misconception. His idea entails a paradigm shift, since the word "quantization" does not make sense anymore. There is no classical world which could be quantized. Exactly the opposite is true: There is only *one fundamental quantum world* described by a state vector of the cosmos in some Hilbert space. Although Everett avoided in this way to introduce a collapse postulate and saved the linearity and unitarity of the theory, it remains to explain why we perceive a world which does not show "macroscopic" quantum correlations and where the probabilistic nature of the measurement process comes from. Everett approached these problems by postulating that each superposition as in (7.3) is realized simultaneously in a "parallel universe" and that the state vector "branches" each time a measurement is performed in one of these parallel worlds.[5] He also ascribed each of those worlds an *objective reality* and not just a hypothetical or potential character in a probabilistic sense. However, he did not solve the problem why we exactly perceive one of the very special and rare worlds where no quantum correlations are present. In fact, the two states in the example where the pointer shows $|D_{\uparrow}\rangle$ or $|D_{\downarrow}\rangle$ are only two very special states in the set of all possible states containing superpositions of $|D_{\uparrow}\rangle$ and $|D_{\downarrow}\rangle$. This problem can be rephrased as the "preferred basis problem". Consider the transition from ρ_F in (7.4) to ρ_R in (7.5). In order to obtain one of the two "classical branches", it seems necessary to associate a definite choice of the basis with a specific branch of the state vector, so that ρ_R is diagonal in that basis. However, it is a legitimate operation to perform a change of basis for $|\Psi\rangle_I$ in (7.3), so that ρ_F is diagonal with respect to this new basis. We notice that this change of basis is not a legitimate operation for the reduced density matrix ρ_R anymore, since by throwing away the non-diagonal terms to arrive from ρ_F at ρ_R, we already had to fix a specific basis and we associated this basis with the "classical branches" where ρ_R is diagonal. Once again, the freedom of basis transformation on the level of the quantum theory just reflects the fact that before the measurement we cannot even predict that the system is either in $|D_{\uparrow}\rangle$ or in $|D_{\downarrow}\rangle$. We only know for sure that the final state will be in *some superposition*. Thus, we need a mechanism which selects a preferred basis, i.e. a basis where we can identify the basis states with the classical branches.

So far, we have only considered an isolated system of two spins. In practice, for most systems it is hard to isolate them from their environment. Following the idea of Everett that there can only be one fundamental quantum state, we have to include the environment $|E\rangle$ in our previous quantum system (system = spin + detector) $|\Phi\rangle := |\Psi\rangle|E\rangle$. In the context of the cosmos one may be tempted to associate the division of system S and environment E with different spatially separated regions in space-time. But this is wrong in general. The division into system and environment is made by associating *abstract degrees of freedom* with the system and the environment respectively. Usually, the degrees of freedom of the environment exceed those of the system by far. If there is an interaction between the system and the environment,[6]

---

[5] Here, we use the terminology of DeWitt and Graham [28] who re-investigated Everett's original work. Everett himself did not use the nomenclature "parallel worlds" but "relative states" [31].

[6] In general, an arbitrary environmental degree of freedom cannot measure aspects of the system. Only those degrees of freedoms which can be modelled by local interaction between the environment and the system do participate in the process of measurement. For example a scattering process of photons with a molecule with centre of mass at a certain point will localize this object due to the electromagnetic interaction. The same setup with neutrinos instead of photons would not lead to such an efficient classical localization of that molecule, since the neutrinos will go through the molecule with almost no interaction.

this can be described by an additional interaction term in $\hat{H} \rightarrow \hat{H}_{\text{tot}} = \hat{H} + \hat{H}_{\text{int}}$. Thus, the unitary evolution including the environment leads to an entanglement with $|E\rangle$

$$\hat{U}_{\text{tot}}|\Phi\rangle = \hat{U}_{\text{tot}}|\Psi\rangle_I|E_0\rangle = \alpha\,|\uparrow\rangle|D_\uparrow\rangle|E_\uparrow\rangle + \beta\,|\downarrow\rangle|D_\downarrow\rangle|E_\downarrow\rangle. \tag{7.6}$$

What did we gain by extending the "von Neumann-measurement-chain" [69]? The crucial point is now that we do not have access to the overwhelming many degrees of the environment. We can express our ignorance about the $|E_i\rangle$; $i = \uparrow, \downarrow$, by limiting ourselves to *local* observables $\hat{O}_{\text{loc}}$, which are "blind" with respect to the environment E. Technically, a local observable can be obtained by a tensor product between a system part $\hat{O}_{\mathcal{S}}$ and a trivial identity operator $\hat{I}_E$, describing the insensitivity of $\hat{O}_{\text{loc}}$ on E [54]:

$$\hat{O}_{\text{loc}} := \hat{O}_{\mathcal{S}} \otimes \hat{I}_E. \tag{7.7}$$

The expectation value $\langle \hat{O} \rangle$ of an (Hermitian) observable $\hat{O}$ with respect to a pure quantum state $|\Phi\rangle$ is defined by

$$\begin{aligned}\langle \hat{O} \rangle = \langle \Phi | \hat{O} | \Phi \rangle &= \sum_i \sum_j \langle o_j |\, \langle \Phi | o_j \rangle\, \hat{O}\, \langle o_i | \Phi \rangle |o_i\rangle = \sum_i o_i |\langle o_i | \Phi \rangle|^2 \\ &= \sum_i \langle o_i |\, |\Phi\rangle \langle \Phi |\, \hat{O} |o_i\rangle = \text{tr}(\rho\, \hat{O}).\end{aligned} \tag{7.8}$$

We have expanded the state $|\Phi\rangle = \sum_i \langle o_i | \Phi \rangle\, |o_i\rangle$ into the eigenstates $|o_i\rangle$ of $\hat{O}$ with eigenvalue o_i.[7] It is exactly at this point where the Born rule enters,[8] by identifying $|\langle o_i | \Phi \rangle|^2$ with the probability of finding a general closed quantum system in the eigenstate $|o_i\rangle$. But what happens if we evaluate the expectation value for our local observable (7.7)? The generalization is obviously given by

$$\begin{aligned}\langle \hat{O}_{\text{loc}} \rangle = \text{tr}(\rho\, \hat{O}_{\text{loc}}) &= \sum_i \sum_j \langle E_i | \langle \Psi_j | \rho\, (\hat{O}_{\mathcal{S}} \otimes \hat{I}_E) |\Psi_j\rangle |E_i\rangle \\ &= \sum_j \langle \Psi_j | (\text{tr}_E(\rho) \hat{O}_{\mathcal{S}} |\Psi_j\rangle = \text{tr}_{\mathcal{S}}(\rho_{\mathcal{S}} \hat{O}_{\mathcal{S}}),\end{aligned} \tag{7.9}$$

where we have defined the *reduced* density matrix of the system $\mathcal{S}$ by the partial trace over the environmental degrees of freedom $\rho_{\mathcal{S}} := \text{tr}_E \rho$. In some sense, this corresponds to performing a kind of quantum average over the environmental degrees of freedom and studying the effective influence this average has on our system. For our model (7.6), this operation yields explicitly

---

[7] The spectral theorem ensures that such an expansion is always possible for Hermitian operators.

[8] Some authors, including Everett himself and Zurek, claim that the Born rule can be *derived* from the theory in terms of "relative frequencies" [31] or "envariance" [75] and is not an independent postulate. However, there are always some additional assumptions needed in the proofs of its derivation, so that it seems fair to say that the final status of this issue remains controversial.

$$
\begin{aligned}
\rho_{\text{red}} &:= \mathrm{tr}_{(E)}\,[\rho_{\text{tot}}] = \sum_i \langle E_i|\Phi\rangle\langle\Phi|E_i\rangle \\
&= \sum_{i=\uparrow,\downarrow} \big(\, |\alpha|^2 \,|\Uparrow\rangle\langle\Uparrow|\; |D_\uparrow\rangle\langle D_\uparrow|\, \langle E_i|E_\uparrow\rangle\langle E_\uparrow|E_i\rangle \\
&\quad + \alpha^*\beta \,|\Downarrow\rangle\langle\Uparrow|\; |D_\downarrow\rangle\langle D_\uparrow|\, \langle E_i|E_\downarrow\rangle\langle E_\uparrow|E_i\rangle \\
&\quad + \beta^*\alpha \,|\Uparrow\rangle\langle\Downarrow|\; |D_\uparrow\rangle\langle D_\downarrow|\, \langle E_i|E_\uparrow\rangle\langle E_\downarrow|E_i\rangle \\
&\quad + |\beta|^2 \,|\Downarrow\rangle\langle\Downarrow|\; |D_\downarrow\rangle\langle D_\downarrow|\, \langle E_i|E_\downarrow\rangle\langle E_\downarrow|E_i\rangle \big).
\end{aligned}
\tag{7.10}
$$

If the environment states $|E_i\rangle$ corresponding to $|D_\uparrow\rangle$ and $|D_\downarrow\rangle$ are orthogonal[9]

$$
\langle E_i|E_j\rangle \approx \delta_{ij}, \tag{7.11}
$$

the reduced density matrix of the system coincides with (7.5)

$$
\rho_{\text{red}} = |\alpha|^2 \,|\Uparrow\rangle\langle\Uparrow|\; |D_\uparrow\rangle\langle D_\uparrow| + |\beta|^2 \,|\Downarrow\rangle\langle\Downarrow|\; |D_\downarrow\rangle\langle D_\downarrow|. \tag{7.12}
$$

This environmental induced destruction of quantum correlations is called decoherence [72]. It also shows how the concept of "classical" probability emerges naturally from this description. It is simply a consequence of our ignorance about the environmental degrees of freedom that carry the information about the interaction with the system. The environment permanently monitors the system and selects a specific pointer basis $|D_\uparrow\rangle$, $|D_\downarrow\rangle$ by conserving (7.11) over the unitary dynamical evolution. Thus, the preferred basis in which the system acquires classical behaviour by the interaction with the environment is selected by the *dynamical* criterion of robustness of the pointer basis. Technically, this means that any mixing $i \neq j$

$$
\langle E_i|E_j\rangle \simeq e^{-\Gamma_{\text{dec}} t} \tag{7.13}
$$

is suppressed exponentially with some decoherence rate Γ_{dec}, also determining the timescale $t \sim 1/\Gamma_{\text{dec}}$ on which decoherence becomes efficient. However, it is important to stress that this relation will never hold *exactly* which is in agreement with the consistent interpretation of the *physical reality* of Everett's "many worlds". If we had chosen a basis different from the pointer basis $|D_\uparrow\rangle$, $|D_\downarrow\rangle$, say for example

$$
|D_\rightarrow\rangle := \frac{1}{\sqrt{2}}(|D_\uparrow\rangle + |D_\downarrow\rangle), \quad |D_\leftarrow\rangle := \frac{1}{\sqrt{2}}(|D_\uparrow\rangle - |D_\downarrow\rangle), \tag{7.14}
$$

so that the pointer shows $|D_\rightarrow\rangle$ and $|D_\leftarrow\rangle$, the corresponding environmental states $|E_k\rangle;\ k = \rightarrow, \leftarrow$ would no longer be orthogonal

$$
\langle E_k|E_l\rangle \not\approx \delta_{kl}, \tag{7.15}
$$

[9] The criterion of mutual orthogonality between two states $|E_i\rangle$, $|E_j\rangle$ is a measure of how well the environment can resolve the pointer positions.

and there would be no suppression of quantum correlations in (7.10). When the reduced density matrix (7.10) becomes diagonal in an environment selected basis $|D_i\rangle$, it follows that an observable $\hat{O}$ whose eigenstates are given by $|D_i\rangle$, will (at least approximately) commute with the interaction Hamiltonian $[\hat{O}, \hat{H}_{\rm int}] \approx 0$ [54]. Thus, $\hat{O}$ is a constant of motion, or in other words: If the system is in an eigenstate of $\hat{O}$, it remains invariant with respect to the action of $\hat{H}_{\rm int}$. Thus, the environment *dynamically* selects the preferred basis that is *stable* or unaffected under the influence of the environment. This explains why we can distinguish different localized objects like a car or a tree on long timescales.[10] The preferred basis in this case would be the position basis and explains why the objects are localized. As an eigenbasis of the observable O, the environmentally induced preferred basis excludes in this way certain "classical" observables. However, in order to decide which one is the preferred basis, we need to know the interaction Hamiltonian $\hat{H}_{\rm int}$ and have to divide the degrees of freedom into system and environment *on the basis of empirical observation.* This subtle point becomes especially interesting in the context of the anthropic principle. Ironically, in a certain sense our ignorance allows for our existence (a stable classical world).[11] The reason why we cannot perceive all degrees of freedom is that we are *local* beings subjected to local interactions, while quantum mechanics is an essentially non-local theory [19]. It is exactly this mismatch between a global state (an abstract vector in Hilbert space which is a functional of the configuration space) and the local interaction operators present in the Hamiltonian, which causes the trouble. The inability to follow all histories (all superpositions) simultaneously introduces the concept of randomness and probability in our world. At the level of the unitary evolution described by the Schrödinger equation, there is no randomness at all. It is a completely deterministic equation.

However, quantum mechanics still incorporates the non-relativistic notion of an external time parameter. The more fundamental quantum gravitational equation used in cosmology is the timeless Wheeler–DeWitt equation [27, 70]. Time does not exist fundamentally but only appears as an emergent semi-classical phenomenon, see e.g. [41]. Thus, only the relation between configuration variables has a fundamental meaning. As a result, the state of the cosmos (the "wave function of the universe" [36]) is timeless and resides in one of the possible configurations which are allowed by the laws of physics. The continuous loss of information which is explained by those degrees of freedom that carry away information about our local system into the environment, can be associated with a locally increasing system entropy and explains our perceived orientation of time.[12] However, we want to emphasize that this process is not irreversible in principle. We could easily imagine a Gedankenexperiment,

---

[10] They do not turn into an oscillating "cree" in the next moment, but we can be confident that the next morning our car still stands on the parking and the tree is still enrooted in the garden.

[11] Although the process of continuous monitoring of the environment would take place just as well without our presence.

[12] The continuous measurement of the system by the environment could perhaps also explain why we perceive time as *continuous and uniform* process. We do not mean the different clock rates due to relativistic effects,but rather the amount of continuous "information flow" carried away from the system into the environment and thereby increasing the entropy.

where we can catch these "escaped" degrees of freedom and transport the amount of information they carry back to the system. Practically, this is of course hard to realize, since we have to perform a simultaneous time reversal of all the degrees of freedom that have escaped into the environment and carry information about the system. This question becomes especially interesting if we consider the decohered branch of our observable universe as a subsystem of the quantum cosmos and ask if it is possible that our classical branch quantum correlates again in the "future" with other branches.[13] Since the robustness of our decohered branch crucially depends on the interaction with the environment, it is possible that it becomes"re-coherent" once the interaction changes. In this picture, it seems possible that our observable universe (with all its history) is just a short "classical" phase of the cosmos which will interfere again and become a superposition of many branches.[14] It is an empirical fact that the quantum-to-classical transition of different degrees of freedom occurs at different timescales. For example, the "classicalization" of the FRW scale factor is needed to allow quantum fluctuations to propagate *on* this "classical background" [44]. Thus, the decoherence rates for different interactions have to be different in order to explain inflation and our world as we observe it.

Finally, we mention that there are more and more refined experiments that have successfully tested different aspects of the decoherence mechanism. In particular, such experiments that are made with mesoscopic objects (C_{70} fullerenes) show that quantum theory is not necessarily restricted to microscopic systems, see e.g. [23]. The non-local aspects of the quantum theory, in turn, are demonstrated by experiments where simple entangled states were constructed over distances of several kilometers [55]. There are also experiments which clearly show that decoherence is a *continuous* process by gradually turning on and off the influence of the environment, see [34]. Moreover, an experimental realization of the Gedankenexperiment mentioned above, would prove that quantum correlations are *always* physically present (at least at the scales of the experiment), supporting the Everett interpretation.

7.1.3 Eternal Inflation and Multiverses

Beside the issue of our classical perception of the world, the following questions still remain: Has there been a moment of "creation" of the cosmos? What happens beyond our cosmological horizon? How and why did inflation start? In order to get an idea of these problems, we can start with the last question. As we have seen in Sect. 2.13, inflation exponentially blows up a tiny patch filled with vacuum energy to a huge space-time region. The physical Hubble horizon stays constant during this inflationary evolution, so that most of the newly produced space-time lies outside our present horizon. If the universe would not be accelerating in the future, we could

---

[13] For example in a recollapsing universe or the "bounce" scenario of Loop Quantum Cosmology.

[14] It could perhaps be interesting to relate the timescale of such a hypothetical "quantum re-coherence cycle" to a quantum version of the classical Poincaré recurrence theorem.

wait in principle for a long time and the structures which may have formed outside our horizon would enter again step by step into our observable region. However, observations have shown that our current universe does expand accelerated [50]. In Sect. 2.13, we have derived constraints for how long inflation has to last in order to solve the horizon problem and to be consistent with CMB observations. However, it seems to be very probable that inflation has already been ongoing before. In Sect. 2.4 we have investigated *small* quantum fluctuations of an inflaton field $\delta\varphi(t,\,\mathbf{x})$, propagating (after decoherence) *on a classical inflating background* $\bar{\varphi}(t)$. Due to the unavoidable zero-point motion, the homogeneous background inflaton field $\bar{\varphi}(t)$ experiences some small random "quantum kicks" $\delta\varphi(t,\,\mathbf{x}) \ll \dot{\bar{\varphi}}(t)$ which constitute the source of the tiny anisotropies in the CMB. In this case, the perturbations $\sim\delta\varphi(t,\,\mathbf{x})$ were much smaller than the "classical" overall slow-roll drift $\sim\dot{\bar{\varphi}}(t)$ down the hill of the potential. However, there might have been some random fluctuations $\delta\varphi \gg \dot{\varphi}$, that are *large* compared to the overall motion $\sim\dot{\bar{\varphi}}(t)$, when the inflaton is on the very flat plateau. These large fluctuations could have pushed the inflaton field also in the opposite direction "upwards the hill". Since these fluctuations depend on the space-time region $(t,\,\mathbf{x})$, they can have different values at different space-time points. Thus, there can be regions where the inflaton stays on the top of the potential for a very long time, while in other regions, the inflaton already rolled down the hill and decayed. While being at the top of the potential, the corresponding space-time regions still undergo inflation and do expand exponentially fast. When the phase transition from the top of the plateau to the valley has occurred, inflation comes to an end. Thus, it seems very probable that, if inflation exists, it already existed before and will still exist in some space-time regions that are unobservable for us. To summarize the situation: We are subjected to two competing processes. On the one hand, we have the exponential expansion of space-time in those regions where the inflaton is still on the top of the potential and on the other hand, there is a certain probability that the inflaton rolls down the hill. If the probability for the transition to the true vacuum is very high, the latter process will dominate and the different regions of true vacuum, the so called "pocket universes", will quickly merge, so that ultimately the whole space-time will be converted into the true vacuum and inflation will come to an end everywhere. If, however, the exponential expansion dominates, the pocket universes will expand by the speed of light in an eternally inflating "sea of space-time" filled with false vacuum.[15] In this sense each pocket universe for itself can be interpreted as an own "parallel-universe". If the height of the local minima of the inflaton potential varies in different regions of space, we are led to the idea of a "landscape" where the different heights of the potential at the different minima correspond to different values of the cosmological constant.[16] We can think of these parallel-universes as being embedded in our decohered semi-classical branch in order to integrate these multiverses in a compatible way with the Everett many worlds. However, if the quantum fluctuations become large compared to the classical

[15] A discussion of the eternally ongoing reproduction of space-time can be found e.g. in [48].

[16] This is a simplified presentation. In the context of string theory [60], the landscape is defined in a parameter-space of the dilaton fields, resulting from compactifications of extra dimensions.

background, as it is the case for eternal inflation, they will potentially lead us out of the realm of our decohered semi-classical branch, and at a certain point, the quantum correlations with other branches must be taken into account again (Fig. 7.1).

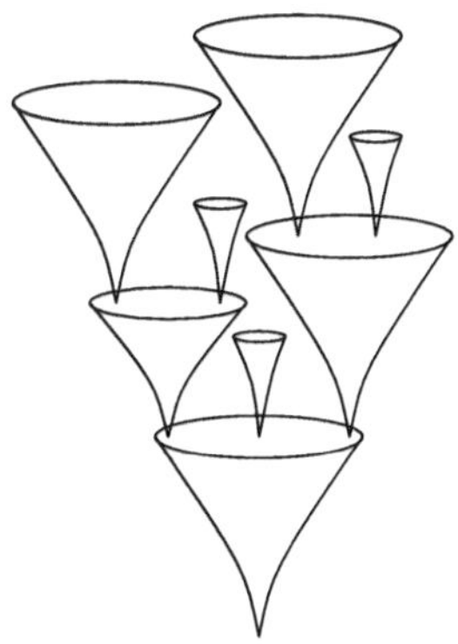

Fig. 7.1 Eternally inflating (self-reproducing) regions of space-time

7.2 Boundary Conditions of the Cosmos

In the previous sections, we have presented some qualitative ideas and conceptual implications about the global structure of the cosmos and our local perception of it. However, all these concepts, like decoherence and inflation, are rather *mechanisms* placed in an already existing cosmos. There has not yet been any explanation of how the cosmos as a whole could have emerged. It could be possible that it is an everlasting ongoing (possibly cyclic) process which has no end and no starting point.

In contrast, following different quantization schemes for gravity, we ultimately obtain a dynamical equation determining the evolution of the state of the cosmos. Since time does not seem to be existent at a fundamental level, but to be rather a derived semi-classical concept, the question for the *initial* conditions of the cosmos reduces to the question of the correct *boundary* conditions for the timeless quantum evolution equation. Regarding the discussion in Sect. 7.1.2, it seems clear that imposing boundary conditions for such a quantum equation is required at a fundamental level. Therefore, we first have to decide which quantum gravitational candidate theory we want to work with, then impose boundary conditions on the resulting dynamical equation and finally calculate the semi-classical limit in order to compare the predictions of this theory with the data of our observed universe today. Unfortunately, this procedure is not unique, even if we had only one candidate for a quantum theory of gravity. There are different boundary conditions discussed in the literature, mainly motivated by cosmological considerations, but the two most influential are certainly the "no-boundary" [37] and the "tunnelling" [63] prescriptions. The hope that the boundary conditions will follow "for free" by the theory itself does not seem to be realized. In the context of the canonical approach to quantum gravity, the dynamical equation is given by the hyperbolic Wheeler–DeWitt equation. The problem of boundary conditions manifests itself already in the symmetry reduced context of minisuperspaces, but it becomes even more difficult to solve at the level of the full Wheeler–DeWitt equation, which should be the natural mathematical environment to setup boundary conditions. However, as cosmologists we have empirical evidence that some homogeneous degrees of freedom, like the scale factor are robust classical quantities, forming a background, while other degrees of freedom can be treated as small perturbations on top of this classical background. We can hope that some boundary conditions do exist in the quantum cosmological context which seem natural by some physical selection criterion. Of course, it remains to show how this generalizes in a theory of full quantum gravity, but it can at least serve as a heuristic starting point. In the following, we will focus on the tunnelling condition which supports the scenario of an inflating universe [67]. For a detailed comparison between different boundary conditions, we refer to [43] and references therein.

7.2.1 Tunnelling Condition in Minisuperspace

In order to present the basic idea behind the tunnelling condition, we consider a simple minisuperspace scenario as e.g. discussed in [67]. Let us consider the Einstein–Hilbert action (2.1) in a spatially closed Friedmann universe $\mathfrak{K} = 1$ with $N = 1$. Using (2.32), we find the Friedmann equation

$$\dot{a}^2 + 1 - H^2\, a^2 = 0\,, \tag{7.16}$$

with the Hubble constant $H^2 = \Lambda/3$. Substituting (2.26) in (2.1) with $\int \mathrm{d}^4x\sqrt{-g} = \int \mathrm{d}t\, \mathcal{S}_3\, a^3$ and the surface $\mathcal{S}_3 = 2\,\pi^2\, R_0^3$ of the unit 3-sphere $R_0 \equiv 1$, we obtain after an integration by parts and $H^2 = \Lambda/3$ for the Lagrangian density

$$\mathcal{L} = \frac{6\,\pi^2}{\kappa}\, a\,(1 - \dot{a}^2 - H^2\, a^2). \tag{7.17}$$

We can interpret this as the Lagrangian density for a "particle" with "coordinate" $a(t)$. In order to simplify the discussion, we will for a moment adopt units $\kappa = 12\,\pi^2$. The momentum associated with a is then given by

$$p_a = \frac{\partial \mathcal{L}}{\partial \dot{a}} = -a\,\dot{a}. \tag{7.18}$$

A Legendre transformation leads to the Hamiltonian density

$$\mathcal{H}(a,\, p_a) = \dot{a}\, p_a - \mathcal{L}(a, p_a) = -\frac{1}{2\,a}(p_a^2 + a^2 - H^2\, a^4). \tag{7.19}$$

The Friedmann Eq. (7.16) implies the vanishing of $\mathcal{H}$

$$\mathcal{H} \approx 0. \tag{7.20}$$

The $\approx$ sign means "weakly equal" in the sense of Dirac [29], i.e. the equality only holds *on* the constraint hypersurface after having used the equations of motion (7.16). In order to quantize this classical system of a particle moving in a potential

$$V(a) := a^2(1 - H^2 a^2) \tag{7.21}$$

in the Schrödinger picture, we have to promote the configuration space variables $a \to \hat{a}$ to operators and substitute the momenta by derivatives with respect to these variables $p_a \to \hat{p}_a = -i\,\frac{\partial}{\partial a}$ which satisfy the commutator relation $[\hat{a},\, \hat{p}_a] = i$. Following the quantization procedure for constraint systems proposed by Dirac [29], we should implement the classical constraint (7.20), by expressing $\mathcal{H}(a,\, p_a) \to \hat{\mathcal{H}} = \mathcal{H}(\hat{a},\, \hat{p}_a)$ in terms of the operators $\hat{a}$ and $\hat{p}_a$[17] and demand

$$\hat{\mathcal{H}}\,\Psi(a) = 0, \tag{7.22}$$

where Ψ is the wave function of the system. Thus, the classical constraint equation is translated in a zero-eigenvalue type stationary Schrödinger equation for a wave function $\Psi(a)$ in some Hilbert space. Omitting the hats from now on and using (7.18) and (7.21) in (7.22), we obtain the Wheeler–DeWitt equation

$$\left[-\frac{\partial^2}{\partial a^2} + V(a)\right]\,\Psi(a) = 0. \tag{7.23}$$

In general, when there are additional matter degrees of freedom present, e.g. a scalar field, it becomes clear that (7.23) has the form of a Klein–Gordon equation rather than that of a Schrödinger equation. It is a hyperbolic equation because the kinetic terms of additional matter fields all enter with a relative minus sign compared to p_a^2. Since an absolute time parameter t is missing, one has to define an intrinsic time in terms of a configuration variable.

Instead of (7.23), let us consider for the moment the equation

$$\left[-\frac{\partial^2}{\partial a^2} + V(a)\right]\,\Psi(a) = \varepsilon\,\Psi(a), \tag{7.24}$$

where we can think of ε as the amount of radiation with the energy density $\rho_r = \varepsilon/a^4$. In order to include this into the original action (2.1), we can just define a total energy density $\rho := \rho_v + \rho_r$ instead of only the vacuum energy density $\rho_v = \Lambda/12\,\pi^2$. We recover our original action in the limit $\varepsilon \to 0$. Following the standard treatment of

[17] In general, a factor ordering problem appears in the transition $\mathcal{H} \to \hat{\mathcal{H}}$, leading to inequivalent quantum theories. We neglect this issue here.

quantum mechanics, we can now consider two different cases: $\varepsilon > V$ and $\varepsilon < V$. In the case $\varepsilon > V$, $\sqrt{\varepsilon - V} \in \mathbb{R}$ and for an approximately constant potential $V(a) \simeq$ const., we obtain the plane wave solution

$$\Psi(a) \propto e^{i\sqrt{\varepsilon - V}\,a} + e^{-i\sqrt{\varepsilon - V}\,a} \propto e^{i\,|p_a|\,a} + e^{-i\,|p_a|\,a}, \tag{7.25}$$

For $\varepsilon < V$, the "particle" is under the potential barrier and $\sqrt{\varepsilon - V}$ becomes imaginary, so that (7.25) becomes the superposition of an exponentially decreasing and increasing solution

$$\Psi(a) \propto e^{-\sqrt{|\varepsilon - V|}\,a} + e^{+\sqrt{|\varepsilon - V|}\,a} \propto e^{-\,|p_a|\,a} + e^{+\,|p_a|\,a}. \tag{7.26}$$

In addition, for the case $\varepsilon < V$, the potential has four distinguished points. Figure 7.2 shows the zeros $V(a_{0_{1/2}}) = 0$ and the intersection points of $V(a)$ with the constant amount of radiation $V(a_{\varepsilon_{1/2}}) = \varepsilon$. Under the condition that a is real and positive, we easily find

$$a_{0_1} = 0 \quad \text{and} \quad a_{0_2} = \frac{1}{H}. \tag{7.27}$$

where the intersection points lie within the interval $0 < a_{\varepsilon_1} < a_{\varepsilon_2} < 1/H$, for $0 < \varepsilon < V$. We subdivide the graph of the potential into three sections (see Fig. 7.2):

$$I : 0 < a < a_{\varepsilon_1}, \quad II : a_{\varepsilon_1} < a < a_{\varepsilon_2}, \quad III : a_{\varepsilon_2} < a. \tag{7.28}$$

Region I and III correspond to the classically allowed sections, whereas region II is classically forbidden. If we calculate the semi-classical WKB approximation,[18] we obtain in region II the wave function (for a general potential)

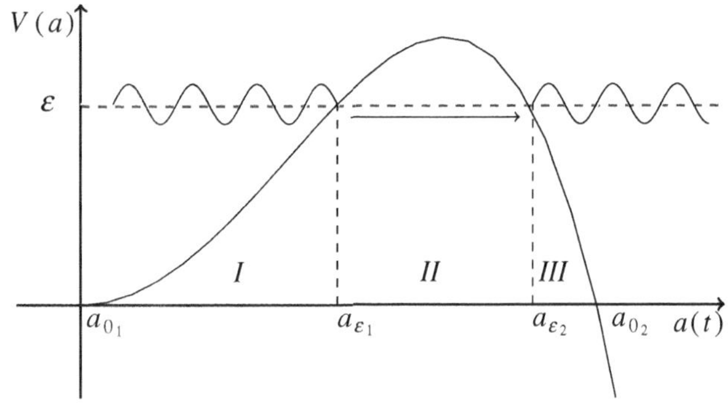

Fig. 7.2 Tunnelling of the wave function through the gravitational potential. Region II corresponds to the classically forbidden sector

[18] See [45] for a systematical expansion of the Wheeler–DeWitt equation in powers of the Planck mass.

$$\Psi_{II}^{\mathrm{WKB}}(a) = \frac{1}{\sqrt{|p_a|}} \left[c_+ \exp\left(\int \mathrm{d}a \, |p_a(a)| \right) + c_- \exp\left(- \int \mathrm{d}a \, |p_a(a)| \right) \right]. \tag{7.29}$$

If the amplitude does not vary quickly compared to the phase, the wave function shows oscillating behaviour with a slowly changing envelope, and we are in a classically allowed region. In the classically forbidden region under the potential barrier, it is just the opposite and the wave function changes its amplitude exponentially while not oscillating. At the classical turning points $V = \varepsilon$, the WKB approximation breaks down and we have to use the matching conditions in order to construct a global solution. The tunneling wave function is then selected by the boundary condition that $\Psi^{\mathrm{WKB}}(a)$ consists only of an *outgoing wave* in region III (for $a \to \infty$). The probability for tunneling from region I to region III then reads

$$\mathcal{P} \propto |\Psi_{\mathrm{tunnel}}^{\mathrm{WKB}}(a)|^2 \propto \exp\left(-2 \int_{a_{\varepsilon_1}}^{a_{\varepsilon_2}} \sqrt{|\varepsilon - V|} \, \mathrm{d}a \right). \tag{7.30}$$

The crucial point is now that this probability is even non-vanishing for the case of no radiation , i.e. $\varepsilon \to 0$. In this limit, the intersection points approach the zeros

$$\lim_{\varepsilon \to 0} a_{\varepsilon_1} = a_{0_1} = 0, \quad \lim_{\varepsilon \to 0} a_{\varepsilon_2} = a_{0_2} = \frac{1}{H}. \tag{7.31}$$

This scenario corresponds to a "tunnelling from nothing" since the classically allowed region I is not present any more. With $\lim_{\varepsilon \to 0} \sqrt{|\varepsilon - V|} = \sqrt{|-V(a)|} = a \sqrt{|H^2 a^2 - 1|}$, the tunnelling probability in the limit $\varepsilon \to 0$ becomes

$$\lim_{\varepsilon \to 0} \mathcal{P} \sim \exp\left(-2 \int_0^{1/H} a \sqrt{|H^2 a^2 - 1|} \, \mathrm{d}a \right) = \exp\left(-\frac{1}{H^2} \int_0^1 \sqrt{|u|} \, \mathrm{d}u \right)$$
$$= \exp\left(-\frac{2}{3 H^2} \right) = \exp\left(-\frac{2}{\Lambda} \right) = \exp\left(-\frac{24 \pi^2 M_{\mathrm{P}}^4}{\rho_v} \right). \tag{7.32}$$

We have substituted $u := H^2 a^2 - 1$ and re-introduced units in terms of the reduced Planck mass $M_{\mathrm{P}}^2 = 1/8 \pi G$ in the last step. We can interpret $\mathcal{P}(\Lambda)$ as the probability distribution for a universe to tunnel "from nothing to existence" with a specific value of the cosmological constant Λ, or, if we describe $\rho_v \sim V(\varphi)$ effectively by a scalar field potential, with a value of φ.

Assuming that our observable universe is by no means distinguished compared to other universes populated with observers, then we should measure the average value of Λ predicted by a peak in the distribution (7.32). Vilenkin denotes this assumption in the context of such a quantum cosmological statistical prediction the "principle of mediocrity". This kind of reasoning goes beyond a purely anthropic explanation of the values of the fundamental constants of nature and corresponds exactly to the situation described in Sect. 7.1.1. Even if we will never be able to measure all the Λ in the different universes, we can still make statistical predictions on the basis of the assumption that our universe is completely average. These predictions can at least be falsified by an experimental observation in our accessible universe.

From the distribution (7.32), it can be seen that the probability for the nucleation is favoured for large values of H^2. This is in accordance with inflation. In contrast, the no-boundary condition would essentially lead to the same distribution but with the opposite sign in the exponential of (7.32).

7.2.2 Beyond the Minisuperspace Treatment

We can now ask if this rather heuristic quantum mechanical derivation for the "tunnelling out of nothing" can be generalized beyond the simplest minisuperspace applications. This should be done within the path integral and the methods of instantons. In the case of the quantum mechanical setup, the WKB method used to calculate the tunnelling effect and the instanton solution do coincide. The advantage of the instanton method is that it can also be applied to field theories.

An instanton solution is defined as a solution of the *Euclidean* equations of motion with a *finite* action. Technically, this solution is obtained by performing a Wick rotation from the Lorentzian to the Euclidean action. This changes the signature of the metric. In contrast to the Lorentzian action, the Euclidean action does not have a different *relative sign* between kinetic and potential term. Note, that rotating to the Euclidean regime corresponds to flipping the potential upside down [25]. Thus, minima turn into maxima and vice versa. A stationary solution extended between two hills of the potential in the Euclidean regime corresponds to a solution under the hill of the potential connecting two valleys. This, in turn, corresponds to the classically forbidden region II of Sect. 7.2.1 and thus to the ordinary tunnelling scenario. It is impossible to find such a solution in a perturbative treatment around the minimum of the potential. An instanton solution corresponds to the saddle-point approximation of the Euclidean path integral, i.e. the dominant contribution of the path integral comes from the neighbourhood around the stationary point. Therefore, we can expand the path integral in the saddle-point approximation up to second order around the stationary point. The constant classical contribution can be pulled out of the integration and the integral quadratic in the deviations from the stationary point is a Gaussian and can be evaluated exactly.

However, naive attempts to go beyond the minisuperspace approximation lead to unnormalizable states in the sector of spatially inhomogeneous degrees of freedom for matter and metric and in particular invalidate the usual Wick rotation from the Lorentzian to the Euclidean space-time. This problem was partly overcome by imposing the normalizability condition on the matter part of the solution of the Wheeler–DeWitt equation [62, 65], but the situation remained controversial for the following reason:

Here, we assume that decoherence, as explained in Sect. 7.1.2, has already taken place and suppressed the quantum interference between the "contracting" and "expanding" branches of the cosmological wave function, an issue discussed for example in [40, 43, 62, 65, 67, 68]. The amplitudes of the no-boundary and tunnelling branches of such a semi-classical solution then take the form

$$\left|\Psi_{\pm}\big(\varphi, \Phi(\mathbf{x})\big)\right| = \exp\left(\mp\frac{1}{2}S_{\mathrm{E}}(\varphi)\right)\left|\Psi_{\mathrm{matter}}\big(\varphi, \Phi(\mathbf{x})\big)\right|, \qquad (7.33)$$

where $\Phi(\mathbf{x})$ is a set of matter fields separate from the spatially homogeneous inflaton φ, and $\Psi_{\mathrm{matter}}\big(\varphi, \Phi(\mathbf{x})\big)$ is their normalizable (quasi-Gaussian) part in the full wave function—in essence representing the Euclidean de Sitter invariant vacuum of linearised fields $\Phi(\mathbf{x})$ on the quasi-de Sitter background with $\Lambda_{\mathrm{eff}} = V(\varphi)/M_{\mathrm{P}}^2$. Quantum averaging over $\Phi(\mathbf{x})$ then leads to the following quantum distribution of the inflaton field

$$\rho_{\pm}^{1-\mathrm{loop}}(\varphi) = \int d[\,\Phi(\mathbf{x})\,]\left|\Psi_{\pm}\big(\varphi, \Phi(\mathbf{x})\big)\right|^2 = \exp\left(\mp S_{\mathrm{E}}(\varphi) - S_{\mathrm{E}}^{1-\mathrm{loop}}(\varphi)\right), \qquad (7.34)$$

where $S_{\mathrm{E}}^{1-\mathrm{loop}}(\varphi) = (1/2)\mathrm{Tr}\ln(\delta^2 S_{\mathrm{E}}[\,\varphi, \Phi\,]/\delta\Phi(x)\,\delta\Phi(y))$ is the contribution of the UV divergent one-loop effective action [5, 6, 9, 14]. By making use of this algorithm, a sharp probability peak was obtained in the *tunnelling* distribution $\rho_{-}^{1-\mathrm{loop}}(\varphi)$ for the model with a strong non-minimal coupling of the inflaton to gravity [9–11]. This peak was interpreted as generating the quantum scale of inflation—the initial condition for its inflationary scenario. Quite remarkably, for accidental reasons this result was free from the usual UV renormalization ambiguity. It did not require the application of the renormalization scheme of absorbing the UV divergences into the redefinition of the coupling constants in the tree-level action $S_{\mathrm{E}}(\varphi)$.

However, beyond the one-loop approximation and for other physical correlation functions the situation changes, and one has to implement a UV renormalization in full. But with the $\mp S_{\mathrm{E}}(\varphi)$ ambiguity in (7.34), this renormalization would be different for the tunnelling and no-boundary states. For instance, an asymptotically free theory in the no-boundary case (associated with the usual Wick rotation to the Euclidean space-time) will not be asymptotically free in the tunnelling case. The tunnelling versus no-boundary gravitational modification of the theory will contradict basic field-theoretical results in flat space-time. This strongly invalidates a naive construction of the tunnelling state of the above type. In particular, it does not allow to go beyond the one-loop approximation in the model of non-minimally coupled inflaton and perform its RG improvement.

Here we suggest a solution of this problem by formulating a new path integral prescription for the tunnelling state of the universe. This formulation is based on a construction of the cosmological density matrix [12, 13] which describes a microcanonical ensemble of cosmological models [7]. The statistical sum of this ensemble was calculated in a spatially closed model with a generic set of scalar, spinor, and vector fields conformally coupled to gravity. It was obtained in the saddle-point approximation dominated by the contribution of the thermal cosmological instantons of topology $S^3 \times S^1$. These instantons also include the vacuum S^4 topology treated as a limiting case of the compactified time dimension S^1 in $S^3 \times S^1$ being ripped in the transition from $S^3 \times S^1$ to S^4. This limiting case exactly recovers the Hartle–Hawking state of [36, 38], so that the whole construction of [7, 12, 13] can be considered as a generalization of the vacuum no-boundary state to the quasi-thermal no-boundary ensemble. The basic physical conclusion for this ensemble was that it exists in a bounded range of values of the effective cosmological constant, that it is capable of generating a big-boost scenario of the cosmological acceleration [8] and that its vacuum Hartle–Hawking member does not really contribute because it is suppressed by the infinite *positive* value of its action. This is a genuine effect of the conformal anomaly of quantum fields [1, 2, 32, 33, 52, 59], which qualitatively changes the tree-level action (7.30).

Below we shall show that the above path integral actually has another saddle-point corresponding to the negative value of the lapse function $N_{\mathrm{E}} < 0$, which is gauge-inequivalent to $N_{\mathrm{E}} > 0$. In the main, this leads to the inversion of the sign of the action in the exponential of the statistical sum and, therefore, deserves the label "tunnelling". In this tunnelling state, the thermal part vanishes and its instanton turns out to be a purely vacuum one. Finally, this construction no longer suffers from the above mentioned controversy with the renormalization. A *full* quantum effective action is supposed to be calculated and renormalized by the usual set of counterterms on the background of a generic metric, and then the result should be analytically continued to $N_{\mathrm{E}} < 0$ and taken at the *tunnelling* saddle-point of the path integral over the lapse function N_{E}.

The purpose of this construction is to extend the results of [15, 16] by suggesting that this model does not only have WMAP-compatible CMB perturbations, but can also generate the initial conditions for the inflationary background upon which these perturbations propagate. These initial conditions are realized in the form of a *sharp probability peak* in the tunnelling distribution function of the inflaton.

7.2.3 Path Integral for the Microcanonical Statistical Sum

The path integral for the microcanonical statistical sum in cosmology [7] can be cast as the integral over a minisuperspace lapse function $N_{\mathrm{E}}(\tau)$ and scale factor $a_{\mathrm{E}}(\tau)$

$$Z := \int \mathcal{D}[a_{\mathrm{E}},\, N_{\mathrm{E}}]\, e^{-S^{\mathrm{eff}}_{\mathrm{E}}[a_{\mathrm{E}}, N_{\mathrm{E}}]}, \tag{7.35}$$

$$e^{-S^{\mathrm{eff}}_{\mathrm{E}}[a_{\mathrm{E}},\, N_{\mathrm{E}}]} := \int \mathcal{D}\,\Phi(x)\; e^{-S_{\mathrm{E}}[a_{\mathrm{E}}, N_{\mathrm{E}};\, \Phi(x)]} \tag{7.36}$$

of a spatially closed Euclidean FRW metric

$$\mathrm{d}s^2 = N_{\mathrm{E}}^2(\tau)\,\mathrm{d}\tau^2 + a_{\mathrm{E}}^2(\tau)\,\mathrm{d}^2\Omega^{(3)}. \tag{7.37}$$

Here, $S_{\mathrm{E}}^{\mathrm{eff}}[\,a_{\mathrm{E}},\, N_{\mathrm{E}}\,]$ is the Euclidean effective action of all inhomogeneous "matter" fields

$$\Phi(x) := \{\phi(x),\ \Psi(x),\ A_{\mu}(x),\ h_{\mu\nu}(x), ...\}, \tag{7.38}$$

containing scalar, spinor, vector gauge fields and tensor fields, including also inhomogeneous metric perturbations $h_{\mu\nu}$. As required by the slow-roll approximation, the homogeneous part of the graviton-inflaton sector of the model $\dot{\varphi}(t) \approx 0$ forms the slowly varying FRW background on which those fluctuations propagate. In contrast, all quantum contributions from the inhomogeneous matter fields running in the quantum loops are integrated out and collected in the effective potential. $S_{\mathrm{E}}[a, N; \Phi(x)]$ is the classical Euclidean action, and the integration runs over periodic fields on the Euclidean space-time with a compactified time τ (of $S^1 \times S^3$ topology).

It is important that the integration over the lapse function N_{E} runs along the imaginary axis from $-\mathrm{i}\infty$ to $+\mathrm{i}\infty$ because this Euclidean path integral represents, in fact, the transformed version of the integral over metrics with Lorentzian signature. This transformation is the usual Wick rotation which can be incorporated by the transition from the Lorentzian lapse function N_{L} to the Euclidean one N_{E} by the relation $N_{\mathrm{L}} = \mathrm{i}N_{\mathrm{E}}$ [7].

It should be mentioned that a full non-perturbative evaluation of the path integral would require a careful inspection of the infinite contours in the complex N_{E}-plane that render the integral convergent, see, for example, [35, 39]. However, such an inspection is not necessary because here we are dealing here with a semi-classical approximation in which only the vicinity of the saddle-point enters.

The convenience of writing the path integral (7.35) in the Euclidean form follows from the needs of the semi-classical approximation. In this approximation, it is dominated by the contribution of a saddle-point, $\Gamma_0 = S_{\mathrm{E}}^{\mathrm{eff}}[\,a_{\mathrm{E}}^0, N_{\mathrm{E}}^0\,]$, where $a_{\mathrm{E}}^0 = a_{\mathrm{E}}^0(\tau)$ and $N_{\mathrm{E}}^0 = N_{\mathrm{E}}^0(\tau)$ solve the equation of motion for $S_{\mathrm{E}}^{\mathrm{eff}}[\,a_{\mathrm{E}}, N_{\mathrm{E}}\,]$ and satisfy periodicity conditions dictated by the definition of the statistical sum. Such periodic solutions exist in the Euclidean domain with real N_{E} rather than in the Lorentzian one with the imaginary lapse. This means that the contour of integration over N_{E} along the imaginary axis should be deformed into the complex plane to traverse the real axis at some $N_{\mathrm{E}}^0 \neq 0$ corresponding to the Euclidean solution of the equations of motion for the minisuperspace action.

The residual one-dimensional diffeomorphism invariance of this action (which is gauged out by the gauge-fixing procedure implicit in the integration measure $D[\,a_{\mathrm{E}}, N_{\mathrm{E}}\,]$) allows to fix the ambiguity in the choice of N_{E}^0. There remains only a double-fold freedom in this choice actually inherited from the sign indefiniteness of the integration range for N_L. This freedom is exhausted by either positive, $N_{\mathrm{E}}^0 > 0$, or negative, $N_{\mathrm{E}}^0 < 0$, values of the lapse, because, on the one hand, all values in each of these equivalence classes are gauge equivalent and, on the other hand, no continuous family of non-degenerate diffeomorphisms exists relating these classes to one another. Without loss of generality, we can choose $N_{\mathrm{E}}^{\pm} = \pm 1$ as representatives of these classes and label the relevant solutions and on-shell actions, respectively, as $a_{\mathrm{E}}^{\pm}(\tau)$ and

$$\Gamma_{\pm} := S_{\rm E}^{\rm eff}[\,a_{\rm E}^{\pm}(\tau), \pm 1\,]\,. \tag{7.39}$$

Gauge inequivalence of these two cases, $\Gamma_- \neq \Gamma_+$, is obvious because, for example, all local contributions to the effective action are odd functionals of $N_{\rm E}$, $S_{\rm E}^{\rm local}[\,a_{\rm E}, N_{\rm E}\,] = -S_{\rm E}^{\rm local}[\,a_{\rm E}, -N_{\rm E}\,]$. Thus, we can heuristically identify the statistical sums $\Gamma_{\pm}$ correspondingly with the "no-boundary" and "tunnelling" prescriptions for the quantum state of the universe,

$$\exp\left(-\Gamma_{\rm no-boundary/tunnel}\right) := e^{-\Gamma_{\pm}}. \tag{7.40}$$

In other words, we use this equation to define "no-boundary" and "tunnelling" in the first place. This result shows that for both prescriptions a *full quantum effective action as a whole* is present in the exponential of the partition function without any splitting into the minisuperspace and matter contributions weighted by different sign factors like in (7.34). This means that the usual renormalization scheme is applicable to the calculation of (7.39). Generally covariant UV counterterms should be calculated on the background of a generic metric and afterwards evaluated at the FRW metric with $N_{\rm E} = \pm 1$, depending on the choice of either the no-boundary or tunnelling prescription. Below, we will demonstrate how this procedure works for the system dominated by quantum fields with heavy masses, whose effective action admits a local expansion in powers of the space-time curvature and matter fields gradients.

We are looking for the cosmological instanton and therefore for finite stationary solutions of the Euclidean effective action

$$S_{\rm E}^{\rm eff}[g_{\mu\nu}] = \int d^4x\, g^{1/2}\left(M_{\rm P}^2\,\Lambda_{\rm eff} - \frac{M_{\rm P}^2}{2}\,R(g_{\mu\nu}) + \ldots\right), \tag{7.41}$$

where we disregard the terms of higher orders in the curvature and derivatives of the mean values of matter fields. Both, the curvature R and $\Lambda_{\rm eff}$, can be seen as constant on the slowly varying background $\phi_{\rm inf} \approx$ const. The slow-roll approximation justifies the omission of the kinetic term $\dot{\phi}^2$ and since the background field $\phi(t)$ is assumed to be homogeneous, it also follows $\nabla\phi = 0$. This effective action does not contain the thermal part characteristic of the statistical ensemble [12, 13] because for heavy quanta the radiation bath is not excited. This is justified by the fact that the effective temperature of this bath turns out to be vanishing (see below).

In order to realize the cosmological setup, we insert the metric (7.37) into (7.41) and use units with $m_{\rm P}^2 := 3\,\pi/4\,G = 6\,\pi^2\,M_{\rm P}^2$. Thus, we can directly use Eqs. (2.21–2.26) with $\mathfrak{K} = 1$ and the volume of the 3-unit sphere $\int {\rm d}^3\Omega_3 = 2\pi^2$ and insert this in (7.41). Integrating by parts, using the derivative $a' := {\rm d}\,a/N\,{\rm d}\tau$ in terms of Euclidean conformal time τ and expressing the cosmological constant $\Lambda_{\rm eff} = 3\,H_{\rm eff}^2$ in terms of the effective Hubble parameter $H_{\rm eff}$, we obtain

$$S_{\rm E}^{\rm eff}[a_{\rm E}, N_{\rm E}] = 2\,m_{\rm P}^2 \int {\rm d}\tau\, N_{\rm E}\,\left(-a_{\rm E}\,a_{\rm E}'^2 - a_{\rm E} + H_{\rm eff}^2\,a_{\rm E}^3\right). \tag{7.42}$$

The saddle-point for the path integral (7.35)—the stationary configuration with respect to variations of the lapse function, $\delta S^{\rm eff}_{\rm E}[\,a_{\rm E}, N_{\rm E}\,]/\delta N_{\rm E} = 0$—satisfies the Euclidean Friedmann equation

$$a'^2_{\rm E} = 1 - H^2_{\rm eff}\, a^2_{\rm E}\,. \tag{7.43}$$

This equation can be solved for $a_{\rm E}$. It has one turning point at $a^+_{\rm E} = 1/H_{\rm eff}$ below which the real solution interpolates between $a^-_{\rm E}(\tau) \equiv a_{\rm E}(0) = 0$ and $a^+_{\rm E}$. In the gauge $N_{\rm E} = \pm 1$ for both no-boundary/tunnelling cases this solution describes the Euclidean de Sitter metric, that is, one hemisphere of S^4,

$$a_{\rm E}(\tau) = \frac{1}{H_{\rm eff}} \sin(H_{\rm eff}\,\tau)\,. \tag{7.44}$$

The Euclidean section is given by half the S^4 sphere from $a^-_{\rm E} = 0$ to the equator $a^+_{\rm E} = \frac{1}{H_{\rm eff}}$. This corresponds to the initial conditions $a_{\rm E} = 0$ and $\frac{{\rm d}a_{\rm E}}{{\rm d}\tau}|_{a_{\rm E}=0} = 1$. We note that there is no ambiguity in sign for the definition of the scale factor, since for both signs of the lapse function $N_{\rm E} \pm 1$ there can be only *one geometrical meaningful positive scale factor*.[19] Thus, this solution is not periodic and in the terminology of [12, 13] describes a purely vacuum contribution to the statistical sum (7.35). As shown in [12, 13], the effective temperature of this state is determined by the inverse of the full period of the instanton solution measured in units of the conformal time η. Therefore, for (7.44) it vanishes because this period between the poles of this spherical instanton is divergent,

$$\eta = 2\int_0^{\pi/2H_{\rm eff}} \frac{d\tau}{a_{\rm E}(\tau)} \to \infty\,. \tag{7.45}$$

This justifies the absence of the thermal part in (7.41). The transition to the Lorentzian regime is achieved by the analytical continuation $\tau \to i\,t$ from the nucleation point $\tau = \frac{\pi}{2\,H_{\rm eff}}$:

$$\begin{aligned} a_L = a_E\left(\frac{\pi}{2\,H_{\rm eff}} + i\,t\right) &= \frac{1}{H_{\rm eff}} \sin\left(\frac{\pi}{2} + i\,H_{\rm eff}\,t\right) \\ = \frac{1}{H_{\rm eff}} \cos\left(i\,H_{\rm eff}\,t\right) &= \frac{1}{H_{\rm eff}} \cosh\left(\,H_{\rm eff}\,t\right) \end{aligned} \tag{7.46}$$

[19] The formal analytic extension from $N^0_{\rm E} = 1$ to $N^0_{\rm E} = -1$ should of course not be applied to $a_{\rm E}(\tau) = \sin(N^0_{\rm E}\, H_{\rm eff}\tau)/H_{\rm eff}$ which would yield a negative $a_{\rm E}(\tau)$ instead of (7.44) because in contrast to the sign-indefinite Lagrange multiplier $N_{\rm E}$ the path integration over $a_{\rm E}(\tau)$ in (7.35) semi-classically always runs in the vicinity of its positive geometrically meaningful value. For this reason, $a_{\rm E}(\tau)$ never brings sign factors into the on-shell action even though it enters the action with odd powers.

According to (7.46), once the nucleation to Lorentzian time has occurred, the scale factor a_L does expand eternally. As already mentioned, the scale factor a must be real at each stage and should be throughout positive to have a geometrical meaningful interpretation. The nucleation from Euclidean space-time to de Sitter space can be visualized by attaching a de Sitter solution at the equator of the half sphere of the Euclidean instanton (Fig. 7.3).

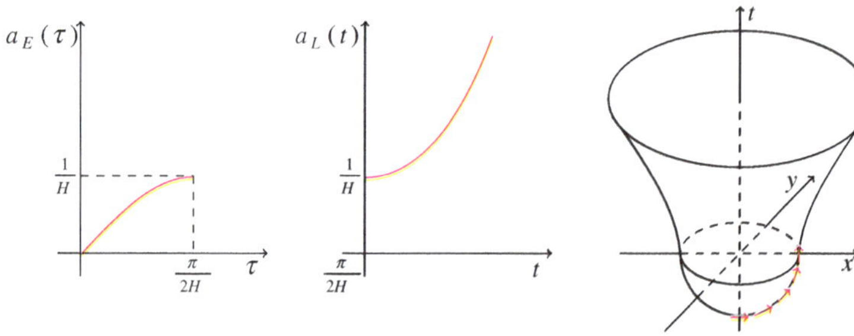

Fig. 7.3 Instanton: Half of the Euclidean four-sphere is attached to expanding de Sitter space. The transition from the Euclidean to the Lorentzian region at the equatorial cross section corresponds to the nucleation of our universe and the emergence of time

As in Fig. 2.10, horizontal slices correspond to a particular "moment of time" or a "snapshot of the universe". We note that in the Euclidean regime, the difference between the time coordinate and the space coordinates becomes blurred and both ultimately condense in a single point, the pole of the Euclidean half sphere. We will now calculate the instanton or saddle-point contribution for both choices of $N_{\mathrm{E}} = \pm 1$ by inserting the equations of motion (7.43) in (7.42) and using the solutions (7.44). Thus, the no-boundary and tunnelling Euclidean effective actions take the form

$$\begin{aligned} S_{\mathrm{E}}^{\mathrm{eff}}[a_{\mathrm{E}}^{\pm}, N_{\mathrm{E}}] &= 2\, m_{\mathrm{P}}^2 \int_{\tau_-}^{\tau_+} \mathrm{d}\tau\, N_{\mathrm{E}} \left(-a_E\, a_E'^2 - a_E + H_{\mathrm{eff}}^2\, a_E^3\right) = \mp 4\, m_{\mathrm{P}}^2 \int_{\tau_-}^{\tau_+} \mathrm{d}\tau\, a_E\, a_E'^2 \\ &= \mp\, 4\, m_{\mathrm{P}}^2 / H_{\mathrm{eff}} \int_0^{\frac{\pi}{2 H_{\mathrm{eff}}}} \mathrm{d}\tau\, \left[\sin(\, H_{\mathrm{eff}}\, \tau)\cos^2(\, H_{\mathrm{eff}}\, \tau)\right] \\ &= \mp\, 4\, \frac{m_{\mathrm{P}}^2}{H_{\mathrm{eff}}^2} \int_0^{\frac{\pi}{2}} \mathrm{d}u\, \left[\sin(u)\cos^2(u)\right] = \mp\, \frac{4}{3}\, \frac{m_{\mathrm{P}}^2}{H_{\mathrm{eff}}^2} = \mp\, 8\, \pi^2\, \frac{M_P^2}{H_{\mathrm{eff}}^2} \end{aligned} \quad (7.47)$$

with $N_{\mathrm{E}} = \pm 1$, $a_{\mathrm{E}}^- := a_{\mathrm{E}}(\tau_-) = a_{\mathrm{E}}(0) = 0$ and $a_{\mathrm{E}}^+ := a(\tau_+) = a(\pi/2\, H_{\mathrm{eff}}) = 1/H_{\mathrm{eff}}$. Thus, in terms of the effective cosmological constant Λ_{eff}, the cosmological instanton is given by

$$\Gamma_{\pm} := S_{\mathrm{E}}^{\mathrm{eff}}[a_{\mathrm{E}}^{\pm},\, \pm 1] = \mp\, 8\, \pi^2\, \frac{M_P^2}{H_{\mathrm{eff}}^2} = \mp \frac{24\, \pi^2\, M_{\mathrm{P}}^2}{\Lambda_{\mathrm{eff}}}. \quad (7.48)$$

The object of major interest here, the tunnelling partition function in the space of positive values of $H_{\text{eff}}^2 = \Lambda_{\text{eff}}/3$, is given by

$$\rho_{\text{tunnel}}(\Lambda_{\text{eff}}) = \exp(-\Gamma_-) = \exp\left(-\frac{24\,\pi^2\,M_{\text{P}}^2}{\Lambda_{\text{eff}}}\right), \quad \Lambda_{\text{eff}} > 0. \tag{7.49}$$

It formally coincides with the semi-classical tunnelling wave function of the universe [46, 47, 53, 73], $|\Psi_{\text{tunnel}}|^2 \simeq \exp(-8\pi^2 M_{\text{P}}^2/H^2)$, derived from the Wheeler–DeWitt equation in the tree-level approximation, however with an *effective* cosmological constant, containing the quantum contributions of the inhomogeneous degrees of freedom.

7.3 Quantum Origin of the Universe in the Higgs Inflation Scenario

The partition function of the type (7.49) can serve as a source of initial conditions for inflation only when the cosmological constant $\Lambda_{\text{eff}} = 3H_{\text{eff}}^2$ becomes a composite field capable of a decay at the exit from inflation. Usually, this is a scalar inflaton field, whose quantum mean value φ is nearly constant in the slow-roll regime, and its effective potential $V_{\text{eff}}(\varphi)$ plays the role of the cosmological constant driving inflation. When the contribution of the inflaton gradients is small, the above formalism remains applicable also with the inclusion of this field whose ultimate effect reduces to the generation of the effective cosmological constant $\Lambda_{\text{eff}} = V_{\text{eff}}(\varphi)/M_{\text{P}}^2$ and the effective Planck mass.

These constants are the coefficients of the zeroth and first order terms of the effective action expanded in powers of the curvature, and they incorporate radiative corrections due to all quantum fields in the path integral (7.36). Now there is no mismatch between the signs of the tree-level and loop parts of the partition function anymore. Therefore, we can apply the usual renormalization and, if necessary, the RG (RG) improvement to obtain the full effective action $S_{\text{E}}^{\text{eff}}[g_{\mu\nu}, \varphi]$ and then repeat the procedure of the previous section.

In the slow-roll approximation, the effective action has the general form

$$S_{\text{E}}^{\text{eff}}[g_{\mu\nu}, \varphi] = \int d^4x\, g^{1/2}\left(V_{\text{eff}}(\varphi) - U_{\text{eff}}(\varphi)\,R(g_{\mu\nu}) + \frac{1}{2}\,G_{\text{eff}}(\varphi)\,(\nabla\varphi)^2 + \dots\right), \tag{7.50}$$

where $V_{\text{eff}}(\varphi)$, $U_{\text{eff}}(\varphi)$ and $G_{\text{eff}}(\varphi)$ are the coefficients of the derivative expansion and we disregard the contribution of higher-derivative operators. In the scenario with the slowly varying inflaton the coefficients $V_{\text{eff}}(\varphi)$ and $U_{\text{eff}}(\varphi)$ play the role of the effective cosmological and Planck mass constants, so that we can identify in (7.41) and (7.50) the effective $M_{\text{P}}^2 = m_{\text{P}}^2/6\,\pi^2$ and H_{eff}^2, respectively, with $2\,U_{\text{eff}}(\varphi)$ and $V_{\text{eff}}(\varphi)/6\,U_{\text{eff}}(\varphi)$, cf. the Jordan frame equations of motion (6.36). Therefore, the tunnelling partition function (7.49) becomes the following distribution of the field φ

$$\rho_{\text{tunnel}}(\varphi) = \exp\left(-\frac{24\pi^2 M_P^4}{\hat{V}_{\text{eff}}(\varphi)}\right), \quad \hat{V}_{\text{eff}}(\varphi) = \left(\frac{M_P^2}{2}\right)^2 \frac{V_{\text{eff}}(\varphi)}{U_{\text{eff}}^2(\varphi)}, \tag{7.51}$$

where $\hat{V}_{\text{eff}}(\varphi)$ in fact coincides with the potential in the Einstein frame of the action (7.50).

So far, the formalism is quite general, but now we want to apply it to the model of inflation driven by the Standard Model Higgs-inflaton. As shown in Chap. 6, the one-loop RG improved action in this model has the form (7.50) for large φ with the running coefficient functions (6.84–6.83) and determined in terms of the running couplings $\lambda(t)$ and $\xi(t)$, and the field renormalization $Z(t)$. They incorporate a summation of powers of logarithms and belong to the solution of the RG Eq. (6.72) which read at the inflationary stage with a large $\varphi \sim M_P/\sqrt{\xi}$ and large $\xi \gg 1$, cf. (6.73) and (6.74)

$$\frac{d\lambda}{dt} = \frac{\mathbf{A}}{16\pi^2}\lambda - 4\gamma\lambda, \tag{7.52}$$

$$\frac{d\xi}{dt} = \frac{6\lambda}{16\pi^2}\xi - 2\gamma\xi. \tag{7.53}$$

In addition, we have the slow running of the field renormalization $dZ/dt = \gamma Z$ with γ being the anomalous dimension of the Higgs field. The running scale $t = \ln(\varphi/M_t)$ is again normalized at the top-quark mass $\mu \equiv M_t$, and $\mathbf{A} = \mathbf{A}(t)$ is the *running anomalous scaling* (6.14).

This quantity and its modification $\mathbf{A_I} = \mathbf{A} - 12\lambda$ due to the RG running of the non-minimal coupling $\xi(t)$ are so important because, for $\xi \gg 1$, they mainly determine the quantum inflationary dynamics [11, 18] and yield the parameters of the CMB generated during inflation [17]. In particular, the value of φ at the beginning of the inflationary stage of duration N in units of the e-folding number has been found in [17] and defined by (6.44) to be

$$\varphi^2 = -\frac{64\pi^2 M_P^2}{\xi \mathbf{A_I}(t_{\text{end}})}(1 - e^{x_{\text{end}}}), \tag{7.54}$$

$$x_{\text{end}} = \frac{N\mathbf{A_I}(t_{\text{end}})}{48\pi^2}. \tag{7.55}$$

We have used the definition (7.55) for the parameter x which directly involves $\mathbf{A_I}(t_{\text{end}})$, evaluated at the end of inflation, $t_{\text{end}} = \ln(\varphi_{\text{end}}/M_t)$, $\varphi_{\text{end}} \simeq 2M_P/\sqrt{3\xi}$. This parameter also enters the quantum corrected formulas for the CMB power spectrum (6.61) and its spectral index (6.63). As shown in Chap. 6, the combination of these parameters with the observational constraints (6.128) lead to the CMB-compatible range of the Higgs mass 135.6 GeV $\lesssim M_H \lesssim$ 184.5 GeV, both bounds being determined by the lower bound on the CMB spectral index.

Now we want to show that, in addition to the good agreement of the spectrum of cosmological perturbations with the CMB data, this model can also describe the mechanism of generating the cosmological *background* itself upon which these perturbations exist. This mechanism consists in the formation of the initial conditions for inflation in the form of a sharp probability peak in the distribution function (7.51) at some appropriate value of the inflaton field φ_0 with which the universe as a whole starts its evolution. The shape and the magnitude of the potential (6.105) is depicted in Fig. 6.8 and clearly indicates the existence of such a peak.

Indeed, the negative of the inverse potential damps the probability of those values φ at which $\hat{V}_{\rm RG}(\varphi) = 0$ to zero after exponentiation and, vice versa, enhances the probability at the positive maxima of the potential. Thus we briefly review the behaviour of the running potential for different Higgs masses: It is known, that the Standard Model has a domain of unstable electroweak vacuum for low $M_{\rm H}$, characterized by negative values of running $\lambda(t)$ at certain energy scales. Thus, we start with the electroweak vacuum instability threshold [30, 58] which exists in this gravitating Standard Model at $M_{\rm H}^{\rm inst} \approx 134.27\,{\rm GeV}$ (cf. Sect. 6.5.2) and which is slightly lower than the CMB compatible range of the Higgs mass ($M_{\rm H}^{\rm inst}$ is chosen for the lowest curve in Fig. 6.8). The potential $\hat{V}_{\rm RG}(\varphi)$ drops to zero at $t_{\rm inst} \simeq 41.6$, $\varphi_{\rm inst} \sim 80 M_{\rm P}$, and forms a false vacuum [15, 16] which is separated from the electroweak vacuum by a large peak at $t \simeq 34$. Correspondingly, the probability of creation of the universe with the initial value of the inflaton field at the electroweak scale $\varphi = \nu$ and at the instability scale $\varphi_{\rm inst}$ is damped to zero, while the most probable value belongs to this peak. The inflationary stage of the formation of the pivotal $N = 60$ CMB perturbation (from the moment $t_{\rm in}$ of the first horizon crossing until the end of inflation $t_{\rm end}$), which is marked by dashed lines in Fig. 6.6, lies to the left of this peak. This conforms to the requirement of the chronological succession of the initial conditions for inflation and the formation of the CMB spectra. The case for $M_{\rm H}^{\rm inst}$ is, however, below the CMB-compatible range of $M_{\rm H}$.[20] An important situation occurs at higher Higgs masses from the lower CMB bound on $M_{\rm H} \simeq 135.6\,{\rm GeV}$ until about 160 GeV. Here we get a family of a metastable vacua with $\hat{V}_{\rm RG} > 0$. An example is the plot for the lower CMB bound $M_{\rm H} = 135.62$ GeV depicted in Fig. 6.7. Despite the shallowness of this vacuum, its small maximum generates via (7.51) a sharp probability peak for the initial inflaton field. This follows from an extremely small value of $\hat{V}_{\rm eff}/M_{\rm P}^4 \sim 10^{-11}$, the reciprocal of which generates a rapidly changing exponential of (7.51). The location of the peak again precedes the inflationary stage for a pivotal $N = 60$ CMB perturbation (also marked by dashed lines in Fig. 6.7).

For even larger $M_{\rm H}$, these metastable vacua get replaced by a negative slope of the potential, see Fig. 6.8. Therefore, for large $M_{\rm H}$ close to the upper CMB bound

[20] Another interesting range of $M_{\rm H}$ is below the instability threshold $M_{\rm H}^{\rm inst}$ where $\hat{V}_{\rm RG}$ becomes negative in the "true" high energy vacuum. As mentioned in the previous section, the tunnelling state rules out aperiodic solutions of effective equations with $H^2 < 0$, which cannot contribute to the quantum ensemble of expanding Lorentzian signature models. Therefore, this range is semi-classically ruled out not only by the instability arguments, but also contradicts the tunnelling prescription.

185 GeV, the probability peak of (7.51) gets separated from the non-perturbative domain of large over-Planckian scales due to a fast drop of $\hat{V} \sim \lambda/\xi^2$ to zero. This, in turn, follows from the fact that $\xi(t)$ grows much faster than $\lambda(t)$ when they both start approaching their Landau pole [15].

The location φ_0 of the probability peak and its quantum width can be found in analytical form. Their derivation shows the crucial role of the running $\mathbf{A_I}(t)$ for the formation of initial conditions for inflation. Indeed, the exponential of the tunnelling distribution (7.51) for $M_P^2/\xi\varphi^2 \ll 1$ reads as

$$\Gamma_-(\varphi) = 24\pi^2 \frac{M_P^4}{\hat{V}_{RG}(\varphi)} \simeq 96\pi^2 \frac{\xi^2}{\lambda}\left(1 + \frac{2M_P^2}{\xi Z^2\varphi^2}\right) \tag{7.56}$$

and in view of the RG Eqs. (7.52) and (7.53) has an extremum satisfying the equation

$$\varphi\frac{\mathrm{d}\Gamma_-}{\mathrm{d}\varphi} = \frac{\mathrm{d}\Gamma_-}{\mathrm{d}t} = -\frac{6\xi^2}{\lambda}\left(\mathbf{A_I} + \frac{64\pi^2 M_P^2}{\xi Z^2\varphi^2}\right) = 0, \tag{7.57}$$

where we again neglect higher order terms in $M_P^2/\xi Z^2\varphi^2$ and $\mathbf{A_I}/64\pi^2$ (extending beyond the one-loop order). Here, $\mathbf{A_I}$ determines the running of the ratio λ/ξ^2

$$16\pi^2 \frac{\mathrm{d}}{\mathrm{d}t}\left(\frac{\lambda}{\xi^2}\right) = \mathbf{A_I}\left(\frac{\lambda}{\xi^2}\right). \tag{7.58}$$

The anomalous scaling $\mathbf{A_I}$ should be negative for the existence of the solution for the probability peak,

$$\varphi_0^2 = -\left.\frac{64\pi^2 M_P^2}{\xi \mathbf{A_I} Z^2}\right|_{t=t_0}. \tag{7.59}$$

As shown in Sect. 6.5.2, this quantity is indeed negative. In the CMB-compatible range of M_H, its running starts from the range $-36 \lesssim \mathbf{A_I}(0) \lesssim -23$ at the electroweak scale and reaches small but still negative values in the range $-11 \lesssim \mathbf{A_I}(t_{end}) \lesssim -2$ at the inflation scale. Moreover, the running of $\mathbf{A_I}(t)$ and $Z(t)$ is very slow—the quantities belonging to the two-loop order—and the duration of inflation is very short $t_0 \sim t_{in} \simeq t_{end} + 2$, see Sect. 6.5.1. Therefore, $\mathbf{A_I}(t_0) \simeq \mathbf{A_I}(t_{end})$, and these estimates also apply to $\mathbf{A_I}(t_0)$. As a result, the second derivative of the tunnelling on-shell action is positive and very large,

$$\frac{\mathrm{d}^2\,\Gamma_-}{\mathrm{d}t^2} \simeq -\frac{12\,\xi^2}{\lambda}\mathbf{A_I} \gg 1, \tag{7.60}$$

which gives an extremely small value of the quantum width of the probability peak,

$$\frac{\Delta\varphi^2}{\varphi_0^2} = -\left.\frac{\lambda}{12\xi^2}\frac{1}{\mathbf{A_I}}\right|_{t=t_0} \sim 10^{-10}. \tag{7.61}$$

This width is about $(24\pi^2/|\mathbf{A_I}|)^{1/2}$ times—one order of magnitude—higher than the CMB perturbation at the pivotal wavelength $k^{-1} = 500\,\text{Mpc}$, which we choose to correspond to $N = 60$. The point φ_{in} of the horizon crossing of this perturbation (and other CMB waves with different N's) easily follows from Eq. (7.54) which in view of $\mathbf{A_I}(t_0) \simeq \mathbf{A_I}(t_{\text{end}})$ takes the form

$$\frac{\varphi_{\text{in}}^2}{\varphi_0^2} = 1 - \exp\left(-N\frac{|\mathbf{A_I}(t_{\text{end}})|}{48\pi^2}\right). \tag{7.62}$$

It indicates that for wavelengths longer than the pivotal one the instant of horizon crossing approaches the moment of "creation" of the universe, but always stays chronologically succeeding it, as it has to be.

7.4 Conclusions

In this chapter we have constructed the tunnelling quantum state of the universe. The corresponding apparent ensemble from the quantum state exists in the unbounded positive range of the effective cosmological constant, unlike the no-boundary state discussed in [7, 12, 13] whose apparent ensemble is bounded by the reciprocated coefficient of the topological term in the overall conformal anomaly. Also, in contrast to the no-boundary case, the tunnelling state turns out to be a radiation-free vacuum one.

The status of the tunnelling versus no-boundary states is rather involved. In fact, the formal Euclidean path integral (7.35) is a transformed version of the microcanonical path integral over Lorentzian metrics, so that its lapse function integration runs along the imaginary axis from $-i\infty$ to $+i\infty$ [7].[21] The absence of periodic solutions for stationary points of (7.35) with the Lorentzian signature makes us to distort the contour of integration over N into a complex plane, so that it traverses the real axis at the points $N = +1$ or $N = -1$ which give rise to no-boundary or tunnelling states. It has been shown that the no-boundary thermal part of the statistical sum of [12, 13] is not analytic in the full complex plane of N. The $N \gtrless 0$ domains are separated by the infinite sequence of its poles densely filling the imaginary axes of N. Therefore, the contour of integration passing through both points $N = \pm 1$ is

[21] This might seem to be equivalent to the tunnelling path integral of [64, 66], but the class of metrics integrated over is very different. We do not impose by hands $a_E^- = 0$ as the boundary condition, but derive it from the saddle-point approximation for the integral over formally periodic configurations. The fact that periodicity gets violated by the boundary condition $a_E^- = 0$ implies that the a priori postulated tunnelling statistical ensemble is exhausted at the dynamical level by the contribution of a pure vacuum state [7, 12, 13].

impossible, and the no-boundary and tunnelling states cannot be obtained by analytic continuation from one another.[22] They represent alternative solutions (quantum states) of the Wheeler–DeWitt equation.

The path-integral formulation of the tunnelling state admits a consistent renormalization scheme and a RG resummation which is very efficient in cosmology according to the series of papers [15, 16, 20, 21, 24, 26]. For this reason we have applied the obtained tunnelling distribution to the model of inflation driven by the SM Higgs boson non-minimally coupled to curvature. In this way a complete cosmological scenario was obtained, embracing the formation of initial conditions for the inflationary background (in the form of a sharp probability peak in the inflaton field distribution) and the ongoing generation of the CMB perturbations on this background. As was shown in [15, 16], the comparison of the CMB amplitude and the spectral index with the WMAP observations impose bounds on the allowed range of the Higgs mass. These bounds turn out to be remarkably consistent with the widely accepted electroweak vacuum stability and perturbation theory restrictions. Interestingly, the behaviour of the running anomalous scaling $\mathbf{A_I}(t) < 0$, being crucially important for these bounds, also guarantees the existence of the obtained probability peak. The quantum width of this peak is one order of magnitude higher than the amplitude of the CMB spectrum at the pivotal wavelength, which could entail interesting observational consequences.

In the sense, discussed in Sect. 7.1.1, the prediction of the initial condition for inflation φ_0 can be tested by observation. If B-polarization signatures of primordial gravitational waves will be found in the CMB, this would offer the possibility to obtain direct information about the energy scale of inflation, cf. (2.141). Thus, if future experiments like the satellite PLANCK will observe primordial gravitational waves, we can compare this value with the prediction $\hat{V}_{\rm RG}(\varphi_0) \propto \Lambda_{\rm eff}(\varphi_0) \propto H^2_{\rm eff}(\varphi_0)$.

Summarizing, the obtained results suggest a convincing unification of quantum cosmology with the particle phenomenology of the Standard Model, inflation theory, and CMB observations. It supports the hypothesis that an appropriately extended Standard Model [3, 4, 57] can be a consistent quantum field theory all the way up to quantum gravity and perhaps explain all major phenomena in early and late cosmology.

References

1. Antoniadis, I., Mazur, P.O., Mottola, E.: Scaling behavior of quantum four-geometries. Phys. Lett. B **323**, 284 (1994)

[22] In the case of the vacuum no-boundary state when the vanishing thermal part of the effective action cannot present an obstacle to analytic continuation in the complex plane of N, the situation stays the same. Indeed, any integration contour from $-i\infty$ to $+i\infty$ crosses the real axes an odd number of times, so that the contribution of only one such crossing survives, because any two (gauge-equivalent) saddle-points traversed in opposite directions give contributions cancelling one another.

2. Antoniadis, I., Mottola, E.: Four-dimensional quantum gravity in the conformal sector. Phys. Rev. D **45**, 2013 (1992)
3. Asaka, T., Blanchet, S., Shaposhnikov, M.: The νMSM, dark matter and neutrino masses. Phys. Lett. B **631**, 151 (2005)
4. Asaka, T., Shaposhnikov, M.: The νMSM, dark matter and baryon asymmetry of the universe. Phys. Lett. B **620**, 17 (2005)
5. Barvinsky, A.O.: Unitarity approach to quantum cosmology. Phys. Rep. **230**, 237 (1993)
6. Barvinsky, A.O.: Reduction methods for functional determinants in quantum gravity and cosmology. Phys. Rev. D **50**, 5115 (1994)
7. Barvinsky, A.O.: Why there is something rather than nothing: cosmological constant from summing over everything in Lorentzian quantum gravity. Phys. Rev. Lett. **99**, 071301 (2007)
8. Barvinsky, A.O., Deffayet, C.: Anomaly driven cosmology: big boost scenario and AdS/CFT correspondence. J. Cosmol. Astropart. Phys. **05**, 020 (2008)
9. Barvinsky, A.O., Kamenshchik, A.Yu.: 1-loop quantum cosmology: the normalizability of the Hartle-Hawking wave function and the probability of inflation. Class. Quantum Grav. **7**, L181 (1990)
10. Barvinsky, A.O., Kamenshchik, A.Yu.: Quantum scale of inflation and particle physics of the early universe. Phys. Lett. B **332**, 270 (1994)
11. Barvinsky, A.O., Kamenshchik, A.Yu.: Effective equations of motion and initial conditions for inflation in quantum cosmology. Nucl. Phys. B **532**, 339 (1998)
12. Barvinsky, A.O., Kamenshchik, A.Yu.: Cosmological landscape from nothing: some like it hot. J. Cosmol. Astropart. Phys. **09**, 014 (2006)
13. Barvinsky, A.O., Kamenshchik, A.Yu.: Thermodynamics via creation from nothing: Limiting the cosmological constant landscape. Phys. Rev. D **74**, 121502 (2006)
14. Barvinsky, A.O., Kamenshchik, A.Yu.: Effective action and decoherence by fermions in quantum cosmology. Nucl. Phys. B **552**, 420 (1999)
15. Barvinsky, A.O., Kamenshchik, A.Yu.: Asymptotic freedom in inflationary cosmology with a non-minimally coupled Higgs field. J. Cosmol. Astropart. Phys. **12**, 003 (2009)
16. Barvinsky, A.O., Kamenshchik, A.Yu.: Higgs boson, renormalization group, and naturalness in cosmology. Eur. Phys. J. C. **72**, 2219 (2012)
17. Barvinsky, A.O., Kamenshchik, A.Yu.: Inflation scenario via the Standard Model Higgs boson and LHC. J. Cosmol. Astropart. Phys. **11**, 021 (2008)
18. Barvinsky, A.O., Nesterov, D.V.: Effective equations in quantum cosmology. Nucl. Phys. B **608**, 333 (2001)
19. Bell, J.S.: On the Einstein-Podolsky-Rosen paradox. Physics **1**, 195 (1964)
20. Bezrukov, F., Shaposhnikov, M.: Standard model Higgs boson mass from inflation: two loop analysis. J. High Energy Phys. **07**, 089 (2009)
21. Bezrukov, F.L., Magnin, A., Shaposhnikov, M.: Standard Model Higgs boson mass from inflation. Phys. Lett. B **675**, 88 (2009)
22. Bohr, N.: Can quantum-mechanical description of physical reality be considered complete? Phys. Rev. **48**, 696 (1935)
23. Brezger, B., Hackmüller, L., Uttenthaler, S., Petschinka, J., Arndt, M., Zeilinger, A.: Matter-wave interferometer for large molecules. Phys. Rev. Lett. **88**, 100404 (2002)
24. Clark, T.E., Liu, B., Love, S.T., ter Veldhuis, T.: Standard model Higgs boson-inflaton and dark matter. Phys. Rev. D **80**, 075019 (2009)
25. Coleman, S.: Aspects of Symmetry. Cambridge University Press, Cambridge (1985)
26. De Simone, A., Hertzberg, M.P., Wilczek, F.: Running inflation in the standard model. Phys. Lett. B **678**, 1 (2009)
27. DeWitt, B.S.: Quantum theory of gravity. I canonical theory. Phys. Rev. **160**, 1113 (1967)
28. DeWitt, B.S., Graham, N.: The Many-Worlds Interpretation of Quantum Mechanics. Princeton University Press, Princeton (1973)
29. Dirac, P.A.M.: Lectures on Quantum Mechanics, 2nd edn. Yeshiva University, New York, Belfer Graduate School of Science (1964)

30. Espinosa, J.R., Giudice, G.F., Riotto, A.: Cosmological implications of the Higgs mass measurement. J. Cosmol. Astropart. Phys. **05**, 002 (2008)
31. Everett, H.: "Relative State" formulation of quantum mechanics. Rev. Mod. Phys. **29**, 454 (1957)
32. Fischetti, M. V., Hartle, J. B., Hu, B. L. Quantum effects in the early universe. I. Influence of trace anomalies on homogeneous, isotropic, classical geometries. Phys. Rev. D **20**, 1757 (1979)
33. Fradkin, E.S., Tseytlin, A.A.: Conformal anomaly in Weyl theory and anomaly free superconformal theories. Phys. Lett. B **134**, 187 (1984)
34. Hackmüller, L., Hornberger, K., Brezger, B., Zeilinger, A., Arndt, M.: Decoherence of matter waves by thermal emission of radiation. Nature **427**, 711 (2004)
35. Halliwell, J. J., Louko, J.: Steepest-descent contours in the path-integral approach to quantum cosmology. III. A general method with applications to anisotropic minisuperspace models. Phys. Rev. D **42**, 3997 (1990)
36. Hartle, J.B., Hawking, S.W.: Wave function of the Universe. Phys. Rev. D **28**, 2960 (1983)
37. Hawking, S.W.: The boundary condition of the universe. Pontificia Acad. Scientarium Scr. Varia **48**, 563 (1982)
38. Hawking, S.W.: The quantum state of the universe. Nucl. Phys. B **239**, 257 (1984)
39. Kiefer, C.: On the meaning of path integrals in quantum cosmology. Ann. Phys. (NY) **207**, 53 (1991)
40. Kiefer, C.: Decoherence in quantum electrodynamics and quantum gravity. Phys. Rev. D **46**, 1658 (1992)
41. Kiefer, C.: Semiclassical gravity and the problem of time. http://arxiv.org/abs/gr-qc/9405039 (1993). Accessed 11 January 2012 (6 pages)
42. Kiefer, C.: Conceptual issues in quantum cosmology. Lect. Notes Phys. **541**, 158 (2000)
43. Kiefer, C.: Quantum Gravity, 2nd edn. Oxford University Press, Oxford (2007)
44. Kiefer, C., Lohmar, I., Polarski, D., Starobinsky, A.A.: Origin of classical structure in the Universe. J. Phys. Conf. Ser. **67**, 012023 (2007)
45. Kiefer, C., Singh, T.P.: Quantum gravitational corrections to the functional Schrödinger equation. Phys. Rev. D **44**, 1067 (1991)
46. Linde, A.D.: Quantum creation of an inflationary universe. Sov. Phys. JETP **60**, 211 (1984)
47. Linde, A.D.: Quantum creation of the inflationary universe. Lett. Nuovo Cimento **39**, 401 (1984)
48. Linde, A.D.: Eternal chaotic inflation. Mod. Phys. Lett. A **1**, 81 (1986)
49. Marolf, D.: Resource letter NSST-1: the nature and status of string theory. Am. J. Phys. **72**, 730 (2004)
50. Perlmutter, S., et al., Supernova Cosmology Project.: Measurements of Ω and Λ from 42 high-redshift supernovae. Astrophys. J. **517**, 565 (1999)
51. Popper, K.: Logik der Forschung. Springer, Vienna (1935)
52. Riegert, R.J.: A nonlocal action for the trace anomaly. Phys. Lett. B **134**, 56 (1984)
53. Rubakov, V.A.: Particle creation in a tunneling universe. JETP Lett. **39**, 107 (1984)
54. Schlosshauer, M.A.: Decoherence and the Quantum-To-Classical Transition. Springer, Berlin (2007)
55. Schmitt-Manderbach, T., et al.: Experimental demonstration of free-space decoy-state quantum key distribution over 144 km. Phys. Rev. Lett. **98**, 010504 (2007)
56. Schrödinger, E.: Die gegenwärtige Situation in der Quantenmechanik. Naturwiss **23**, 807 (1935)
57. Shaposhnikov, M., Zenhausern, D.: Quantum scale invariance, cosmological constant and hierarchy problem. Phys. Lett. B **671**, 162 (2009)
58. Sher, M.: Electroweak Higgs potentials and vacuum stability. Phys. Rep. **179**, 273 (1989)
59. Starobinsky, A.A.: A new type of isotropic cosmological models without singularity. Phys. Lett. B **91**, 99 (1980)
60. Susskind, L.: The anthropic landscape of string theory. http://arxiv.org/abs/hep-th/0302219 (2003). Accessed 11 January 2012 (22 pages)

61. Tegmark, M.: The multiverse hierarchy. http://arxiv.org/abs/0905.1283 (2007). Accessed 11 January 2012 (15 pages)
62. Vachaspati, T., Vilenkin, A.: Uniqueness of the tunneling wave function of the universe. Phys. Rev. D **37**, 898 (1988)
63. Vilenkin, A.: Creation of universes from nothing. Phys. Lett. B **117**, 25 (1982)
64. Vilenkin, A.: Quantum creation of universes. Phys. Rev. D **30**, 509 (1984)
65. Vilenkin, A.: Quantum cosmology and the initial state of the universe. Phys. Rev. D **37**, 888 (1988)
66. Vilenkin, A.: Approaches to quantum cosmology. Phys. Rev. D **50**, 2581 (1994)
67. Vilenkin, A.: The quantum cosmology debate. http://arxiv.org/abs/gr-qc/9812027 (1998). Accessed 11 January 2012 (9 pages)
68. Vilenkin, A.: The Wave function discord. Phys. Rev. D **58**, 067301 (1998)
69. von Neumann, J.: Mathematische Grundlagen der Quantenmechanik. Springer, Berlin (1932)
70. Wheeler, J.A.: Superspace and the nature of quantum geometrodynamics. In: DeWitt, C.M., Wheeler, J.A. (eds.) Battelle Rencontres, pp. 242–307. Benjamin, New York (1968)
71. Wigner, E.P.: Remarks on the mind-body question. In: Good, I.J. (ed.) The Scientist Speculates: An Anthology of Partly-Baked Ideas, pp. 284–302. Heinemann, London (1961)
72. Zeh, H.D.: On the interpretation of measurement in quantum theory. Found. Phys. **1**, 69 (1970)
73. Zeldovich, Y.B., Starobinsky, A.A.: Quantum creation of a universe in a nontrivial topology. Sov. Astron. Lett. **10**, 135 (1984)
74. Zurek, W. H.: Decoherence and the transition from quantum to classical. Phys. Today **44N10**, 36 (1991)
75. Zurek, W.H.: Probabilities from entanglement, Born's rule $p_k = |\psi_k|^2$ from envariance. Phys. Rev. A **71**, 052105 (2005)

Chapter 8
Summary and Outlook

In this thesis, we have investigated the model of non-minimal Higgs inflation. The unified description of the Standard Model Higgs boson and the cosmological inflaton required a strong non-minimal coupling to gravity in order to ensure consistency with the data of the cosmic microwave background radiation. We have derived a numerical constraint for the mass of the Higgs-inflaton to lie within the interval $135.6\ \text{GeV} \leq M_{\text{H}} \leq 184.5\ \text{GeV}$. These numerically predicted values are close to the expected Higgs mass range. During the process of writing up this thesis there has been an official CERN press release, claiming that experiments at the LHC show a strong evidence for a detection of the Higgs particle with a mass around 125 GeV. It remains to be seen if this announcement will ultimately be confirmed. However, even if this first indication is correct, it does not mean that we have to abandon the non-minimal Higgs inflation scenario. First of all, numerical test runs and the results of [2, 5] have clearly shown a generic tendency that the inclusion of two-loop quantum corrections will improve the situation and shift the Higgs mass interval to lower values, very close to the announced value of 125 GeV. Moreover, if more accurate measurements would result in a lower value for the top-quark mass $M_{\text{t}} < 171\ \text{GeV}$, this would also shift the Higgs mass range to lower values. Aside from these numerical uncertainties, there still remain unresolved theoretical problems like the gauge and reparametrization dependence of the results that could also affect the quantitative predictions. However, most of the general properties of the model seem to be robust. We have revealed the qualitative key features of non-minimal Higgs inflation: a suppression mechanism of Higgs propagators that leads to a phenomenon similar to asymptotic freedom. In addition, we have shown that the Goldstone bosons account for significant quantum contributions that are crucial for the determination of the renormalization group flow from the electroweak scale up to inflationary energies. In contrast to the Higgs propagator, the Goldstone propagators are not suppressed for high energies, which makes them especially important in the inflationary analysis. Moreover, contrary to the criticism of [1, 3], we have shown that there is no violation of unitarity in this model for the high energies during inflation. All higher dimensional operators are reliably suppressed by a field-dependent cut-off.

C. F. Steinwachs, *Non-minimal Higgs Inflation and Frame Dependence in Cosmology*, Springer Theses, DOI: 10.1007/978-3-319-01842-3_8,
© Springer International Publishing Switzerland 2014

We have used the results of the non-minimal Higgs inflation scenario in a quantum cosmological framework. Within a path integral formulation, we have calculated the probability distribution of the universe to tunnel from "nothing into existence". We have found a sharp peak in this distribution, corresponding to a specific value of the inflaton field that can be interpreted as the initial condition for inflation. Thus, the non-minimal Higgs inflation is not only in good agreement with the power spectrum of primordial fluctuations but can also explain the generation of the inflationary background on which these fluctuations propagate. Moreover, this quantum cosmological aspect of the model can be tested by the analysis of primordial gravitational waves that leave their imprint in the B-polarization of the cosmic microwave background.

The non-minimal Higgs inflation model required the calculation of the divergent part of the one-loop off-shell effective action for the complex Higgs doublet with a non-minimal curvature coupling. We have performed this calculation in the Jordan frame for an even more general setup. This included arbitrary functional couplings and N scalar fields obeying an internal $O(N)$ symmetry. We have presented the general one-loop results in a closed form. For a specific choice of the tree-level couplings and $N = 4$, we have recovered the results relevant for the application within the non-minimal Higgs inflation model. However, due to its universality the result can serve as a reference for many different cosmological models.

This result also provided the basis for an investigation of the equivalence between the Jordan and Einstein frame at the quantum level. We have constructed an explicit transformation rule for a transition between the Jordan and Einstein frame for the model of the $O(N)$-symmetric scalar multiplet non-minimally coupled to gravity. However due to a non-trivial integrability condition this transition is not possible in general. This is in agreement with the results obtained in [4, 6, 7]. By individually calculating and comparing the off-shell effective actions in both parametrizations for a single scalar field, we have found that the quantum corrections are indeed frame-dependent. We have provided a natural explanation for the failure of off-shell equivalence within a generalized geometric framework developed in [8]. Ultimately, the parametrization dependence can be traced back to the lack of covariance of the conventional formalism which is used to calculate the effective action. Without being able to present a final resolution of this problem, this important insight nevertheless entails immediate consequences for the long-lasting and ongoing cosmological debate about the equivalence between Jordan and Einstein frame. It shows that this cosmological problem is just a very special aspect of a more general problem and further reveals that hitherto the cosmological debate has simply missed this point. In view of the *physical* requirement of reparametrization invariance of configuration space variables, it seems to be a meaningless undertaking to collect physical arguments either in favour of or against one or the other frame. Rather, the solution of this problem requires the development of a theoretical framework that ensures the covariance at *all levels of the formalism* along the lines of [8].

References

1. Barbon, J.L.F., Espinosa, J.R.: On the naturalness of Higgs inflation. Phys. Rev. D **79**, 081302 (2009)
2. Bezrukov, F., Shaposhnikov, M.: Standard model Higgs boson mass from inflation: two loop analysis. J. High Energy Phys. **07**, 089 (2009)
3. Burgess, C.P., Lee, H.M., Trott, M.: Power-counting and the validity of the classical approximation during inflation. J. High Energy Phys. **09**, 103 (2009)
4. Burgess, C.P., Lee, H.M., Trott, M.: On Higgs inflation and naturalness. J. High Energy Phys. **07**, 007 (2010)
5. De Simone, A., Hertzberg, M.P., Wilczek, F.: Running inflation in the standard model. Phys. Lett. B **678**, 1 (2009)
6. Hertzberg, M.P.: On inflation with non-minimal coupling. J. High Energy Phys. **11**, 023 (2010)
7. Kaiser, D.I.: Conformal transformations with multiple scalar fields. Phys. Rev. D **81**, 084044 (2010)
8. Vilkovisky, G.A.: The unique effective action in quantum field theory. Nucl. Phys. B **234**, 125 (1984)

Appendix A
Riemann Curvature Tensor

We will often use the notation $(...)_{,\,\mu}$ and $(...)_{;\,\mu}$ for a partial and covariant derivative respectively, and the [anti]-symmetrization brackets defined here by

$$T_{[ab]} := \frac{1}{2}\left(T_{ab} - T_{ba}\right) \quad \text{and} \quad T_{(ab)} := \frac{1}{2}\left(T_{ab} + T_{ba}\right), \tag{A.1}$$

for a tensor of second rank T_{ab}, but easily generalized for tensors of arbitrary rank.

We can define the Riemannian curvature tensor in coordinate representation by the action of the commutator of two covariant derivatives on a vector field v^{α}

$$[\nabla_{\mu}, \nabla_{\nu}]\, v^{\rho} = R^{\rho}{}_{\sigma\mu\nu}\, v^{\sigma}, \tag{A.2}$$

with the explicit formula in terms of the symmetric Christoffel symbols $\Gamma^{\rho}_{\mu\nu}$

$$R^{\rho}{}_{\sigma\mu\nu} = \partial_{\mu}\Gamma^{\rho}_{\sigma\nu} - \partial_{\nu}\Gamma^{\rho}_{\sigma\mu} + \Gamma^{\rho}_{\alpha\mu}\Gamma^{\alpha}_{\sigma\nu} - \Gamma^{\rho}_{\alpha\nu}\Gamma^{\alpha}_{\sigma\mu}. \tag{A.3}$$

From this definition it is obvious that $R^{\rho}{}_{\sigma\mu\nu}$ possesses the following symmetries

$$R_{\alpha\beta\gamma\delta} = -R_{\beta\alpha\gamma\delta} = -R_{\alpha\beta\delta\gamma} = R_{\gamma\delta\alpha\beta}. \tag{A.4}$$

In addition, there is an identity for the cyclic permutation of the last three indices

$$R_{\alpha[\beta\gamma\delta]} = \frac{1}{3}(R_{\alpha\beta\gamma\delta} + R_{\alpha\delta\beta\gamma} + R_{\alpha\gamma\delta\beta}) = 0. \tag{A.5}$$

An arbitrary tensor of fourth rank in d dimension has d^4 independent components. Since a tensor is built from tensor products, we can think of $R_{\alpha\beta\gamma\delta}$ as being composed of two $(d \times d)$ matrices $A^1_{\alpha\beta}$ and $A^2_{\gamma\delta}$. From (A.4), it follows that A^1 and A^2 are antisymmetric, each having $n(n-1)/2$ independent components. Writing $R_{\alpha\beta\gamma\delta} \equiv R_{A^1A^2}$, where we have collected the index pairs $(\alpha\beta) \rightarrow A^1$ and $(\gamma\delta) \rightarrow A^2$, this corresponds to a matrix, in which each index A^1 and A^2 labels $d(d-1)/2$

C. F. Steinwachs, *Non-minimal Higgs Inflation and Frame Dependence in Cosmology*, Springer Theses, DOI: 10.1007/978-3-319-01842-3,
© Springer International Publishing Switzerland 2014

components. Taking into account the symmetry under pairwise exchange of $(\alpha\beta)$ and $(\gamma\delta)$, we can consider $R_{A^1 A^2}$ as a symmetric matrix in A^1 and A^2, having $A^1(A^1+1)/2$ independent components. Altogether, we find

$$\frac{1}{2}\left[\frac{1}{2}d\,(d-1)\right]\left\{\left[\frac{1}{2}d\,(d-1)\right]+1\right\}=\frac{1}{8}d^4-\frac{1}{4}d^3+\frac{3}{8}d^2-\frac{1}{4}d \tag{A.6}$$

independent components. We still need to figure out how many components are related by the cyclic identity (A.5) in d dimensions. We can artificially write

$$R_{\alpha\beta\gamma\delta}=\frac{1}{8}\left[R_{\alpha\beta\gamma\delta}-R_{\alpha\beta\delta\gamma}+R_{\beta\alpha\delta\gamma}-R_{\beta\alpha\gamma\delta}+(\alpha\beta)\leftrightarrow(\gamma\delta)\right] \tag{A.7}$$

and similar expressions for $R_{\alpha\gamma\delta\beta}$ and $R_{\alpha\delta\beta\gamma}$. Inserting these expressions into (A.5) leads to the condition $R_{[\alpha\beta\gamma\delta]}=0$. This totally antisymmetric object is identical zero for two identical indices but gives one additional constraint for each choice of four distinct, orderless indices, reducing the number of independent components by one respectively. In d dimensions, the number of additional constraints corresponds to "choose 4 out of d". For integers d, n there exists a product formula to calculate the binomial

$$\binom{d}{n}=\prod_{k=1}^{n}\frac{d-n+k}{k}. \tag{A.8}$$

Inserting $n=4$, we obtain

$$\binom{d}{4}=\frac{1}{24}\left[d\,(d-1)\,(d-2)\,(d-3)\right]=\frac{1}{24}d^4-\frac{1}{4}d^3+\frac{11}{24}d^2-\frac{1}{4}d. \tag{A.9}$$

Thus, the number of independent components $I_{\text{Riem}}(d)$ of $R_{\alpha\beta\gamma\delta}$ is given by

$$I_{\text{Riem}}(d):=\frac{1}{4}d\,(d-1)\left[\frac{1}{2}d\,(d-1)+1\right]-\binom{d}{4}=\frac{1}{12}d^2\,(d^2-1). \tag{A.10}$$

We consider $I_{\text{Riem}}(d)$ for different dimensions d:

$$I_{\text{Riem}}(1)=0,\quad I_{\text{Riem}}(2)=1,\quad I_{\text{Riem}}(3)=6,\quad I_{\text{Riem}}(4)=20. \tag{A.11}$$

This shows that gravity in one dimension is trivial, since there is no dynamical degree of freedom. It further shows that gravity in two dimensions can be described by the Ricci scalar R and in three dimensions by $R^{(3)}_{\mu\nu}=R^{(3)}_{\nu\mu}$. In general, the d-dimensional Ricci tensor $R_{\mu\nu}:=g^{\gamma\delta}R_{\gamma\mu\delta\nu}$ has $I_{\text{Ric}}(d):=\frac{1}{2}d\,(d+1)$ independent components. In particular $I_{\text{Ric}}(4)=10$. Thus, ten components of $R_{\alpha\beta\gamma\delta}$ not contained in $R_{\mu\nu}$ still remain. They are contained in the Weyl tensor

$$C_{\alpha\beta\gamma\delta} := R_{\alpha\beta\gamma\delta} - \frac{2}{d-2}\left(g_{\alpha[\gamma}R_{\delta]\beta} - g_{\beta[\gamma}R_{\delta]\alpha}\right) + \frac{2}{(d-1)\,(d-2)}\, g_{\alpha[\gamma}g_{\delta]\beta}\, R. \tag{A.12}$$

It follows that the Weyl tensor has (for $d > 3$)

$$I_{\text{Weyl}}(d) = I_{\text{Riem}}(d) - I_{\text{Ric}}(d) = \frac{1}{12}\, d\,(d+1)\,(d+2)\,(d-3) \tag{A.13}$$

independent components, and in particular $I_{\text{Weyl}}(4) = 10$.

Appendix B
Variations and Derivatives

B.1 Covariant Differentiation in General Relativity

The action of the metric compatible covariant derivative $\nabla_\mu\, g_{\alpha\beta} = 0$ with respect to the Christoffel symbol $\Gamma^\rho_{\mu\nu}$ on a general tensor is defined as

$$\nabla_\mu\, T^{\alpha\ldots\gamma}{}_{\beta\ldots\delta} = \partial_\mu\, T^{\alpha\ldots\gamma}{}_{\beta\ldots\delta} + \Gamma^\alpha_{\mu\rho}\, T^{\rho\ldots\gamma}{}_{\beta\ldots\delta} + \ldots + \Gamma^\gamma_{\mu\rho}\, T^{\alpha\ldots\rho}{}_{\beta\ldots\delta} \\ - \Gamma^\rho_{\mu\beta}\, T^{\alpha\ldots\gamma}{}_{\rho\ldots\delta} - \ldots - \Gamma^\rho_{\mu\delta}\, T^{\alpha\ldots\gamma}{}_{\beta\ldots\rho}, \tag{B.1}$$

with the symmetric Christoffel symbol

$$\Gamma^\alpha_{\mu\nu}(g) := \frac{1}{2}\, g^{\alpha\rho}\left(\partial_\nu\, g_{\mu\rho} + \partial_\mu\, g_{\nu\rho} - \partial_\rho\, g_{\mu\nu}\right) = \Gamma^\alpha_{\nu\mu}(g). \tag{B.2}$$

For scalar functions $\nabla_\mu\varphi = \partial_\mu\varphi$. For vector fields v^μ, a useful formula is

$$\nabla_\mu\, v^\mu = \frac{1}{\sqrt{g}}\, \partial_\mu\left(\sqrt{g}\, v^\mu\right) \quad \text{or} \quad \sqrt{g}\, \nabla_\mu\, v^\mu = \partial_\mu\left(\sqrt{g}\, v^\mu\right). \tag{B.3}$$

B.2 Functional Derivative

We use again the condensed DeWitt notation for a generalized field $\phi^i = \phi^A(x)$, introduced in Sect. 4.4. We want to emphasize that an object with two DeWitt indices like $\Omega_{ij'}$ corresponds to a *two-point function* or a generalized *bi-tensor* $\Omega_{AB}(x, x')$ in conventional notation. The primed indices like j' refer to the space-time point x'. This means we can construct objects which behave as tensors of different rank at different space-time points. A particular case of a generalized bi-tensor is a bi-scalar which has no indices $A, B, ..$ at all. A special bi-scalar, in turn, is the Dirac Delta-Distribution $\delta(x,\, x')$ which is defined in a 2ω-dimensional curved space-time with metric $g_{\mu\nu}(x)$ for a general test field $\Psi^A(x)$ by the equation

C. F. Steinwachs, *Non-minimal Higgs Inflation and Frame Dependence in Cosmology*, Springer Theses, DOI: 10.1007/978-3-319-01842-3,
© Springer International Publishing Switzerland 2014

$$\Psi^A(x) = \int \mathrm{d}^{2\omega}x' \, |g(x')|^{1/2} \, \delta(x, x') \, \Psi^A(x'), \tag{B.4}$$

For simplicity, we specify $2\omega = 4$. In the condensed notation (B.4) takes the form $\Psi^i = \delta^i{}_{j'} \Psi^{j'}$ with

$$\delta^i_{j'} = |g(x')|^{1/2} \, \delta^A{}_B \, \delta(x, x') = \delta^A{}_B \, \tilde{\delta}(x, x') \quad . \tag{B.5}$$

The quantity $\tilde{\delta}(x, x') := |g(x')|^{1/2} \delta(x, x')$ is no longer a bi-scalar, since it transforms as a scalar at the point x, but as a scalar density at the point x'. The expansion of $Z[\Psi]$ up to linear order in condensed notation is defined by

$$Z[\Psi + \zeta] =: Z[\Psi] + Z_{,i}[\Psi] \, \zeta^i, \tag{B.6}$$

with $\zeta^i := \delta \psi^i$ and

$$Z_{,i}[\Psi] \, \zeta^i = \int d^4x' \, \frac{\delta Z[\Psi(x)]}{\delta \Psi^A(x')} \, \zeta^A(x') \tag{B.7}$$

in conventional notation. Therefore, it seems obvious to define the functional derivative $Z_{,i} = \delta Z[\Psi(x)]/\delta \Psi^A(x')$ by the variation [3]

$$\lim_{\epsilon \to 0} \frac{1}{\epsilon} \left(Z[\Psi + \epsilon \zeta] - Z[\Psi] \right) =: \int d^4x' \, \frac{\delta Z[\Psi(x)]}{\delta \Psi^A(x')} \, \zeta^A(x'). \tag{B.8}$$

For the special case $Z = \mathrm{Id}$, we obtain from (B.8)

$$\zeta^A = \int d^4x' \, \frac{\delta \Psi^A(x)}{\delta \Psi^B(x')} \, \zeta^B(x') \quad \text{or} \quad \frac{\delta \Psi^A(x)}{\delta \Psi^B(x')} = \delta^A{}_B \, \tilde{\delta}(x, x'). \tag{B.9}$$

B.3 Lie Derivative

The Lie derivative $\mathcal{L}_\xi$ of an arbitrary tensor $T^{\gamma \ldots \delta}_{\alpha \ldots \beta}$ in the direction of the vector ξ^μ is a measure of the difference between $T^{\gamma' \ldots \delta'}_{\alpha' \ldots \beta'}(x')$ "dragged" along the integral curve γ_ξ of ξ^μ compared to the tensor $T^{\gamma \ldots \delta}_{\alpha \ldots \beta}(x')$ at this point. Calculating the infinitesimal flow of $T^{\gamma' \ldots \delta'}_{\alpha' \ldots \beta'}(x')$ along γ_ξ back from x' to x with the infinitesimal coordinate transformation matrix $\partial x^{\mu'}/\partial x^\nu = \delta^\mu_\nu + \epsilon \, \xi^\mu_{,\nu}$ yields

$$T^{\gamma' \ldots \delta'}_{\alpha' \ldots \beta'}(x') = T^{\gamma \ldots \delta}_{\alpha \ldots \beta} + \epsilon \left(\xi^\gamma_{,\rho} T^{\rho \ldots \delta}_{\alpha \ldots \beta} + \ldots + \xi^\delta_{,\rho} T^{\gamma \ldots \rho}_{\alpha \ldots \beta} - \xi_{\alpha,}{}^\rho T^{\gamma \ldots \delta}_{\rho \ldots \beta} - \ldots - \xi_{\beta,}{}^\rho T^{\gamma \ldots \delta}_{\alpha \ldots \rho} \right). \tag{B.10}$$

Taylor expansion of $T^{\gamma \ldots \delta}_{\alpha \ldots \beta}(x')$ around x yields

$$T^{\gamma\ldots\delta}_{\alpha\ldots\beta}(x+\epsilon\,\xi) = T^{\gamma\ldots\delta}_{\alpha\ldots\beta}(x) + \epsilon\, T^{\gamma\ldots\delta}_{\alpha\ldots\beta,\,\rho}\,\xi^{\rho}. \tag{B.11}$$

Since the right-hand sides of (B.10) and (B.11) involve only quantities at x, we can compare these two objects at x and define the Lie derivative as the difference

$$\begin{aligned}\left(\mathcal{L}_{\xi}\,T\right)^{\gamma\ldots\delta}_{\alpha\ldots\beta}(x) &:= \lim_{\epsilon\to 0}\frac{T^{\gamma\ldots\delta}_{\alpha\ldots\beta}(x') - T'^{\gamma'\ldots\delta'}_{\alpha'\ldots\beta'}(x')}{\epsilon} \\ &= T^{\gamma\ldots\delta}_{\alpha\ldots\beta,\,\rho}\,\xi^{\rho} - T^{\rho\ldots\delta}_{\alpha\ldots\beta}\,\xi^{\gamma}_{,\rho} - \ldots - T^{\gamma\ldots\rho}_{\alpha\ldots\beta}\,\xi^{\delta}_{,\rho} \\ &\quad + T^{\gamma\ldots\delta}_{\rho\ldots\beta}\,\xi_{\alpha,}^{\ \rho} + \ldots + T^{\gamma\ldots\delta}_{\alpha\ldots\rho}\,\xi_{\beta,}^{\ \rho}.\end{aligned} \tag{B.12}$$

A special case is the Lie derivative of the metric tensor. The vanishing of the Lie derivative $\left(\mathcal{L}_{\xi}\,g\right)_{\mu\nu} = 0$ in direction ξ^{μ} signalizes a symmetry of the space-time manifold and means that the vector ξ^{μ} is a generator of an isometry

$$\left(\mathcal{L}_{\xi}\,g\right)_{\mu\nu} = g_{\mu\nu,\,\rho}\,\xi^{\rho} + g_{\rho\nu}\,\xi^{\rho}_{,\,\mu} + g_{\nu\rho}\,\xi^{\rho}_{,\,\nu} = \xi_{\mu;\,\nu} + \xi_{\nu;\,\mu} = 0. \tag{B.13}$$

The vector fields ξ^{μ} obeying (B.13) are called Killing vector fields.

B.4 Variation of Metric Quantities

In the case of pure gravity, $\Psi^{i} = g_{\mu\nu}(x)$ and $\zeta^{i} = h_{\mu\nu}(x)$, we find from (B.9)

$$\frac{\delta g_{\mu\nu}(x)}{\delta g_{\alpha\beta}(x')} = \delta^{\alpha\beta}_{\mu\nu}\,\tilde{\delta}(x,x') \quad \text{with} \quad \delta^{\alpha\beta}_{\mu\nu} := \delta^{\alpha}_{(\mu}\delta^{\beta}_{\nu)} = \frac{1}{2}\left(\delta^{\alpha}_{\mu}\delta^{\beta}_{\nu} + \delta^{\alpha}_{\nu}\delta^{\beta}_{\mu}\right). \tag{B.14}$$

By using (B.8) with $F(\Psi) = F(g) = (g_{\mu\nu})^{-1}$ and $g^{\mu\nu} := (g_{\mu\nu})^{-1}$, it follows

$$\delta_{g}\,g^{\mu\nu} = -g^{\mu(\alpha}\,g^{\beta)\nu}\,h_{\alpha\beta}. \tag{B.15}$$

By using (B.8) with $F(g) = g^{1/2}$ and the identity $\det(g) = \exp\left\{\operatorname{tr}[\log(g_{\mu\nu})]\right\}$, we find

$$\delta_{g}\,g^{1/2} = \frac{1}{2}\,g^{1/2}\,g^{\alpha\beta}\,h_{\alpha\beta}. \tag{B.16}$$

By using (B.4) with an *arbitrary* test function $t(x) \neq 0$ that does not depend on $g_{\mu\nu}$ and differentiating both sides, we find

$$\int \mathrm{d}^4x \left[\frac{\delta}{\delta g_{\mu\nu}(y)}\,\tilde{\delta}(x,x')\right] t(x') = 0 \quad \text{or} \quad \frac{\delta}{\delta g_{\mu\nu}(y)}\,\tilde{\delta}(x,x') = 0. \tag{B.17}$$

Combining (B.14) with (B.17) leads to

$$\frac{\delta^2 g_{\mu\nu}(x)}{\delta g_{\alpha\beta}(y)\, \delta g_{\gamma\delta}(z)} = 0 \quad \text{or} \quad \delta_g^2\, g_{\mu\nu} = \delta_g\, h_{\mu\nu} = 0. \tag{B.18}$$

With the basic results (B.15), (B.16) and (B.18), we can construct the variation of more complicated objects. Using the fact that the operations of variation and partial differentiation commute, we calculate the variation of the Christoffel symbol

$$\begin{aligned}\delta_g \Gamma^{\rho}_{\mu\nu} = & -\frac{1}{2} g^{\rho\alpha} g^{\delta\beta} \left(\partial_\nu\, g_{\mu\delta} + \partial_\mu\, g_{\nu\delta} - \partial_\delta\, g_{\mu\nu}\right)\, h_{\alpha\beta} \\ & + \frac{1}{2}\, g^{\rho\delta} \left(\partial_\nu\, h_{\mu\delta} + \partial_\mu\, h_{\nu\delta} - \partial_\delta\, h_{\mu\nu}\right).\end{aligned} \tag{B.19}$$

Using (B.1), we rewrite the ∂_μ's acting on the $h_{\mu\nu}$'s in terms of the ∇_μ's

$$\begin{aligned}\delta_g \Gamma^{\rho}_{\mu\nu} = & -g^{\rho\alpha}\, \Gamma^{\beta}_{\mu\nu}\, h_{\alpha\beta} + \frac{1}{2}\, g^{\rho\delta} \Big(\nabla_\nu\, h_{\mu\delta} + \nabla_\mu\, h_{\nu\delta} - \nabla_\delta\, h_{\mu\nu} \\ & + \Gamma^{\epsilon}_{\nu\mu} h_{\epsilon\delta} + \Gamma^{\epsilon}_{\nu\delta} h_{\mu\epsilon} + \Gamma^{\epsilon}_{\mu\nu} h_{\epsilon\delta} + \Gamma^{\epsilon}_{\mu\rho} h_{\nu\epsilon} - \Gamma^{\epsilon}_{\mu\rho} h_{\epsilon\nu} - \Gamma^{\epsilon}_{\nu\rho} h_{\mu\epsilon}\Big) \\ = & -g^{\rho\alpha}\, \Gamma^{\beta}_{\mu\nu}\, h_{\alpha\beta} + \frac{1}{2}\, g^{\rho\delta} \Big(\nabla_\nu\, h_{\mu\delta} + \nabla_\mu\, h_{\nu\delta} - \nabla_\delta\, h_{\mu\nu} + 2\Gamma^{\epsilon}_{\nu\mu} h_{\epsilon\delta}\Big) \\ = & \frac{1}{2}\, g^{\rho\delta} \left(\nabla_\nu\, h_{\mu\delta} + \nabla_\mu\, h_{\nu\delta} - \nabla_\delta\, h_{\mu\nu}\right) = \frac{1}{2}\, g^{\rho\delta} \left(h_{\mu\delta;\,\nu} + h_{\nu\delta;\,\mu} - h_{\mu\nu;\,\delta}\right).\end{aligned} \tag{B.20}$$

Then, the variation of the Riemann tensor yields

$$\begin{aligned}\delta_g R^{\rho}{}_{\sigma\mu\nu} & = \delta_g \left(\partial_\mu \Gamma^{\rho}_{\nu\sigma} - \partial_\nu \Gamma^{\rho}_{\mu\sigma} + \Gamma^{\rho}_{\mu\alpha} \Gamma^{\alpha}_{\nu\sigma} - \Gamma^{\rho}_{\nu\alpha} \Gamma^{\alpha}_{\mu\sigma}\right) \\ & = (\delta_g \Gamma^{\rho}_{\nu\sigma})_{,\,\mu} - (\delta_g \Gamma^{\rho}_{\mu\sigma})_{,\,\nu} + (\delta_g \Gamma^{\rho}_{\mu\alpha}) \Gamma^{\alpha}_{\nu\sigma} + (\delta_g \Gamma^{\alpha}_{\nu\sigma}) \Gamma^{\rho}_{\mu\alpha} \\ & \quad - (\delta_g \Gamma^{\rho}_{\nu\alpha}) \Gamma^{\alpha}_{\mu\sigma} - (\delta_g \Gamma^{\alpha}_{\mu\sigma}) \Gamma^{\rho}_{\nu\alpha}.\end{aligned} \tag{B.21}$$

Using (B.1), we can again rewrite the partial derivative in terms of a covariant derivative plus terms proportional to the connection

$$(\delta_g \Gamma^{\rho}_{\nu\sigma})_{,\,\mu} = (\delta_g \Gamma^{\rho}_{\nu\sigma})_{;\,\mu} - (\delta_g \Gamma^{\alpha}_{\nu\sigma}) \Gamma^{\rho}_{\alpha\mu} + (\delta_g \Gamma^{\rho}_{\alpha\sigma}) \Gamma^{\alpha}_{\nu\mu} + (\delta_g \Gamma^{\rho}_{\nu\alpha}) \Gamma^{\alpha}_{\sigma\mu}. \tag{B.22}$$

Substituting this expression (and the same term interchanging μ and ν) in (B.21), we find that all terms proportional to $(\delta_g \Gamma)\Gamma$ cancel identically. Thus, the variation of the Riemann tensor yields

$$\delta_g R^{\rho}{}_{\sigma\mu\nu} = (\delta_g \Gamma^{\rho}_{\nu\sigma})_{;\,\mu} - (\delta_g \Gamma^{\rho}_{\mu\sigma})_{;\,\nu}. \tag{B.23}$$

For the variation of the Ricci tensor, we contract (B.23) over ρ and μ

$$\delta^\mu_\rho (\delta_g R^\rho{}_{\sigma\mu\nu}) = \delta_g (\delta^\mu_\rho R^\rho{}_{\sigma\mu\nu}) = \delta_g R^\rho{}_{\sigma\rho\nu} = (\delta_g \Gamma^\rho_{\nu\sigma})_{;\,\rho} - (\delta_g \Gamma^\rho_{\rho\sigma})_{;\,\nu}. \tag{B.24}$$

Substituting the variation (B.20), we obtain

$$\delta_g R_{\mu\nu} = (\delta_g \Gamma^\rho_{\nu\mu})_{;\,\rho} - (\delta_g \Gamma^\rho_{\mu\rho})_{;\,\nu} = \frac{1}{2} g^{\rho\delta} \left(h_{\mu\rho;\,\nu\delta} + h_{\nu\delta;\,\mu\rho} - h_{\mu\nu;\,\rho\delta} - h_{\rho\delta;\,\mu\nu} \right). \tag{B.25}$$

Finally, the variation of the Ricci scalar can ultimately be obtained from (B.15) and (B.25) by the relation

$$\delta_g R = (\delta_g g^{\mu\nu}) R_{\mu\nu} + g^{\mu\nu} (\delta_g R_{\mu\nu}). \tag{B.26}$$

Appendix C
Young Tableaux for $SU(3)$

The most efficient way to calculate direct products of different representations and their dimensionality is to use the technique of Young tableaux. This is an abstract symbolical notation for projecting out the symmetric and antisymmetric components of a tensor. A linear representation ρg of a group element $g \in G$ can be written as a multilinear form (a tensor). Each index runs over the dimension of the group parameters from $1, ..., N$. Everything we will work out can easily be generalized to $SU(N)$, but we are mainly interested in $SU(3)$. We have to specify several conditions and rules which define a valid diagram and its combination with other diagrams. The fundamental representation $\mathbf{3}$ is represented as a single box

$$\mathbf{3} = \square. \tag{C.1}$$

This corresponds to a tensor A^i with one contravariant index (for $SU(3)$ all indices run from $1, ..., 3$). The conjugated representation $\bar{\mathbf{3}}$ corresponds to a tensor A_i with one covariant index. To obtain the corresponding diagram, we have to raise this index with the three-dimensional epsilon tensor ϵ^{ijk}. This gives a completely antisymmetric contravariant tensor A^{jk} of second rank. Anti-symmetrization is diagrammatically indicated by vertical boxes, one box for each index and symmetrization by boxes in a row. Thus, we obtain the diagram for $\bar{\mathbf{3}}$

$$\bar{\mathbf{3}} = \begin{array}{|c|}\hline \\ \hline \\ \hline \end{array}. \tag{C.2}$$

For a valid Young diagram The numbers of boxes in a corresponding row, starting from the first row, has to be smaller or equal to the number of boxes in the previous row. For $SU(N)$ the number of vertical boxes can never exceed N because otherwise the procedure of anti-symmetrization of $N + 1$ indices, each index running from $1, ..., N$ would result in zero. We write tensor product between two fundamental representations $\mathbf{3}$ as

C. F. Steinwachs, *Non-minimal Higgs Inflation and Frame Dependence in Cosmology*, Springer Theses, DOI: 10.1007/978-3-319-01842-3,
© Springer International Publishing Switzerland 2014

$$\mathbf{3}\otimes\mathbf{3} = \square \otimes \boxed{a} = \square\boxed{a} \oplus \begin{array}{|c|}\hline \\ \hline a \\ \hline\end{array} = \mathbf{6}\oplus\bar{\mathbf{3}}. \tag{C.3}$$

The algorithm of the calculation can be summarized in different steps:

1. First, we fill all boxes in the first row of the right diagram with a, all boxes in the second row of the right diagram with b etc., until we have filled all rows of the second diagram in that way.
2. Second, we take all boxes filled with a and attach them to the first diagram in all possible ways to construct a valid Young diagram. After having attached the blocks containing the a's, we do the same procedure successively for the b's, the c's etc., until all boxes of the original right diagram are attached to the left diagram. This will result in a big number of different diagrams (two diagrams are considered equal only if they have the same shape and the same a's, b', etc. in the same boxes).
3. Third, there is one more rule which guarantees the avoidance of double counting. We will write a sequence of letters for each diagram in the following way. We start with the last row and write down all a's, b's etc. in the order they appear reading the row from left to right. We continue writing the sequence generated by the letters from the last row by attaching a sequence of letters generated from the above row in the same way. We do this for all rows ending with the first row. We finally have a string of letters for each diagram. We keep only those diagrams in which for an arbitrary position in the string the number of a's right to that position is bigger or equal than the number of b's right to this position. We require the same for the number of b's with respect to the number of c's and so on. We throw away all diagrams which do not satisfy this constraint. The resulting diagrams correspond to the decomposition into a direct sum of irreducible representations.

We finally have to estimate the dimensionality of the representation corresponding to a diagram. We do this again in several steps.

1. First, we duplicate the corresponding diagram.
2. In one diagram we fill the boxes according to the following rule: Begin with the box of the first row in the right corner. Insert the dimension of the Group (in case of SU(3), a 3). Fill the remaining boxes in the same row by the value of the foregoing box and add 1. Fill the first box of the second row with the value of the box above it and subtract one. Fill the remaining boxes in the same row according to the same rule like for the first row. Proceed in this way until you have filled all boxes. Calculate the product of the numbers contained in all boxes of the diagram. Take the result as the value of the numerator of the number for the dimensionality.
3. Fill the copy of the diagram accordingly to the following rule: For each box count the number of boxes to the right and the number of boxes below and add one. Fill this value into the box. Fill all boxes in this way and calculate again the product of the numbers of all boxes. The result is the denominator of the dimensionality.

4. The fraction of these two numbers gives the dimensionality of the representation corresponding to each diagram in the direct sum.

The dimensionality for the simple example above can be calculated by

$$\frac{\begin{array}{|c|c|}\hline 3&4\\\hline\end{array}}{\begin{array}{|c|c|}\hline 2&1\\\hline\end{array}} = \frac{3\cdot 4}{2\cdot 1} = \frac{12}{2} = 6. \tag{C.4}$$

Now we calculate the product between the fundamental representation and its conjugated. It is easier to reverse the order of multiplication because the right-hand side has fewer boxes, reducing the amount of calculations.

$$\mathbf{\bar{3}}\otimes\mathbf{3} = \begin{array}{|c|}\hline \;\\\hline \;\\\hline\end{array} \otimes \begin{array}{|c|}\hline a\\\hline\end{array} = \begin{array}{|c|c|}\hline \; & a\\\hline \; & \multicolumn{1}{c}{}\\\cline{1-1}\end{array} \oplus \begin{array}{|c|}\hline \;\\\hline \;\\\hline a\\\hline\end{array} = \mathbf{8}\oplus\mathbf{1}, \tag{C.5}$$

where we have calculated the dimensionality by

$$\frac{\begin{array}{|c|c|}\hline 3&4\\\hline 2&\multicolumn{1}{c}{}\\\cline{1-1}\end{array}}{\begin{array}{|c|c|}\hline 3&1\\\hline 1&\multicolumn{1}{c}{}\\\cline{1-1}\end{array}} = \frac{4\cdot 3\cdot 2}{3\cdot 1\cdot 1} = 8, \quad \frac{\begin{array}{|c|}\hline 3\\\hline 2\\\hline 1\\\hline\end{array}}{\begin{array}{|c|}\hline 3\\\hline 2\\\hline 1\\\hline\end{array}} = 1. \tag{C.6}$$

The product of three quarks can be calculated by

$$\begin{aligned}\mathbf{3}\otimes\mathbf{3}\otimes\mathbf{3} &= (\mathbf{6}\oplus\mathbf{\bar{3}})\otimes\mathbf{3} = \left(\begin{array}{|c|c|}\hline \;&\;\\\hline\end{array} \oplus \begin{array}{|c|}\hline \;\\\hline \;\\\hline\end{array}\right)\otimes\begin{array}{|c|}\hline a\\\hline\end{array}\\ &= \begin{array}{|c|c|c|}\hline \;&\;&a\\\hline\end{array} \oplus \begin{array}{|c|c|}\hline \;&\;\\\hline a&\multicolumn{1}{c}{}\\\cline{1-1}\end{array} \oplus \begin{array}{|c|c|}\hline \;&a\\\hline \;&\multicolumn{1}{c}{}\\\cline{1-1}\end{array} \oplus \begin{array}{|c|}\hline \;\\\hline \;\\\hline a\\\hline\end{array} = \mathbf{10}\oplus\mathbf{8}\oplus\mathbf{8}\oplus\mathbf{1}\end{aligned} \tag{C.7}$$

Finally, we have to figure out how the product of two gluons transforms. Since the gluon itself transforms in the adjoint representation **8** we have to calculate

$$\begin{aligned}\mathbf{8}\otimes\mathbf{8} &= \begin{array}{|c|c|}\hline \;&\;\\\hline \;&\multicolumn{1}{c}{}\\\cline{1-1}\end{array} \otimes \begin{array}{|c|c|}\hline a&a\\\hline b&\multicolumn{1}{c}{}\\\cline{1-1}\end{array} = \left(\begin{array}{|c|c|c|c|}\hline \;&\;&a&a\\\hline \;&\multicolumn{3}{c}{}\\\cline{1-1}\end{array} \oplus \begin{array}{|c|c|c|}\hline \;&\;&a\\\hline \;&a&\multicolumn{1}{c}{}\\\cline{1-2}\end{array} \oplus \begin{array}{|c|c|c|}\hline \;&\;&a\\\hline \;&\multicolumn{2}{c}{}\\\cline{1-1} a&\multicolumn{2}{c}{}\\\cline{1-1}\end{array} \oplus \begin{array}{|c|c|}\hline \;&\;\\\hline \;&a\\\hline a&\multicolumn{1}{c}{}\\\cline{1-1}\end{array} \oplus \cancel{\begin{array}{|c|c|}\hline \;&\;\\\hline \;&\multicolumn{1}{c}{}\\\cline{1-1} a&\multicolumn{1}{c}{}\\\cline{1-1} a&\multicolumn{1}{c}{}\\\cline{1-1}\end{array}}\right)\otimes\begin{array}{|c|}\hline b\\\hline\end{array}\\ &= \cancel{\begin{array}{|c|c|c|c|c|}\hline \;&\;&a&a&b\\\hline \;&\multicolumn{4}{c}{}\\\cline{1-1}\end{array}} \oplus \begin{array}{|c|c|c|c|}\hline \;&\;&a&a\\\hline \;&b&\multicolumn{2}{c}{}\\\cline{1-2}\end{array} \oplus \begin{array}{|c|c|c|c|}\hline \;&\;&a&a\\\hline \;&\multicolumn{3}{c}{}\\\cline{1-1} b&\multicolumn{3}{c}{}\\\cline{1-1}\end{array} \oplus \cancel{\begin{array}{|c|c|c|c|}\hline \;&\;&a&b\\\hline \;&a&\multicolumn{2}{c}{}\\\cline{1-2}\end{array}} \oplus \begin{array}{|c|c|c|}\hline \;&\;&a\\\hline \;&a&b\\\hline\end{array}\end{aligned}$$

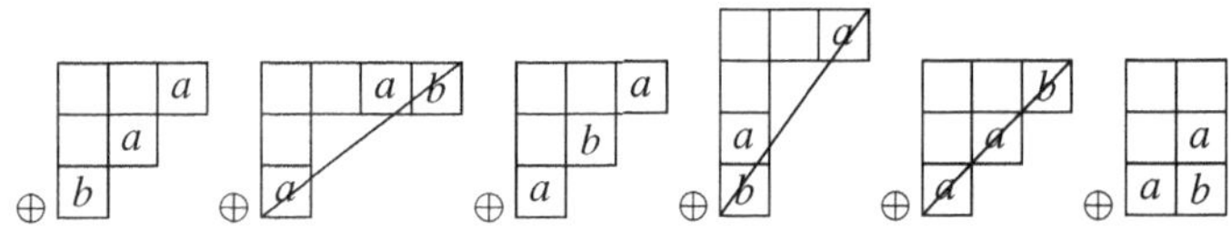

$$= \mathbf{27} \oplus \mathbf{10} \oplus \mathbf{10} \oplus \mathbf{8} \oplus \mathbf{8} \oplus \mathbf{1} \tag{C.8}$$

We have calculated all relevant products of representations of $SU(3)$ needed for the application in the Standard Model.

Appendix D
Synge's World Function

The notation "world function" dates back to the Irish relativist Synge [7] and describes a non-local extension $\sigma(x, x')$ of the space-time metric $g_{\mu\nu}(x)$. It is defined by connecting the two space-time points $\{x = z(s_1), x' = z(s_0)\} \in \mathcal{M}, \quad s \in \mathbb{R}$ by the unique geodesic $\gamma \equiv z(s)$ parametrized through s

$$\sigma(x, x') := \frac{1}{2}(s_1 - s_0) \int_{s_0}^{s_1} g_{\mu\nu}(z(s))\, t^\mu t^\nu \, \mathrm{d}s. \tag{D.1}$$

The restriction to unique geodesics γ excludes conjugate points on γ. This is always satisfied for sufficient small distances between x and x' and it is necessary in order to define the derivative, the normalized tangential vector

$$t^\mu := \frac{\mathrm{d}z^\mu(s)}{\mathrm{d}s} \in \mathrm{T}\mathcal{M}. \tag{D.2}$$

The world function is a *bi-scalar* that transforms *independently* at x and x' as a scalar. The tangential vectors are parallelly transported along the geodesic γ

$$\frac{\mathrm{D}t^\mu}{\mathrm{D}s} := \frac{\mathrm{d}^2 z^\mu(s)}{\mathrm{d}s^2} + \Gamma^\mu_{\alpha\beta} \frac{\mathrm{d}z^\alpha(s)}{\mathrm{d}s} \frac{\mathrm{d}z^\beta(s)}{\mathrm{d}s} = 0. \tag{D.3}$$

Thus, the scalar product and in particular the norm are conserved on γ

$$g_{\mu\nu}(z)\, t^\mu t^\nu := N = \text{const.} \tag{D.4}$$

This allows us to write the world function with $\Delta s := (s_1 - s_0)$ as

$$\sigma(x, x') = \frac{1}{2}(s_1 - s_0)^2\, g_{\mu\nu}(z)\, t^\mu t^\nu = \frac{N}{2}\, (\Delta s)^2. \tag{D.5}$$

C. F. Steinwachs, *Non-minimal Higgs Inflation and Frame Dependence in Cosmology*, Springer Theses, DOI: 10.1007/978-3-319-01842-3,
© Springer International Publishing Switzerland 2014

The sign of N fixes the causal nature of the geodesic, i.e. $N = (-1, +1, 0)$ corresponds to γ =(time-like, space-like, light-like). From (D.5) it is clear that $\sigma(x, x') = \sigma(x', x)$ is symmetric under the exchange $x \leftrightarrow x'$ For Minkowski space-time ($g_{\mu\nu} = \eta_{\mu\nu}$), geodesics are straight lines and (D.1) reduces to

$$\sigma(x,\, x') = \frac{1}{2}\eta_{\mu\nu}(x^\mu - x'^\mu)(x^\nu - x'^\nu). \tag{D.6}$$

Without loss of generality we can fix one point, say $x' = 0$ and regard $\sigma(x,\, 0)$ as an ordinary scalar function at x. The light cone structure emerges from

$$\sigma(x, 0) = -\frac{(x^0)^2}{(\sqrt{2})^2} + \frac{(x^1)^2}{(\sqrt{2})^2} + \frac{(x^2)^2}{(\sqrt{2})^2} + \frac{(x^3)^2}{(\sqrt{2})^2} = 0. \tag{D.7}$$

D.1 Derivatives of the World Function

By repeated differentiation of the world function, we can generate *bi-tensors* of different rank at x and x'. The covariant derivative of $\sigma(x,\, x')$ at the point x is defined as $\nabla_\alpha\sigma := \sigma_{;\,\alpha} = \sigma_{,\,\alpha}$. Correspondingly, the derivative of σ at the point x' is defined as $\sigma_{;\,\alpha'}$. The world function and its derivatives transform independently at x and x'. For example, this means $\sigma_{;\,\alpha'\beta'\gamma}$ transforms like a tensor of second rank at x' and like a covariant vector at x. This implies that the order of primed and unprimed indices can be changed arbitrarily for any general bi-tensor $A_{...;\,\beta'\alpha} = A_{...;\,\alpha\,\beta'}$. In particular, we can freely permute primed and unprimed derivatives of σ as long as the order among the primed and the order among the unprimed indices remain maintained, e.g.

$$\sigma_{;\,\alpha'\beta'\gamma} = \sigma_{;\,\alpha'\gamma\beta'} = \sigma_{;\,\gamma\alpha'\beta'} \neq \sigma_{;\,\beta'\alpha'\gamma}. \tag{D.8}$$

For an explicit expression of the first derivative of σ at x, we fix the point x' and vary x, i.e. $\delta z(s_0) := \delta x' = 0$ and $\delta z(s_1) := \delta x$, see [4]

$$\delta\sigma = \sigma(x + \delta x,\, x') - \sigma(x,\, x') = \Delta s\, g_{\mu\nu}\, t^\mu\, \delta x^\nu. \tag{D.9}$$

Comparison with the result of the chain rule $\delta\sigma = \sigma_{,\,\mu}\delta x^\mu$ yields

$$\sigma_{;\,\mu} = \Delta s\, (g_{\mu\nu}\, t^\nu)|_{s=s_1} \quad \text{or} \quad \sigma^{;\,\mu} = \Delta s\, t^\mu, \tag{D.10}$$

which shows that $\sigma_{;\,\mu}$ is proportional to the tangential vector t_μ with length of the geodesic distance between x and x' and pointing in direction $x' \to x$. The derivative of σ at x' can be obtained in an analogue manner

$$\sigma_{;\,\mu'} = -\Delta s\, (g_{\mu'\nu'}\, t^{\nu'})|_{s=s_0} \quad \text{or} \quad \sigma^{;\,\mu'} = -\Delta s\, t^{\mu'}. \tag{D.11}$$

It has the same properties as (D.10) but points in the opposite direction $x \to x'$. With (D.10) we can derive an important identity for the norm of $\sigma_{;\,\mu}$

$$g^{\mu\nu}\,\sigma_{,\mu}\sigma_{;\,\nu} = g_{\mu\nu}\,(\Delta s)^2\,t^{\mu}t^{\nu} = (\Delta s)^2\,N = 2\sigma \quad \text{or} \quad \sigma = \frac{1}{2}\sigma_{;\,\mu}\sigma^{;\,\mu}. \tag{D.12}$$

D.2 Densities and Determinants

For any ordinary tensor $A^{b_1\ldots b_n}_{a_1\ldots a_n}$ a tensor density $\mathcal{A}^{b_1\ldots b_n}_{a_1\ldots a_n}$ of weight w is defined by its transformation behaviour under general coordinate transformations $x^{\mu} \to \tilde{x}^{\mu}$

$$\tilde{\mathcal{A}}^{a_1\ldots a_n}_{b_1\ldots b_n} = \det\left(\frac{\partial x_c}{\partial \tilde{x}^d}\right)^w \frac{\partial \tilde{x}^{a_1}}{\partial x^{a_1'}}\cdots\frac{\partial \tilde{x}^{a_n}}{\partial x^{a_n'}}\frac{\partial x^{b_1'}}{\partial \tilde{x}^{b_1}}\cdots\frac{\partial x^{b_n'}}{\partial \tilde{x}^{b_n}}\mathcal{A}^{a_1'\ldots a_n'}_{b_1'\ldots b_n'}, \tag{D.13}$$

where $g = \det(g_{\mu\nu}(x)) = \frac{\partial(\tilde{x}_{\mu})}{\partial(x^{\nu})}$ is the Jacobian that takes care of a coordinate transformation from x^{μ} to $\tilde{x}^{\nu}$. Since we deal with bi-tensors, we can generalize this for the transformation of the two variables (x^{μ}, x'^{ρ}) that characterize the geodesic γ by its endpoints to the variables $(\sigma_{;\,\nu'}, x'_{\sigma})$ that characterize γ by one endpoint and the tangent at this point. This Jacobian is given by

$$\left|\frac{\partial\,(\sigma_{;\,\nu'},\, x'_{\sigma})}{\partial\,(x^{\mu},\, x'^{\rho})}\right| = \left|\frac{\partial\,(\sigma_{;\,\nu'})}{\partial\,(x^{\mu})}\right| = \det(\sigma_{;\,\mu\nu'}) = \det(-D_{\mu\nu'}). \tag{D.14}$$

Following [2], we have defined the matrix $D_{\mu\nu'} := -\sigma_{;\,\mu\nu'}$ and its determinant as $D(x, x') := -\det(D_{\mu\nu'})$. The convention of the minus sign takes care of the Lorentzian signature of space-time. The square root of this determinant can be thought of as the generalize bi-volume factor $D^{1/2}(x, x')$ replacing $g^{1/2}(x)$. Differentiating (D.12) twice with respect to x^{μ} and x'^{ν}, we obtain the relation

$$D_{\mu\nu'} = \sigma_{;}{}^{\rho}{}_{\mu}\,D_{\rho\nu'} + \sigma_{;}{}^{\rho}\,D_{\mu\nu';\,\rho}. \tag{D.15}$$

Multiplication with $(D^{-1})^{\mu\nu'}$, and using Jacobi's formula yields

$$D^{-1}\,(D\,\sigma_{;}{}^{\rho})_{;\,\rho} = d. \tag{D.16}$$

with $d = 4$ in four space-time dimensions. Finally, we introduce the Van-Fleck determinate $\Delta(x, x')$ by

$$\Delta := g^{-1/2}\,D\,g'^{-1/2}. \tag{D.17}$$

Since $(g^{1/2})_{;\,\mu} = 0$, we obtain relation (D.16) in terms of Δ

$$\Delta^{-1}\,(\Delta\,\sigma_{;}{}^{\rho})_{;\,\rho} = d. \tag{D.18}$$

D.3 The Parallel Propagator

We further need the geometrical concept of parallel transport. Following [4], we define the parallel propagator as the tensor product of two tetrads. We introduce an orthonormal tetrad $e_I^{\mu}(z)$ which satisfies

$$g_{\mu\nu}\, e_I^{\mu}\, e_J^{\nu} = \eta_{IJ}, \tag{D.19}$$

with $(I, J, ... = 0, 1, 2, 3)$ being Lorentz-indices of the frame with the Minkowski metric $\eta_{IJ} = \mathrm{diag}(-1, 1, 1, 1)$. We define the dual triad

$$e_{\mu}^{I} := \eta^{IJ}\, g_{\mu\nu}\, e_J^{\nu} \tag{D.20}$$

and via the completeness relation

$$g^{\mu\nu} = \eta^{IJ}\, e_I^{\mu} e_J^{\nu} \quad \text{it follows} \quad e_I^{\mu} e_{\nu}^{I} = \delta_{\nu}^{\mu} \quad \text{and} \quad e_I^{\mu} e_{\mu}^{J} = \delta_I^{J}. \tag{D.21}$$

The tetrad basis $e_I^{\mu}(z)$ and its dual $e_{\mu}^{I}(z)$ are parallelly transported along $\gamma = z(s)$

$$\frac{\mathrm{D}e_I^{\mu}}{\mathrm{D}s} = 0 \quad \text{and} \quad \frac{\mathrm{D}e_{\mu}^{I}}{\mathrm{D}s} = 0. \tag{D.22}$$

We can expand each contravariant vector field $v^{\nu}(z)$ into this orthonormal basis

$$v^{\mu} = v^{I}\, e_I^{\mu}, \tag{D.23}$$

whereas the Lorentz components of v can be expanded in the dual basis

$$v^{I} = v^{\nu} e_{\nu}^{I}. \tag{D.24}$$

If the vector v^{μ} is parallelly transported along γ, then the frame coefficients v^{I} are constant since e_I^{μ} is also parallelly transported along γ. Since the v^{I} are constant, we can express them as $v^{I} = v^{\nu'} e_{\nu'}^{I}$ where the vector $v^{\nu'}$ and the tetrad $e_{\nu'}^{I}$ are evaluated at x'. Putting all this together, v^{μ} at x can be expressed by

$$v^{\mu}(x) = (v^{\nu'}\, e_{\nu'}^{I})\, e_I^{\mu} \quad \text{or} \quad v^{\mu} = g^{\mu}{}_{\mu'}\, v^{\mu'}, \tag{D.25}$$

with $g^{\mu}{}_{\mu'}$ being the parallel propagator $g^{\mu}{}_{\mu'} := e_I^{\mu}\, e_{\mu'}^{I}$. It is a bi-vector that transports the vector $v^{\mu'}$ from the point $x' = z(s_1)$ to the vector v^{μ} at the point $x = z(s_0)$ along the unique geodesic $\gamma = z(s)$. The inverse parallel transporter from x to x' is given by $g^{\mu'}{}_{\mu} = e_I^{\mu'}\, e_{\mu}^{I}$. It is clear that the parallel transporter $g_{\mu\nu'}$ satisfies the boundary condition

$$\lim_{x'\to x} g^{\mu}{}_{\nu'} = \delta_{\nu}^{\mu} \quad \text{or} \quad \lim_{x'\to x} g_{\mu\nu'} = g_{\mu\nu}. \tag{D.26}$$

The parallel transport easily generalizes to tensors of arbitrary rank $T^{\mu\ldots\nu}_{\rho\ldots\sigma}$. Since a tensor is a multilinear form we need one (inverse) parallel propagator for each (covariant) contravariant index to transport it along γ from x' to x.

$$T^{\mu\ldots\nu}_{\rho\ldots\sigma} = g^{\mu}{}_{\mu'}\cdots g^{\nu}{}_{\nu'}\, g^{\rho'}{}_{\rho}\cdots g^{\sigma'}{}_{\sigma}\, T^{\mu'\ldots\nu'}_{\rho'\ldots\sigma'}\,. \tag{D.27}$$

Since the tetrads e^{μ}_{I} are parallelly propagated along γ this means

$$\frac{\mathrm{D}e^{\mu}_{I}(z(s))}{\mathrm{D}s} = e^{\mu}_{I\,;\,\nu}\frac{\mathrm{d}z^{\nu}(s)}{\mathrm{d}s} = \Delta s\; e^{\mu}_{I\,;\,\nu}\sigma^{\nu} = 0. \tag{D.28}$$

From (D.28) it follows $e^{\mu}_{I\,;\,\nu}\,\sigma_{;}{}^{\nu} = 0$ and $e^{I}_{\mu'\,;\,\nu}\,\sigma_{;}{}^{\nu} = 0$. This, in turn, implies

$$g^{\mu}{}_{\mu'\,;\,\nu}\,\sigma_{;}{}^{\nu} = 0 \quad\text{and}\quad g^{\mu}{}_{\mu'\,;\,\nu'}\,\sigma_{;}{}^{\nu'} = 0. \tag{D.29}$$

Since $\sigma_{;\,\mu} \propto -t_{\mu}$, it is automatically parallelly transported along γ and thus

$$\sigma_{;\,\mu} = -g^{\mu'}{}_{\mu}\,\sigma_{;\,\mu'} \quad\text{and}\quad \sigma_{;\,\mu'} = -g^{\mu}{}_{\mu'}\,\sigma_{;\,\mu}. \tag{D.30}$$

From the determinants of the tetrad and its dual

$$\det(e^{\mu}_{I}) = g^{-1/2} \quad\text{and}\quad \det(e^{I}_{\mu}) = g^{1/2}, \tag{D.31}$$

we find for the determinant of the parallel propagator

$$\begin{aligned}\det(-g_{\mu\nu'}) &= \det(-g_{\mu\nu}g^{\nu}{}_{\nu'}) = \det(-g_{\mu\nu}e^{\nu}_{I}e^{I}_{\nu'})\\ &= \det(-g_{\mu\nu})\det(e^{\nu}_{I})\det(e^{I}_{\nu'}) = g^{1/2}g'^{1/2}.\end{aligned} \tag{D.32}$$

D.4 Field Curvature

We assume that we have a set of fields collected in the generalized field $\phi^{A}(x)$. Each component of the generalized field consists of some fundamental field present in the action of the theory together forming the field space $\mathcal{C}$. If we consider $\mathcal{C}$ as a manifold, $\phi^{A}(x)$ corresponds to a coordinate of a point in this manifold. The fields collected in $\phi^{A}(x)$ do have their own symmetries. In order to define invariant differentiation, we need to introduce a covariant derivative. We introduce an abbreviation by suppressing the capital index and write instead $\phi := \phi^{A}$ and a hat for operators $\hat{O} := O_{A}{}^{B}$. The law of covariant differentiation is not commutative but defines the field (bundle) curvature $\hat{\mathcal{R}}_{\mu\nu}$

$$[\hat{\nabla}_{\mu},\, \hat{\nabla}_{\nu}]\,\phi =: \hat{\mathcal{R}}_{\mu\nu}\,\phi. \tag{D.33}$$

To clarify the role of ϕ, we will consider two examples. We assume $\phi = \phi^A \equiv q^a$ consists of a single field q^a with an internal gauge symmetry described by the Lie group G and the group index a. Covariant differentiation of q^a means

$$\hat{\nabla}_\mu q^a \equiv \partial_\mu q^a + A_\mu q^a \tag{D.34}$$

with the gauge vector field $A_\mu := A^a_\mu G_a$ and the generators G_a of the group G. The commutator of covariant derivatives then yields

$$[\hat{\nabla}_\mu, \hat{\nabla}_\nu]\phi \equiv (A^a_{\mu,\,\nu} - A^a_{\nu,\,\mu} + c^a_{\ bc}\, A^b_\mu\, A^c_\nu)\,\phi_a = F^a_{\mu\nu}\,\phi_a \tag{D.35}$$

with the field strength tensor

$$\hat{\mathcal{R}}_{\mu\nu} \equiv F^a_{\mu\nu} = A^a_{\mu,\,\nu} - A^a_{\nu,\,\mu} + c^a_{bc}\, A^b_\mu\, A^c_\nu \tag{D.36}$$

and the structure constants c^a_{bc} of the group G. If the field ϕ instead only consists of a symmetric tensor field $\phi \equiv q_{\mu\nu} = q_{\nu\mu}$ with ordinary space-time indices, the covariant derivative is defined with respect to the usual Christoffel symbol $\hat{\nabla}_\mu q_{\alpha\beta} \equiv \nabla_\mu q_{\alpha\beta} = \partial_\mu q_{\alpha\beta} - \Gamma^\gamma_{\ \alpha\mu} q_{\gamma\beta} - \Gamma^\gamma_{\ \beta\mu} q_{\alpha\gamma}$ and we obtain

$$[\hat{\nabla}_\mu, \hat{\nabla}_\nu]\phi \equiv (\nabla_\mu\nabla_\nu - \nabla_\nu\nabla_\mu)\, q_{\alpha\beta} = -R^\gamma_{\ \alpha\mu\nu} q_{\gamma\beta} - R^\gamma_{\ \beta\mu\nu} q_{\alpha\gamma} \tag{D.37}$$

where we have used $u_{\alpha;\,\beta\gamma} - u_{\alpha;\,\gamma\beta} = R^\delta_{\ \alpha\beta\gamma} u_\delta$ and have written $q_{\alpha\beta}$ without loss of generality as the tensor product $q_{\alpha\beta} := u_\alpha \otimes v_\beta$. This can also be written as

$$[\hat{\nabla}_\mu, \hat{\nabla}_\nu]\phi \equiv (\nabla_\mu\nabla_\nu - \nabla_\nu\nabla_\mu)\, q_{\alpha\beta} = -2\,\delta^{(\gamma}_{(\alpha}\, R^{\delta)}_{\ \beta)\,\mu\nu}\, q_{\gamma\delta} \tag{D.38}$$

and shows that in this case the field curvature is given by the operator

$$\hat{\mathcal{R}}_{\mu\nu} \equiv -2\,\delta^{(\gamma}_{(\alpha}\, R^{\delta)}_{\ \beta)\,\mu\nu}, \tag{D.39}$$

acting on $q_{\gamma\delta}$. In general, ϕ is a collection of many different fields. We can therefore define a parallel displacement bi-matrix $\hat{I}(x,\, x')$, which parallelly transports $\phi(x)$ to $\phi(x')$ along the geodesic γ. It is a matrix in field space and from the general properties of a parallel propagator, it follows as in (D.29) and (D.26) that $\hat{I}(x,\, x')$ must satisfy the relations

$$\sigma^{;\,\mu}\,\hat{I}_{;\,\mu} = 0 \quad , \quad \sigma^{;\,\mu'}\,\hat{I}_{;\,\mu'} = 0 \quad \text{and} \quad \lim_{x\to x'} \hat{I}(x,\, x') = \delta^A_B(x) := \hat{\mathbf{1}}, \tag{D.40}$$

where $\hat{\mathbf{1}}$ is the unit matrix at the point x.

D.5 Coincidence Limits

Since $\sigma(x, x')$ is a bi-scalar and derivatives of σ are bi-tensors, we are interested in the value of these objects in the coincidence limit $x \to x'$. For any bi-tensor $A^{\nu\dots}_{\mu\dots}(x, x')$ we define the limit $x \to x'$ by the symbolic bracket-notation

$$[\, A^{\dots}_{\dots}(x, x')\,] := \lim_{x' \to x} A^{\dots}_{\dots}(x, x'). \tag{D.41}$$

The result of this limit is an ordinary tensor at x'. There is a relation between coincidence limits of primed and unprimed indices, denoted Synge's rule [4]

$$[A_{\dots\alpha'}] = [A_{\dots}]_{;\,\alpha'} - [A_{\dots\alpha}]. \tag{D.42}$$

By repeated differentiation of (D.12) we have a recursive algorithm to systematically calculate the coincidence limit of higher derivatives $[\sigma_{;\,\alpha_1,\dots,\alpha_N}]$.

$$\sigma = \frac{1}{2}\sigma^{;\,\lambda}\sigma_{,\lambda}, \tag{D.43}$$

$$\sigma_{;\,\alpha} = \sigma_{;\ \ \alpha}^{\ \lambda}\sigma_{\lambda}, \tag{D.44}$$

$$\sigma_{;\,\alpha\beta} = \sigma_{;\ \ \alpha\beta}^{\ \lambda}\sigma_{;\,\lambda} + \sigma_{;\ \ \alpha}^{\ \lambda}\sigma_{;\,\lambda\beta}, \tag{D.45}$$

$$\sigma_{;\,\alpha\beta\gamma} = \sigma_{;\ \ \alpha\beta\gamma}^{\ \lambda}\sigma_{;\,\lambda} + \sigma_{;\ \ \alpha\beta}^{\ \lambda}\sigma_{;\,\lambda\gamma} + \sigma_{;\ \ \alpha\gamma}^{\ \lambda}\sigma_{;\,\lambda\beta} + \sigma_{;\ \ \alpha}^{\ \lambda}\sigma_{;\,\lambda\beta\gamma}, \tag{D.46}$$

$$\begin{aligned}\sigma_{;\,\alpha\beta\gamma\delta} &= \sigma_{;\ \ \alpha\beta\gamma\delta}^{\ \lambda}\sigma_{;\,\lambda} + \sigma_{;\ \ \alpha\beta\gamma}^{\ \lambda}\sigma_{;\,\lambda\delta} + \sigma_{;\ \ \alpha\beta\delta}^{\ \lambda}\sigma_{;\,\lambda\gamma} + \sigma_{;\ \ \alpha\beta}^{\ \lambda}\sigma_{;\,\lambda\gamma\delta}, \\ &\quad + \sigma_{;\ \ \alpha\gamma\delta}^{\ \lambda}\sigma_{;\,\lambda\beta} + \sigma_{;\ \ \alpha\gamma}^{\ \lambda}\sigma_{;\,\lambda\beta\delta} + \sigma_{;\ \ \alpha\delta}^{\ \lambda}\sigma_{;\,\lambda\beta\gamma} + \sigma_{;\ \ \alpha}^{\ \lambda}\sigma_{;\,\lambda\beta\gamma\delta} \\ &\ \ \vdots\end{aligned} \tag{D.47}$$

From (D.5) and (D.10), it follows directly that $[\sigma] = 0$ and $[\sigma_{;\,\mu}] = 0$. From (D.45) it is clear that $[\sigma_{;\,\mu\nu}] = g_{\mu\nu}$. The remaining coincidence limits for (D.46) and (D.47) are obtained recursively. By commuting covariant derivatives we generate curvature expressions. Using the symmetries of the Riemann tensor, we obtain

$$[\,\sigma_{;\,\alpha\beta\gamma}\,] = 0 \quad \text{and} \quad [\,\sigma_{;\,\alpha\beta\gamma\delta}\,] = \frac{1}{3}\left(R_{\alpha\gamma\beta\delta} + R_{\alpha\delta\beta\gamma}\right). \tag{D.48}$$

For further application we will also need two additional coincidence limits. Contracting indices in (D.47) and differentiating twice, we obtain after some algebra

$$\left[\sigma_{;\ \mu\ \nu\ \alpha}^{\ \ \mu\ \ \nu}\right] = R_{;\,\alpha}, \tag{D.49}$$

$$\left[\sigma_{;\ \mu\ \nu\ \alpha}^{\ \ \mu\ \ \nu\ \ \alpha}\right] = \frac{8}{5}\,R_{;\ \ \alpha}^{\ \alpha} + \frac{4}{15}\,R^{\alpha\beta}R_{\alpha\beta} - \frac{4}{15}\,R^{\alpha\beta\gamma\delta}R_{\alpha\beta\gamma\delta}. \tag{D.50}$$

We are ultimately interested in the coincidence limit of the coefficient matrix $\hat{a}_2$. As shown in Chap. 4, $\hat{a}_2$ can be calculated by the recursion relation (4.109). The coincidence limits can then be again obtained recursively. In order to calculate $[\,\hat{a}_1\,]$ and $[\,\hat{a}_2\,]$, we must calculate several further coincidence limits. It is easy to see that

$$[\,D_{\mu\nu'}\,] = g_{\mu\nu}, \quad [\,D\,] = g \quad \text{and} \quad [\,\Delta\,] = 1. \tag{D.51}$$

For the coincidence limits of higher derivatives of the field parallel propagator matrix $\hat{I}$, we differentiate (D.40) and use (D.33) in order to obtain

$$\left[\,\hat{I}_{;\,\mu}\,\right] = 0, \quad \left[\,\hat{I}_{;\,\mu\nu}\,\right] = -\frac{1}{2}\,\hat{\mathcal{R}}_{\mu\nu}, \tag{D.52}$$

$$\left[\,\hat{I}_{;\,\mu\;\;\nu}^{\;\;\;\mu}\,\right] = \frac{1}{3}\,\hat{\mathcal{R}}_{\mu\nu;}^{\;\;\;\;\;\mu}, \quad \left[\,\hat{I}_{;\,\mu\;\;\nu}^{\;\;\;\mu\;\;\nu}\,\right] = \frac{1}{2}\,\hat{\mathcal{R}}_{\mu\nu}\hat{\mathcal{R}}^{\mu\nu}. \tag{D.53}$$

Proceeding in a similar manner, we differentiate (D.18) in order to obtain recursively the coincidence limits of the derivatives of the Van-Fleck determinant

$$\left[\,\Delta_{;\,\mu}^{1/2}\,\right] = 0, \quad \left[\,\Delta_{;\,\mu\nu}^{1/2}\,\right] = -\frac{1}{6}\,R_{\mu\nu}, \tag{D.54}$$

$$\left[\,\Delta_{;\,\mu\nu}^{-1/2}\,\right] = \frac{1}{6}\,R_{\mu\nu}, \quad \left[\,\Delta^{1/2}{}_{;\,\mu\;\;\nu}^{\;\;\;\mu}\,\right] = -\frac{1}{6}\,R_{;\,\nu} \tag{D.55}$$

$$\left[\,\Delta^{1/2}{}_{;\,\mu\;\;\nu}^{\;\;\;\mu\;\;\nu}\,\right] = -\frac{1}{5}\,R_{;\;\;\alpha}^{\;\;\alpha} + \frac{1}{36}\,R^2 - \frac{1}{30}\,R^{\alpha\beta}R_{\alpha\beta} + \frac{1}{30}\,R^{\alpha\beta\gamma\delta}R_{\alpha\beta\gamma\delta}. \tag{D.56}$$

Appendix E
Corrections to the Initial Values of λ and y_t

The pole mass matching scheme used to correct the initial values for λ and $y_{\rm t}$ was developed in [6, 8]. We have used these results at the one-loop level. The correction function for the Higgs mass is

$$\Delta_{\rm H} = \frac{G_{\rm F}}{\sqrt{2}}\frac{M_{\rm Z}^2}{16\pi^2}\left[\zeta\, f_1(\zeta) + f_0(\zeta) + \zeta^{-1} f_{-1}(\zeta)\right] \tag{E.1}$$

with

$$f_1(\zeta) = 6\ln\frac{M_{\rm t}^2}{M_{\rm H}^2} + \frac{3}{2}\ln\zeta - \frac{1}{2}Z\left[\frac{1}{\zeta}\right] - Z\left[\frac{c_{\rm W}^2}{\zeta}\right] - \ln c_{\rm W}^2 + \frac{9}{2}\left(\frac{25}{9} - \frac{\pi}{\sqrt{3}}\right), \tag{E.2}$$

$$\begin{aligned} f_0(\zeta) = &-6\ln\frac{M_{\rm t}^2}{M_{\rm Z}^2}\left(1 + 2c_{\rm W}^2 - 2\frac{M_{\rm t}^2}{M_{\rm Z}^2}\right) + \frac{3c_{\rm W}^2\zeta}{\zeta - c_{\rm W}^2}\ln\frac{\zeta}{c_{\rm W}^2} + 2Z\left[\frac{1}{\zeta}\right] \\ &+ 4c_{\rm W}^2 Z\left[\frac{c_{\rm W}^2}{\zeta}\right] + \left(\frac{3c_{\rm W}^2}{s_{\rm W}^2} + 12c_{\rm W}^2\right)\ln c_{\rm W}^2 - \frac{15}{2}(1 + 2c_{\rm W}^2) \\ &- 3\frac{M_{\rm t}^2}{M_{\rm Z}^2}\left(2Z\left[\frac{M_{\rm t}^2}{M_{\rm Z}^2\zeta}\right] + 4\ln\frac{M_{\rm t}^2}{M_{\rm Z}^2} - 5\right), \end{aligned} \tag{E.3}$$

$$\begin{aligned} f_{-1}(\zeta) = &\,6\ln\frac{M_{\rm t}^2}{M_{\rm Z}^2}\left(1 + 2c_{\rm W}^2 - 4\frac{M_{\rm t}^2}{M_{\rm Z}^2}\right) - 6Z\left[\frac{1}{\zeta}\right] - 12c_{\rm W}^4 Z\left[\frac{c_{\rm W}^2}{\zeta}\right] - 12c_{\rm W}^4\ln c_{\rm W}^2 \\ &+ 8(1 + 2c_{\rm W}^4) + 24\frac{M_{\rm t}^4}{M_{\rm Z}^4}\left(\ln\frac{M_{\rm t}^2}{M_{\rm Z}^2} - 2 + Z\left[\frac{M_{\rm t}^2}{M_{\rm Z}^2\zeta}\right]\right), \end{aligned} \tag{E.4}$$

the Fermi constant $G_{\rm F}$ and $\zeta := M_{\rm H}^2/M_{\rm Z}^2$, $s_{\rm W}^2 := \sin^2\theta_{\rm W}$, $c_{\rm W}^2 := \cos^2\theta_{\rm W}$. The function $Z[z]$ is defined with respect to its domains depending on z

$$Z[z] := \begin{cases} 2A\arctan\frac{1}{A}, & (z > 1/4), \\ A\ln\frac{1+A}{1-A}, & (z < 1/4), \end{cases} \qquad A := \sqrt{|1 - 4z|}. \tag{E.5}$$

C. F. Steinwachs, *Non-minimal Higgs Inflation and Frame Dependence in Cosmology*, Springer Theses, DOI: 10.1007/978-3-319-01842-3,
© Springer International Publishing Switzerland 2014

The correction for the top-quark mass is $\Delta_{\mathrm{t}}(M_{\mathrm{H}})$ and reads

$$\Delta_{\mathrm{t}}(M_{\mathrm{H}}) = \delta_{\mathrm{t}}^{\mathrm{QCD}} + \delta_{\mathrm{t}}^{\mathrm{W}} + \delta_{t}^{\mathrm{QED}}, \quad \delta_{\mathrm{t}}^{\mathrm{QCD}} = -\frac{4\,\alpha_s(M_{\mathrm{t}})}{3\pi}, \quad \delta_{\mathrm{t}}^{\mathrm{QED}} + \delta_{\mathrm{t}}^{\mathrm{W}} = -\frac{4\,\alpha(M_{\mathrm{t}})}{9\,\pi}. \tag{E.6}$$

We have neglected two-loop corrections. The numerical values of α and α_s can be obtained from [1].

Appendix F
Transfer Equations

Not all structures appearing in the divergent part of the one-loop effective action are independent. Neglecting surface terms and making use of the Bianchi identities, we can convert certain structures into others via an integration by parts. In this way, we can reduce the number of different structures in (5.102) to a minimum. The "transfer equations" below describe explicitly how the contributions of the dependent structures are distributed among the minimal set of independent structures:

$$F\,\Phi^{a\ \ \mu}_{;\mu}\,n_a \rightarrow\ -F'\,\Gamma_4 - \frac{F}{\varphi}\,(\Gamma_3 - \Gamma_4), \tag{F.1}$$

$$\begin{aligned} F\,\Phi^a_{;\,\mu\nu}\Phi^{;\,\mu\nu}_a \rightarrow\ & F'\,\Gamma_{21} - F(\Gamma_7 - \Gamma_{17}) + \frac{F'}{\varphi}\,\Gamma_{15} \\ & - \frac{1}{2}\left(\frac{F'}{\varphi} - F''\right)\Gamma_{14} + \frac{1}{2}F'\,\Gamma_{19} - \frac{1}{2}\frac{F'}{\varphi}\,\Gamma_{12}, \end{aligned} \tag{F.2}$$

$$\begin{aligned} F\,\Phi^a_{;\,\mu\nu}\,n_a\Phi^{b;\,\mu\nu}\,n_b \rightarrow\ & \left(4\frac{F}{\varphi^2} - \frac{5}{2}\frac{F'}{\varphi} + \frac{1}{2}F''\right)\Gamma_{13} + \left(2\frac{F'}{\varphi} - 4\frac{F}{\varphi^2}\right)\Gamma_{16} \\ & + \left(\frac{3}{2}F' - 3\frac{F}{\varphi}\right)\Gamma_{20} + \left(\frac{1}{2}\frac{F'}{\varphi} - \frac{F}{\varphi^2}\right)\Gamma_{14} \\ & + 2\frac{F}{\varphi}\,\Gamma_{21} - F\Gamma_8 + \frac{F}{\varphi}\,\Gamma_{19} + F\,\Gamma_{18} + \frac{F}{\varphi^2}\,\Gamma_{15}, \end{aligned} \tag{F.3}$$

$$\begin{aligned} F\,\Phi^a_{;\,\mu\nu}\,n_a\,\Phi^{b,\,\mu}\,\Phi^{;\,\nu}_b \rightarrow\ & -\frac{1}{2}\left(\frac{F}{\varphi} - F'\right)\Gamma_{14} + \left(\frac{F}{\varphi} - F'\right)\Gamma_{16} \\ & - F\,\Gamma_{21} + \frac{1}{2}\,F\,\Gamma_{19} + \frac{1}{2}\frac{F}{\varphi}\Gamma_{12} - \frac{F}{\varphi}\Gamma_{15}, \end{aligned} \tag{F.4}$$

$$F\,\Phi^a_{;\,\mu\nu}\Phi^{;\,\mu}_a\,\Phi^{;\,\nu}_b\,n^b \rightarrow \frac{1}{2}\left(\frac{F}{\varphi} - F'\right)\Gamma_{14} - \frac{F}{2}\,\Gamma_{19} - \frac{1}{2}\frac{F}{\varphi}\,\Gamma_{12}, \tag{F.5}$$

C. F. Steinwachs, *Non-minimal Higgs Inflation and Frame Dependence in Cosmology*, Springer Theses, DOI: 10.1007/978-3-319-01842-3,
© Springer International Publishing Switzerland 2014

$$F\,\Phi^a_{;\,\mu\nu}\,n_a\,\Phi_b^{;\,\mu}\,n^b\,\Phi_c^{;\,\nu}\,n^c \to \frac{1}{2}\left(3\,\frac{F}{\varphi} - F'\right)\Gamma_{13} - \frac{F}{\varphi}\,\Gamma_{16} - \frac{1}{2}\,F\,\Gamma_{20} - \frac{1}{2}\,\frac{F}{\varphi}\,\Gamma_{14}, \tag{F.6}$$

$$F\,R^{\mu\nu}\,\Phi^a_{;\,\mu\nu}\,n_a \to \left(\frac{F}{\varphi} - F'\right)\Gamma_8 - \frac{1}{2}\left(\frac{F}{\varphi} - F'\right)\Gamma_{10} - \frac{F}{\varphi}\,\Gamma_7 + \frac{1}{2}\,F\,\Gamma_{11} + \frac{1}{2}\,\frac{F}{\varphi}\,\Gamma_9, \tag{F.7}$$

$$F\,R^{\mu\nu\rho\sigma}R_{\mu\nu\rho\sigma} \to 4\,F\,\Gamma_5 - F\,\Gamma_6. \tag{F.8}$$

The last equation is a topological invariant, the Gauss–Bonnet identity. In the single field limit, cf. Table 5.1, the other seven transfer equations reduce to the three reduction formulas given in [5]. In principle, even more scalar invariants composed of different scalar contractions between derivatives ∇_μ and field variables $(g_{\mu\nu},\ \Phi^a)$, containing up to four derivatives, can appear at the one-loop level, but they do not occur in our calculations.

Appendix G
Gradient Structures

As already explained in Sect. 4.5.2, the coefficients α_i, $i = 12, ..., 21$ do all correspond to gradient structures, symbolically denoted as $\partial^4\Phi\Phi\Phi\Phi$. In the cosmological model of Higgs inflation, discussed in Chap. 6, these structures are additionally suppressed compared to (5.108)–(5.118) and thus less important from the viewpoint of an effective field theory. For the sake of completeness we list their coefficients in a closed form.

The coefficient α_{12} belonging to the structure $(\Phi^a_{\ ,\mu}\Phi_a^{;\mu})^2$:

$$\alpha_{12} = s^2\Bigg(\frac{25\,(U')^8}{48G^2\varphi^4U^4} + \frac{3\,(U')^8}{16U^6} - \frac{(U')^7}{4G\varphi^3U^4} - \frac{3\,(U')^7}{8\varphi U^5} + \frac{5G'\,(U')^6}{4G^2\varphi^3U^3} - \frac{11\,(U')^6}{72G\varphi^4U^3}$$
$$- \frac{9\,(U')^6}{8\varphi^2U^4} - \frac{13G'\,(U')^5}{12G\varphi^2U^3} - \frac{G'\,(U')^5}{2U^4} - \frac{3U''\,(U')^5}{4G\varphi^3U^3} + \frac{35\,(U')^5}{12\varphi^3U^3}$$
$$+ \frac{13\,(G')^2\,(U')^4}{24G^2\varphi^2U^2} - \frac{G'\,(U')^4}{G\varphi^3U^2} + \frac{15G'\,(U')^4}{8\varphi U^3} + \frac{3G''\,(U')^4}{8U^3} - \frac{695\,(U')^4}{432\varphi^4U^2}$$
$$+ \frac{(G')^2\,(U')^3}{8G\varphi U^2} - \frac{47G'\,(U')^3}{18\varphi^2U^2} - \frac{3G''\,(U')^3}{4\varphi U^2} + \frac{3G'U''\,(U')^3}{4G\varphi^2U^2} - \frac{U''\,(U')^3}{4\varphi^3U^2}$$
$$- \frac{(G')^3\,(U')^2}{4G^2\varphi U} + \frac{139\,(G')^2\,(U')^2}{72G\varphi^2U} + \frac{(G')^2\,(U')^2}{8U^2} + \frac{19G'\,(U')^2}{36\varphi^3U}$$
$$+ \frac{3G'G''\,(U')^2}{4G\varphi U} + \frac{(G')^3\,U'}{6GU} - \frac{(G')^2\,U'}{12\varphi U} - \frac{G'G''U'}{4U} + \frac{G'U''U'}{4\varphi^2U}$$
$$+ \frac{(G')^4}{48G^2} - \frac{7\,(G')^3}{24G\varphi} + \frac{11\,(G')^2}{24\varphi^2} + \frac{(G'')^2}{8} - \frac{(G')^2\,G''}{8G} + \frac{G'G''}{2\varphi}\Bigg)$$
$$+ s\Bigg(-\frac{25\,(U')^6}{72G^2\varphi^4U^3} - \frac{(U')^6}{8U^5} + \frac{(U')^5}{6G\varphi^3U^3} - \frac{37\,(U')^5}{8\varphi U^4} + \frac{5G'\,(U')^4}{12G^2\varphi^3U^2}$$
$$- \frac{43\,(U')^4}{216G\varphi^4U^2} + \frac{21\,(U')^4}{4\varphi^2U^3} + \frac{13G'\,(U')^3}{36G\varphi^2U^2} + \frac{11G'\,(U')^3}{12U^3} + \frac{U''\,(U')^3}{2G\varphi^3U^2}$$

C. F. Steinwachs, *Non-minimal Higgs Inflation and Frame Dependence in Cosmology*, Springer Theses, DOI: 10.1007/978-3-319-01842-3,
© Springer International Publishing Switzerland 2014

$$+\frac{3U''(U')^3}{2\varphi U^3}-\frac{107(U')^3}{36\varphi^3U^2}+\frac{77(G')^2(U')^2}{72G^2\varphi^2U}+\frac{15G'(U')^2}{4G\varphi^3U}-\frac{15G'(U')^2}{8\varphi U^2}$$
$$-\frac{G''(U')^2}{8U^2}-\frac{3U''(U')^2}{\varphi^2U^2}-\frac{(U')^2}{12\varphi^4U}+\frac{5(G')^2U'}{6G\varphi U}+\frac{13G'U'}{6\varphi^2U}-\frac{G'U''U'}{4G\varphi^2U}$$
$$+\frac{U''U'}{12\varphi^3U}-\frac{(G')^3}{6G^2\varphi}-\frac{13(G')^2}{12G\varphi^2}-\frac{G'G''}{4G\varphi}+\frac{G'U''}{2\varphi U}\Bigg)+\frac{25(U')^4}{432G^2\varphi^4U^2}$$
$$+\frac{(U')^4}{48U^4}+\frac{59(U')^3}{36G\varphi^3U^2}+\frac{19(U')^3}{12\varphi U^3}-\frac{G'(U')^2}{9G^2\varphi^3U}+\frac{5(U')^2}{12G\varphi^4U}+\frac{17(U')^2}{3\varphi^2U^2}$$
$$-\frac{4G'U'}{3G\varphi^2U}-\frac{G'U'}{4U^2}-\frac{5U''U'}{12G\varphi^3U}-\frac{U''U'}{2\varphi U^2}-\frac{5U'}{3\varphi^3U}+\frac{13GU'}{6\varphi U^2}$$
$$+\frac{(N-1)(G')^2}{2G^2\varphi^2}-\frac{5(G')^2}{12G^2\varphi^2}-\frac{7G'}{6\varphi U}+\frac{5U''}{3\varphi^2U}-\frac{G^2}{4U^2} \tag{G.1}$$

The coefficient α_{13} belonging to the structure $(\Phi^a_{,\mu}n_a\Phi^{b,\mu}n_b)^2$:

$$\alpha_{13}=s^4\Bigg(\frac{81(U')^{12}}{32U^8}-\frac{81U''(U')^{10}}{4U^7}-\frac{27G'(U')^9}{8U^6}+\frac{243(U'')^2(U')^8}{4U^6}$$
$$+\frac{81G'U''(U')^7}{4U^5}-\frac{81(U'')^3(U')^6}{U^5}+\frac{27(G')^2(U')^6}{16U^4}-\frac{81G'(U'')^2(U')^5}{2U^4}$$
$$+\frac{81(U'')^4(U')^4}{2U^4}-\frac{27(G')^2U''(U')^4}{4U^3}-\frac{3(G')^3(U')^3}{8U^2}$$
$$+\frac{27G'(U'')^3(U')^3}{U^3}+\frac{27(G')^2(U'')^2(U')^2}{4U^2}+\frac{3(G')^3U''U'}{4U}+\frac{(G')^4}{32}\Bigg)$$
$$+s^3\Bigg(-\frac{3(U')^{10}}{2G\varphi^2U^6}+\frac{45(U')^{10}}{8U^7}-\frac{81(U')^9}{8\varphi U^6}-\frac{9G'(U')^8}{4G\varphi U^5}+\frac{19U''(U')^8}{4G\varphi^2U^5}$$
$$+\frac{27U''(U')^8}{8U^6}+\frac{25(U')^8}{4\varphi^2U^5}+\frac{G'(U')^7}{G\varphi^2U^4}-\frac{15G'(U')^7}{4U^5}+\frac{81U''(U')^7}{2\varphi U^5}$$
$$+\frac{5(G')^2(U')^6}{8GU^4}+\frac{155(U'')^2(U')^6}{8G\varphi^2U^4}-\frac{351(U'')^2(U')^6}{4U^5}+\frac{39G'(U')^6}{8\varphi U^4}$$
$$+\frac{9G''(U')^6}{8U^4}+\frac{3G'U''(U')^6}{G\varphi U^4}-\frac{305U''(U')^6}{12\varphi^2U^4}+\frac{3(G')^2(U')^5}{2G\varphi U^3}$$
$$-\frac{81(U'')^2(U')^5}{2\varphi U^4}-\frac{25G'(U')^5}{6\varphi^2U^3}+\frac{15G'U''(U')^5}{4G\varphi^2U^3}-\frac{39G'U''(U')^5}{4U^4}$$
$$+\frac{261(U'')^3(U')^4}{2U^4}+\frac{29(G')^2(U')^4}{24U^3}+\frac{93G'(U'')^2(U')^4}{4G\varphi U^3}+\frac{803(U'')^2(U')^4}{24\varphi^2U^3}$$
$$+\frac{(G')^2U''(U')^4}{GU^3}-\frac{8G'U''(U')^4}{\varphi U^3}-\frac{9G''U''(U')^4}{2U^3}+\frac{(G')^2(U')^3}{8\varphi U^2}$$
$$+\frac{39G'(U'')^2(U')^3}{U^3}-\frac{3G'G''(U')^3}{4U^2}+\frac{9(G')^2U''(U')^3}{G\varphi U^2}+\frac{41G'U''(U')^3}{4\varphi^2U^2}$$
$$-\frac{27(U'')^4(U')^2}{U^3}+\frac{3(G')^3(U')^2}{4G\varphi U}+\frac{3(G')^2(U')^2}{4\varphi^2U}-\frac{31(G')^2(U'')^2(U')^2}{8GU^2}$$

$$
\begin{aligned}
&+\frac{13G'(U'')^2(U')^2}{4\varphi U^2}+\frac{9G''(U'')^2(U')^2}{2U^2}+\frac{41(G')^2U''(U')^2}{24U^2}-\frac{(G')^3U'}{4U}\\
&-\frac{9G'(U'')^3U'}{U^2}-\frac{3(G')^3U''U'}{2GU}+\frac{3(G')^2U''U'}{2\varphi U}+\frac{3G'G''U''U'}{2U}-\frac{(G')^4}{8G}+\frac{(G')^3}{8\varphi}\\
&-\frac{13(G')^2(U'')^2}{24U}+\frac{1}{8}(G')^2G''\Bigg)+s^2\Bigg(\frac{25(U')^8}{16G^2\varphi^4U^4}-\frac{19(U')^8}{6G\varphi^2U^5}-\frac{189(U')^8}{16U^6}\\
&-\frac{5(U')^7}{4G\varphi^3U^4}-\frac{123(U')^7}{8\varphi U^5}+\frac{15G'(U')^6}{4G^2\varphi^3U^3}+\frac{G'(U')^6}{4G\varphi U^4}-\frac{U''(U')^6}{6G\varphi^2U^4}+\frac{273U''(U')^6}{8U^5}\\
&+\frac{133(U')^6}{24G\varphi^4U^3}-\frac{223(U')^6}{72\varphi^2U^4}-\frac{25G'(U')^5}{24G\varphi^2U^3}+\frac{85G'(U')^5}{8U^4}-\frac{5G''(U')^5}{4G\varphi U^3}-\frac{15G'U''(U')^5}{4G^2\varphi^2U^3}\\
&-\frac{33U''(U')^5}{4G\varphi^3U^3}+\frac{87U''(U')^5}{4\varphi U^4}+\frac{3U'''(U')^5}{4G\varphi^2U^3}-\frac{3U'''(U')^5}{2U^4}-\frac{113(U')^5}{12\varphi^3U^3}+\frac{(G')^2(U')^4}{G^2\varphi^2U^2}\\
&-\frac{4(G')^2(U')^4}{3GU^3}-\frac{55(U'')^2(U')^4}{6G\varphi^2U^3}-\frac{81(U'')^2(U')^4}{8U^4}+\frac{5G'(U')^4}{2G\varphi^3U^2}+\frac{167G'(U')^4}{24\varphi U^3}\\
&-\frac{7G''(U')^4}{8U^3}+\frac{45G'U''(U')^4}{4G\varphi U^3}+\frac{881U''(U')^4}{36\varphi^2U^3}+\frac{889(U')^4}{144\varphi^4U^2}-\frac{7(G')^2(U')^3}{8G\varphi U^2}\\
&+\frac{9(U'')^2(U')^3}{2\varphi U^3}+\frac{325G'(U')^3}{72\varphi^2U^2}-\frac{8G''(U')^3}{3\varphi U^2}-\frac{9(G')^2U''(U')^3}{2G^2\varphi U^2}-\frac{31G'U''(U')^3}{2G\varphi^2U^2}\\
&-\frac{25G'U''(U')^3}{U^3}-\frac{13G''U''(U')^3}{4G\varphi U^2}-\frac{11U''(U')^3}{4\varphi^3U^2}-\frac{3G'U'''(U')^3}{4G\varphi U^2}-\frac{6U''U'''(U')^3}{U^3}\\
&+\frac{U'''(U')^3}{4\varphi^2U^2}-\frac{3(G')^3(U')^2}{2G^2\varphi U}-\frac{39(U'')^3(U')^2}{2U^3}-\frac{8(G')^2(U')^2}{3G\varphi^2U}-\frac{19(G')^2(U')^2}{8U^2}\\
&-\frac{43G'(U'')^2(U')^2}{4G\varphi U^2}-\frac{713(U'')^2(U')^2}{72\varphi^2U^2}-\frac{31G'(U')^2}{12\varphi^3U}-\frac{125(G')^2U''(U')^2}{24GU^2}\\
&-\frac{2G'U''(U')^2}{3\varphi U^2}+\frac{7G''U''(U')^2}{4U^2}-\frac{G'U'''(U')^2}{2U^2}-\frac{(G')^3U'}{8GU}-\frac{23(G')^2U'}{12\varphi U}\\
&+\frac{17G'(U'')^2U'}{2U^2}-\frac{G'G''U'}{4U}+\frac{3(G')^3U''U'}{4G^2U}-\frac{4G'U''U'}{3\varphi^2U}+\frac{3G'G''U''U'}{2GU}\\
&+\frac{23G''U''U'}{12\varphi U}-\frac{G'U'''U'}{4\varphi U}+\frac{3(G')^4}{16G^2}+\frac{9(U'')^4}{2U^2}+\frac{(G')^3}{8G\varphi}+\frac{5(G')^2}{8\varphi^2}+\frac{(G'')^2}{8}\\
&+\frac{13(G')^2(U'')^2}{24GU}-\frac{5G'(U'')^2}{2\varphi U}-\frac{3G''(U'')^2}{2U}+\frac{(G')^2G''}{8G}+\frac{3G'G''}{4\varphi}+\frac{(G')^2U''}{2U}\Bigg)\\
&+s\Bigg(-\frac{25(U')^6}{24G^2\varphi^4U^3}-\frac{35(U')^6}{9G\varphi^2U^4}+\frac{61(U')^6}{8U^5}+\frac{7G'(U')^5}{8G^2\varphi^2U^3}-\frac{37(U')^5}{6G\varphi^3U^3}+\frac{189(U')^5}{8\varphi U^4}\\
&+\frac{5(G')^2(U')^4}{8G^3\varphi^2U^2}-\frac{3G'(U')^4}{8G^2\varphi^3U^2}-\frac{11G'(U')^4}{12G\varphi U^3}-\frac{5G''(U')^4}{8G^2\varphi^2U^2}+\frac{16U''(U')^4}{9G\varphi^2U^3}-\frac{227U''(U')^4}{8U^4}\\
&-\frac{79(U')^4}{72G\varphi^4U^2}+\frac{521(U')^4}{27\varphi^2U^3}-\frac{5(G')^2(U')^3}{2G^2\varphi U^2}-\frac{56G'(U')^3}{9G\varphi^2U^2}-\frac{13G'(U')^3}{2U^3}-\frac{19G''(U')^3}{12G\varphi U^2}\\
&+\frac{G'U''(U')^3}{G^2\varphi^2U^2}+\frac{3U''(U')^3}{2G\varphi^3U^2}-\frac{95U''(U')^3}{2\varphi U^3}-\frac{U'''(U')^3}{2G\varphi^2U^2}+\frac{3U'''(U')^3}{U^3}+\frac{101(U')^3}{36\varphi^3U^2}\\
&+\frac{3(G')^3(U')^2}{4G^3\varphi U}+\frac{17(G')^2(U')^2}{8G^2\varphi^2U}+\frac{(G')^2(U')^2}{4GU^2}+\frac{101(U'')^2(U')^2}{72G\varphi^2U^2}+\frac{53(U'')^2(U')^2}{2U^3}
\end{aligned}
$$

$$
\begin{aligned}
&+\frac{11G'\left(U'\right)^2}{8G\varphi^3U}-\frac{433G'\left(U'\right)^2}{36\varphi U^2}-\frac{41G''\left(U'\right)^2}{24G\varphi^2U}+\frac{G''\left(U'\right)^2}{U^2}-\frac{17G'U''\left(U'\right)^2}{6G\varphi U^2}-\frac{21U''\left(U'\right)^2}{4\varphi^2U^2}\\
&+\frac{3U'''\left(U'\right)^2}{\varphi U^2}+\frac{\left(U'\right)^2}{4\varphi^4U}+\frac{3\left(G'\right)^3U'}{8G^2U}+\frac{17\left(G'\right)^2U'}{12G\varphi U}+\frac{3\left(U''\right)^2U'}{2\varphi U^2}-\frac{17G'U'}{12\varphi^2U}+\frac{G'G''U'}{4GU}\\
&-\frac{G''U'}{2\varphi U}+\frac{3\left(G'\right)^2U''U'}{2G^2\varphi U}+\frac{17G'U''U'}{4G\varphi^2U}+\frac{27G'U''U'}{2U^2}+\frac{13G''U''U'}{12G\varphi U}-\frac{5U''U'}{12\varphi^3U}\\
&+\frac{G'U'''U'}{4G\varphi U}-\frac{U''U'''U'}{U^2}-\frac{U'''U'}{12\varphi^2U}-\frac{\left(G'\right)^4}{8G^3}-\frac{\left(G'\right)^3}{8G^2\varphi}-\frac{\left(U''\right)^3}{2U^2}+\frac{23\left(G'\right)^2}{24U}-\frac{\left(G'\right)^2}{4G\varphi^2}\\
&-\frac{\left(G''\right)^2}{4G}+\frac{G'\left(U''\right)^2}{G\varphi U}+\frac{\left(U''\right)^2}{6\varphi^2U}-\frac{\left(G'\right)^2G''}{8G^2}-\frac{G'G''}{2G\varphi}+\frac{\left(G'\right)^2U''}{2GU}+\frac{25G'U''}{12\varphi U}-\frac{G'U'''}{U}\Bigg)\\
&+\frac{25\left(U'\right)^4}{144G^2\varphi^4U^2}+\frac{46\left(U'\right)^4}{27G\varphi^2U^3}+\frac{85\left(U'\right)^4}{8U^4}-\frac{\left(G'\right)^3}{4G^3\varphi}-\frac{7G'\left(U'\right)^3}{24G^2\varphi^2U^2}+\frac{133\left(U'\right)^3}{36G\varphi^3U^2}+\frac{3\left(U'\right)^3}{4\varphi U^3}\\
&-\frac{23\left(G'\right)^2}{24GU}-\frac{\left(G'\right)^2}{2G^2\varphi^2}-\frac{5\left(G'\right)^2\left(U'\right)^2}{24G^3\varphi^2U}-\frac{7G'\left(U'\right)^2}{24G^2\varphi^3U}-\frac{89G'\left(U'\right)^2}{36G\varphi U^2}-\frac{\left(U'\right)^2}{4G\varphi^4U}+\frac{3\left(U'\right)^2}{2\varphi^2U^2}\\
&-\frac{\left(U''\right)^2}{6G\varphi^2U}+\frac{9\left(U''\right)^2}{2U^2}-\frac{\left(G'\right)^2U'}{2G^2\varphi U}-\frac{19G'U'}{12G\varphi^2U}+\frac{5\left(U'\right)^2G''}{24G^2\varphi^2U}+\frac{U'G''}{2G\varphi U}-\frac{9\left(U'\right)^2U''}{4G\varphi^2U^2}\\
&-\frac{3\left(U'\right)^2U''}{4U^3}+\frac{11G'U''}{12G\varphi U}+\frac{G'U'U''}{12G^2\varphi^2U}+\frac{5U'U''}{12G\varphi^3U}-\frac{6U'U''}{\varphi U^2}+\frac{U'U'''}{12G\varphi^2U}-\frac{U'U'''}{U^2}\\
&+\frac{(N-1)\left(G'\right)^2}{8G^2\varphi^2}+\frac{(N-1)\left(G'\right)^3}{8G^3\varphi}+\frac{(N-1)\left(G'\right)^4}{32G^4}-\frac{(N-1)G'G''}{4G^2\varphi}\\
&-\frac{(N-1)\left(G'\right)^2G''}{8G^3}+\frac{(N-1)\left(G''\right)^2}{8G^2}
\end{aligned}
\tag{G.2}
$$

The coefficient α_{14} belonging to the structure $(\Phi^a_{,\mu}\Phi_a^{;\mu})\,(\Phi^c_{,\nu}n_c\Phi^{d,\nu}n_d)$:

$$
\begin{aligned}
\alpha_{14}=s^3\Bigg(&\frac{\left(U'\right)^{10}}{2G\varphi^2U^6}-\frac{9\left(U'\right)^{10}}{8U^7}+\frac{9\left(U'\right)^9}{8\varphi U^6}+\frac{3G'\left(U'\right)^8}{4G\varphi U^5}-\frac{19U''\left(U'\right)^8}{12G\varphi^2U^5}+\frac{9U''\left(U'\right)^8}{2U^6}\\
&+\frac{29\left(U'\right)^8}{12\varphi^2U^5}-\frac{G'\left(U'\right)^7}{3G\varphi^2U^4}+\frac{9G'\left(U'\right)^7}{4U^5}-\frac{9U''\left(U'\right)^7}{2\varphi U^5}-\frac{5\left(G'\right)^2\left(U'\right)^6}{24GU^4}\\
&-\frac{155\left(U''\right)^2\left(U'\right)^6}{24G\varphi^2U^4}-\frac{9\left(U''\right)^2\left(U'\right)^6}{2U^5}-\frac{19G'\left(U'\right)^6}{8\varphi U^4}-\frac{9G''\left(U'\right)^6}{8U^4}-\frac{G'U''\left(U'\right)^6}{G\varphi U^4}\\
&-\frac{343U''\left(U'\right)^6}{36\varphi^2U^4}-\frac{\left(G'\right)^2\left(U'\right)^5}{2G\varphi U^3}+\frac{9\left(U''\right)^2\left(U'\right)^5}{2\varphi U^4}-\frac{29G'\left(U'\right)^5}{18\varphi^2U^3}-\frac{5G'U''\left(U'\right)^5}{4G\varphi^2U^3}\\
&-\frac{15G'U''\left(U'\right)^5}{2U^4}-\frac{95\left(G'\right)^2\left(U'\right)^4}{72U^3}-\frac{31G'\left(U''\right)^2\left(U'\right)^4}{4G\varphi U^3}+\frac{493\left(U''\right)^2\left(U'\right)^4}{72\varphi^2U^3}\\
&-\frac{\left(G'\right)^2U''\left(U'\right)^4}{3GU^3}+\frac{26G'U''\left(U'\right)^4}{3\varphi U^3}+\frac{9G''U''\left(U'\right)^4}{2U^3}+\frac{29\left(G'\right)^2\left(U'\right)^3}{24\varphi U^2}\\
&+\frac{6G'\left(U''\right)^2\left(U'\right)^3}{U^3}+\frac{3G'G''\left(U'\right)^3}{4U^2}-\frac{3\left(G'\right)^2U''\left(U'\right)^3}{G\varphi U^2}+\frac{31G'U''\left(U'\right)^3}{12\varphi^2U^2}\\
&-\frac{\left(G'\right)^3\left(U'\right)^2}{4G\varphi U}+\frac{\left(G'\right)^2\left(U'\right)^2}{4\varphi^2U}+\frac{31\left(G'\right)^2\left(U''\right)^2\left(U'\right)^2}{24GU^2}-\frac{121G'\left(U''\right)^2\left(U'\right)^2}{12\varphi U^2}
\end{aligned}
$$

$$- \frac{9G''(U'')^2(U')^2}{2U^2} + \frac{43(G')^2U''(U')^2}{18U^2} + \frac{(G')^3U'}{4U} + \frac{(G')^3U''U'}{2GU}$$

$$- \frac{7(G')^2U''U'}{2\varphi U} - \frac{3G'G''U''U'}{2U} + \frac{(G')^4}{24G} - \frac{7(G')^3}{24\varphi} - \frac{5(G')^2(U'')^2}{72U}$$

$$- \frac{1}{8}(G')^2G''\bigg) + s^2\bigg(- \frac{25(U')^8}{24G^2\varphi^4U^4} - \frac{13(U')^8}{9G\varphi^2U^5} - \frac{15(U')^8}{2U^6} + \frac{(U')^7}{2G\varphi^3U^4}$$

$$+ \frac{39(U')^7}{4\varphi U^5} - \frac{5G'(U')^6}{2G^2\varphi^3U^3} + \frac{7G'(U')^6}{12G\varphi U^4} - \frac{7U''(U')^6}{9G\varphi^2U^4} + \frac{141U''(U')^6}{8U^5}$$

$$+ \frac{11(U')^6}{36G\varphi^4U^3} - \frac{185(U')^6}{54\varphi^2U^4} + \frac{127G'(U')^5}{72G\varphi^2U^3} + \frac{11G'(U')^5}{8U^4} - \frac{5G''(U')^5}{12G\varphi U^3}$$

$$+ \frac{5G'U''(U')^5}{4G^2\varphi^2U^3} + \frac{3U''(U')^5}{2G\varphi^3U^3} - \frac{111U''(U')^5}{4\varphi U^4} - \frac{3U'''(U')^5}{4G\varphi^2U^3} - \frac{35(U')^5}{6\varphi^3U^3}$$

$$- \frac{7(G')^2(U')^4}{8G^2\varphi^2U^2} + \frac{17(G')^2(U')^4}{18GU^3} + \frac{55(U'')^2(U')^4}{18G\varphi^2U^3} - \frac{27(U'')^2(U')^4}{4U^4}$$

$$+ \frac{2G'(U')^4}{G\varphi^3U^2} - \frac{47G'(U')^4}{9\varphi U^3} + \frac{G''(U')^4}{U^3} - \frac{131G'U''(U')^4}{12G\varphi U^3} + \frac{1249U''(U')^4}{108\varphi^2U^3}$$

$$+ \frac{695(U')^4}{216\varphi^4U^2} - \frac{(G')^2(U')^3}{4G\varphi U^2} + \frac{39(U'')^2(U')^3}{2\varphi U^3} + \frac{1307G'(U')^3}{216\varphi^2U^2} + \frac{31G''(U')^3}{36\varphi U^2}$$

$$+ \frac{3(G')^2U''(U')^3}{2G^2\varphi U^2} - \frac{11G'U''(U')^3}{12G\varphi^2U^2} + \frac{29G'U''(U')^3}{4U^3} - \frac{13G''U''(U')^3}{12G\varphi U^2}$$

$$+ \frac{U''(U')^3}{2\varphi^3U^2} + \frac{3G'U'''(U')^3}{4G\varphi U^2} - \frac{U'''(U')^3}{4\varphi^2U^2} + \frac{3(G')^3(U')^2}{4G^2\varphi U} - \frac{97(G')^2(U')^2}{24G\varphi^2U}$$

$$+ \frac{2(G')^2(U')^2}{3U^2} + \frac{25G'(U'')^2(U')^2}{4G\varphi U^2} - \frac{583(U'')^2(U')^2}{216\varphi^2U^2} - \frac{19G'(U')^2}{18\varphi^3U} - \frac{7G'G''(U')^2}{4G\varphi U}$$

$$+ \frac{143(G')^2U''(U')^2}{72GU^2} - \frac{6G'U''(U')^2}{\varphi U^2} - \frac{17G''U''(U')^2}{4U^2} - \frac{11(G')^3U'}{24GU} - \frac{2(G')^2U'}{9\varphi U}$$

$$- \frac{13G'(U'')^2U'}{2U^2} + \frac{G'G''U'}{2U} - \frac{(G')^3U''U'}{4G^2U} + \frac{2(G')^2U''U'}{G\varphi U} + \frac{5G'U''U'}{12\varphi^2U} + \frac{G'G''U''U'}{2GU}$$

$$+ \frac{23G''U''U'}{36\varphi U} + \frac{G'U'''U'}{4\varphi U} - \frac{(G')^4}{12G^2} + \frac{5(G')^3}{6G\varphi} - \frac{3(G')^2}{4\varphi^2} - \frac{(G'')^2}{4} - \frac{49(G')^2(U'')^2}{72GU}$$

$$+ \frac{49G'(U'')^2}{18\varphi U} + \frac{3G''(U'')^2}{2U} + \frac{(G')^2G''}{3G} - \frac{11G'G''}{12\varphi} - \frac{2(G')^2U''}{3U}\bigg) + s\bigg(\frac{25(U')^6}{36G^2\varphi^4U^3}$$

$$+ \frac{43(U')^6}{54G\varphi^2U^4} + \frac{12(U')^6}{U^5} + \frac{5G'(U')^5}{24G^2\varphi^2U^3} - \frac{(U')^5}{3G\varphi^3U^3} - \frac{7(U')^5}{\varphi U^4} - \frac{5(G')^2(U')^4}{24G^3\varphi^2U^2}$$

$$- \frac{5G'(U')^4}{24G^2\varphi^3U^2} - \frac{47G'(U')^4}{36G\varphi U^3} + \frac{5G''(U')^4}{8G^2\varphi^2U^2} + \frac{79U''(U')^4}{54G\varphi^2U^3} - \frac{157U''(U')^4}{8U^4} + \frac{43(U')^4}{108G\varphi^4U^2}$$

$$- \frac{3281(U')^4}{324\varphi^2U^3} - \frac{(G')^2(U')^3}{G^2\varphi U^2} + \frac{137G'(U')^3}{108G\varphi^2U^2} - \frac{4G'(U')^3}{3U^3} + \frac{41G''(U')^3}{36G\varphi U^2} - \frac{5G'U''(U')^3}{6G^2\varphi^2U^2}$$

$$- \frac{U''(U')^3}{G\varphi^3U^2} + \frac{23U''(U')^3}{\varphi U^3} + \frac{U'''(U')^3}{2G\varphi^2U^2} + \frac{3U'''(U')^3}{2U^3} + \frac{107(U')^3}{18\varphi^3U^2} - \frac{(G')^3(U')^2}{4G^3\varphi U}$$

$$- \frac{25(G')^2(U')^2}{9G^2\varphi^2 U} - \frac{11(G')^2(U')^2}{12GU^2} - \frac{65(U'')^2(U')^2}{216G\varphi^2U^2} + \frac{15(U'')^2(U')^2}{4U^3} - \frac{139G'(U')^2}{24G\varphi^3 U}$$
$$+ \frac{1231G'(U')^2}{216\varphi U^2} + \frac{G'G''(U')^2}{2G^2\varphi U} + \frac{41G''(U')^2}{24G\varphi^2 U} - \frac{G''(U')^2}{24U^2} + \frac{35G'U''(U')^2}{6G\varphi U^2}$$
$$+ \frac{145U''(U')^2}{108\varphi^2U^2} - \frac{3U'''(U')^2}{\varphi U^2} + \frac{(U')^2}{6\varphi^4 U} - \frac{(G')^3U'}{8G^2U} - \frac{79(G')^2U'}{36G\varphi U} - \frac{11(U'')^2U'}{2\varphi U^2}$$
$$- \frac{41G'U'}{12\varphi^2 U} + \frac{G'G''U'}{4GU} + \frac{7G''U'}{6\varphi U} - \frac{(G')^2U''U'}{2G^2\varphi U} + \frac{4G'U''U'}{9G\varphi^2U} - \frac{2G'U''U'}{U^2} + \frac{13G''U''U'}{36G\varphi U}$$
$$- \frac{U''U'}{6\varphi^3 U} - \frac{G'U'''U'}{4G\varphi U} + \frac{U'''U'}{12\varphi^2 U} + \frac{(G')^4}{24G^3} + \frac{11(G')^3}{24G^2\varphi} - \frac{19(G')^2}{72U} + \frac{3(G')^2}{2G\varphi^2}$$
$$- \frac{(G'')^2}{12G} - \frac{11G'(U'')^2}{9G\varphi U} - \frac{(G')^2G''}{24G^2} - \frac{G'G''}{12G\varphi} - \frac{(G')^2U''}{2GU} + \frac{7G'U''}{36\varphi U} + \frac{2G''U''}{3U}$$
$$+ \frac{G'U'''}{2U}\Bigg) - \frac{25(U')^4}{216G^2\varphi^4U^2} - \frac{10(U')^4}{81G\varphi^2U^3} - \frac{25(U')^4}{8U^4} - \frac{5G'(U')^3}{72G^2\varphi^2U^2} - \frac{53(U')^3}{18G\varphi^3U^2}$$
$$- \frac{6(U')^3}{\varphi U^3} + \frac{5(G')^2(U')^2}{72G^3\varphi^2U} + \frac{25G'(U')^2}{72G^2\varphi^3U} + \frac{55G'(U')^2}{108G\varphi U^2} - \frac{5G''(U')^2}{24G^2\varphi^2U}; + \frac{125U''(U')^2}{108G\varphi^2U^2}$$
$$+ \frac{53U''(U')^2}{12U^3} - \frac{(U')^2}{6G\varphi^4U} - \frac{27(U')^2}{2\varphi^2U^2} - \frac{3G(U')^2}{U^3} + \frac{(G')^2U'}{2G^2\varphi U} + \frac{35G'U'}{12G\varphi^2U}$$
$$+ \frac{15G'U'}{4U^2} - \frac{G''U'}{2G\varphi U} + \frac{5G'U''U'}{36G^2\varphi^2U} + \frac{U''U'}{6G\varphi^3U} + \frac{83U''U'}{6\varphi U^2} - \frac{U'''U'}{12G\varphi^2U} - \frac{U'''U'}{2U^2}$$
$$- \frac{2GU'}{\varphi U^2} + \frac{(G')^3}{12G^3\varphi} + \frac{55(G')^2}{72GU} + \frac{3(G')^2}{4G^2\varphi^2} - \frac{(U'')^2}{3U^2} + \frac{G'}{\varphi U} - \frac{G'G''}{6G^2\varphi} - \frac{G''}{U}$$
$$- \frac{37G'U''}{36G\varphi U} + \frac{2GU''}{U^2} - \frac{(N-1)(G')^2}{2G^2\varphi^2} - \frac{(N-1)(G')^3}{4G^3\varphi} + \frac{(N-1)G'G''}{2G^2\varphi} \tag{G.3}$$

The coefficient α_{15} belonging to the structure $\Phi^a_{,\mu}\Phi_{a,\nu}\Phi^{b,\mu}\Phi_b^{;\nu}$:

$$\alpha_{15} = s^2\Bigg(\frac{25(U')^8}{24G^2\varphi^4U^4} + \frac{3(U')^8}{8U^6} - \frac{3(U')^7}{\varphi U^5} + \frac{5G'(U')^6}{2G^2\varphi^3U^3} + \frac{205(U')^6}{36G\varphi^4U^3} + \frac{9(U')^6}{\varphi^2U^4}$$
$$+ \frac{5G'(U')^5}{6G\varphi^2U^3} + \frac{G'(U')^5}{2U^4} - \frac{12(U')^5}{\varphi^3U^3} + \frac{13(G')^2(U')^4}{12G^2\varphi^2U^2} + \frac{6G'(U')^4}{G\varphi^3U^2} - \frac{3G'(U')^4}{\varphi U^3}$$
$$+ \frac{1681(U')^4}{216\varphi^4U^2} + \frac{(G')^2(U')^3}{G\varphi U^2} + \frac{113G'(U')^3}{18\varphi^2U^2} - \frac{(G')^3(U')^2}{2G^2\varphi U} - \frac{77(G')^2(U')^2}{36G\varphi^2U}$$
$$+ \frac{(G')^2(U')^2}{4U^2} - \frac{41G'(U')^2}{18\varphi^3U} - \frac{(G')^3U'}{6GU} - \frac{2(G')^2U'}{3\varphi U} + \frac{(G')^4}{24G^2} + \frac{(G')^3}{6G\varphi} + \frac{(G')^2}{6\varphi^2}\Bigg)$$
$$+ s\Bigg(-\frac{25(U')^6}{36G^2\varphi^4U^3} + \frac{2(U')^6}{U^5} - \frac{5(U')^5}{2G\varphi^3U^3} + \frac{(U')^5}{2\varphi U^4} - \frac{5G'(U')^4}{3G^2\varphi^3U^2} - \frac{205(U')^4}{108G\varphi^4U^2}$$
$$+ \frac{7(U')^4}{\varphi^2U^3} - \frac{16G'(U')^3}{9G\varphi^2U^2} - \frac{G'(U')^3}{6U^3} + \frac{19(U')^3}{6\varphi^3U^2} - \frac{31(G')^2(U')^2}{36G^2\varphi^2U} - \frac{2G'(U')^2}{G\varphi^3U}$$
$$- \frac{5G'(U')^2}{2\varphi U^2} - \frac{(G')^2U'}{3G\varphi U} - \frac{11G'U'}{6\varphi^2U} + \frac{(G')^3}{6G^2\varphi} + \frac{(G')^2}{4U} + \frac{(G')^2}{3G\varphi^2}\Bigg) + \frac{25(U')^4}{216G^2\varphi^4U^2}$$

$$
\begin{aligned}
&-\frac{17\left(U'\right)^4}{24U^4}+\frac{2\left(U'\right)^3}{3G\varphi^3U^2}+\frac{\left(U'\right)^3}{6\varphi U^3}+\frac{G'\left(U'\right)^2}{9G^2\varphi^3U}-\frac{\left(U'\right)^2}{3G\varphi^4U}-\frac{5\left(U'\right)^2}{3\varphi^2U^2}+\frac{G\left(U'\right)^2}{4U^3}\\
&+\frac{G'U'}{2G\varphi^2U}+\frac{U''U'}{3G\varphi^3U}+\frac{5U'}{3\varphi^3U}-\frac{GU'}{6\varphi U^2}+\frac{\left(G'\right)^2}{6G^2\varphi^2}+\frac{7G'}{6\varphi U}-\frac{5U''}{3\varphi^2U}+\frac{3G^2}{2U^2}
\end{aligned}
\tag{G.4}
$$

The coefficient α_{16} belonging to the structure $\Phi^a_{,\mu}n_a\Phi^b_{,\nu}n_b\Phi^{c,\mu}\Phi_c^{;\nu}$:

$$
\begin{aligned}
\alpha_{16}=s^3\Bigg(&\frac{\left(U'\right)^{10}}{G\varphi^2U^6}-\frac{9\left(U'\right)^{10}}{4U^7}+\frac{9\left(U'\right)^9}{\varphi U^6}+\frac{3G'\left(U'\right)^8}{2G\varphi U^5}-\frac{19U''\left(U'\right)^8}{6G\varphi^2U^5}+\frac{9U''\left(U'\right)^8}{U^6}\\
&-\frac{26\left(U'\right)^8}{3\varphi^2U^5}-\frac{2G'\left(U'\right)^7}{3G\varphi^2U^4}-\frac{36U''\left(U'\right)^7}{\varphi U^5}-\frac{5\left(G'\right)^2\left(U'\right)^6}{12GU^4}-\frac{155\left(U''\right)^2\left(U'\right)^6}{12G\varphi^2U^4}\\
&-\frac{9\left(U''\right)^2\left(U'\right)^6}{U^5}-\frac{5G'\left(U'\right)^6}{2\varphi U^4}-\frac{2G'U''\left(U'\right)^6}{G\varphi U^4}+\frac{629U''\left(U'\right)^6}{18\varphi^2U^4}-\frac{\left(G'\right)^2\left(U'\right)^5}{G\varphi U^3}\\
&+\frac{36\left(U''\right)^2\left(U'\right)^5}{\varphi U^4}+\frac{52G'\left(U'\right)^5}{9\varphi^2U^3}-\frac{5G'U''\left(U'\right)^5}{2G\varphi^2U^3}+\frac{3G'U''\left(U'\right)^5}{U^4}+\frac{13\left(G'\right)^2\left(U'\right)^4}{36U^3}\\
&-\frac{31G'\left(U''\right)^2\left(U'\right)^4}{2G\varphi U^3}-\frac{1451\left(U''\right)^2\left(U'\right)^4}{36\varphi^2U^3}-\frac{2\left(G'\right)^2U''\left(U'\right)^4}{3GU^3}-\frac{2G'U''\left(U'\right)^4}{3\varphi U^3}\\
&-\frac{4\left(G'\right)^2\left(U'\right)^3}{3\varphi U^2}-\frac{6G'\left(U''\right)^2\left(U'\right)^3}{U^3}-\frac{6\left(G'\right)^2U''\left(U'\right)^3}{G\varphi U^2}-\frac{77G'U''\left(U'\right)^3}{6\varphi^2U^2}\\
&-\frac{\left(G'\right)^3\left(U'\right)^2}{2G\varphi U}-\frac{\left(G'\right)^2\left(U'\right)^2}{\varphi^2U}+\frac{31\left(G'\right)^2\left(U''\right)^2\left(U'\right)^2}{12GU^2}+\frac{41G'\left(U''\right)^2\left(U'\right)^2}{6\varphi U^2}\\
&-\frac{11\left(G'\right)^2U''\left(U'\right)^2}{9U^2}+\frac{\left(G'\right)^3U''U'}{GU}+\frac{2\left(G'\right)^2U''U'}{\varphi U}+\frac{\left(G'\right)^4}{12G}+\frac{\left(G'\right)^3}{6\varphi}-\frac{5\left(G'\right)^2\left(U''\right)^2}{36U}\Bigg)\\
+s^2\Bigg(&-\frac{25\left(U'\right)^8}{12G^2\varphi^4U^4}+\frac{83\left(U'\right)^8}{18G\varphi^2U^5}+\frac{15\left(U'\right)^8}{2U^6}+\frac{\left(U'\right)^7}{G\varphi^3U^4}+\frac{9\left(U'\right)^7}{\varphi U^5}-\frac{5G'\left(U'\right)^6}{G^2\varphi^3U^3}\\
&-\frac{5G'\left(U'\right)^6}{6G\varphi U^4}+\frac{17U''\left(U'\right)^6}{18G\varphi^2U^4}-\frac{33U''\left(U'\right)^6}{2U^5}-\frac{205\left(U'\right)^6}{18G\varphi^4U^3}-\frac{73\left(U'\right)^6}{54\varphi^2U^4}-\frac{17G'\left(U'\right)^5}{36G\varphi^2U^3}\\
&-\frac{19G'\left(U'\right)^5}{4U^4}+\frac{5G''\left(U'\right)^5}{3G\varphi U^3}+\frac{5G'U''\left(U'\right)^5}{2G^2\varphi^2U^3}+\frac{15U''\left(U'\right)^5}{2G\varphi^3U^3}+\frac{6U''\left(U'\right)^5}{\varphi U^4}+\frac{73\left(U'\right)^5}{3\varphi^3U^3}\\
&-\frac{7\left(G'\right)^2\left(U'\right)^4}{4G^2\varphi^2U^2}+\frac{7\left(G'\right)^2\left(U'\right)^4}{18GU^3}+\frac{55\left(U''\right)^2\left(U'\right)^4}{9G\varphi^2U^3}-\frac{19G'\left(U'\right)^4}{2G\varphi^3U^2}-\frac{11G'\left(U'\right)^4}{18\varphi U^3}\\
&-\frac{G''\left(U'\right)^4}{U^3}-\frac{G'U''\left(U'\right)^4}{3G\varphi U^3}-\frac{973U''\left(U'\right)^4}{27\varphi^2U^3}-\frac{1681\left(U'\right)^4}{108\varphi^4U^2}-\frac{24\left(U''\right)^2\left(U'\right)^3}{\varphi U^3}\\
&-\frac{1537G'\left(U'\right)^3}{108\varphi^2U^2}+\frac{23G''\left(U'\right)^3}{9\varphi U^2}+\frac{3\left(G'\right)^2U''\left(U'\right)^3}{G^2\varphi U^2}+\frac{47G'U''\left(U'\right)^3}{3G\varphi^2U^2}+\frac{G'U''\left(U'\right)^3}{U^3}\\
&+\frac{13G''U''\left(U'\right)^3}{3G\varphi U^2}+\frac{5U''\left(U'\right)^3}{2\varphi^3U^2}+\frac{3\left(G'\right)^3\left(U'\right)^2}{2G^2\varphi U}+\frac{83\left(G'\right)^2\left(U'\right)^2}{12G\varphi^2U}+\frac{\left(G'\right)^2\left(U'\right)^2}{3U^2}\\
&+\frac{9G'\left(U''\right)^2\left(U'\right)^2}{2G\varphi U^2}+\frac{1361\left(U''\right)^2\left(U'\right)^2}{108\varphi^2U^2}+\frac{97G'\left(U'\right)^2}{18\varphi^3U}+\frac{G'G''\left(U'\right)^2}{G\varphi U}\\
&+\frac{29\left(G'\right)^2U''\left(U'\right)^2}{9GU^2}+\frac{20G'U''\left(U'\right)^2}{3\varphi U^2}+\frac{2G''U''\left(U'\right)^2}{U^2}+\frac{7\left(G'\right)^3U'}{12GU}+\frac{26\left(G'\right)^2U'}{9\varphi U}
\end{aligned}
$$

$$
\begin{aligned}
&+\frac{2G'(U'')^2U'}{U^2}-\frac{(G')^3U''U'}{2G^2U}-\frac{2(G')^2U''U'}{G\varphi U}+\frac{2G'U''U'}{3\varphi^2U}-\frac{2G'G''U''U'}{GU}\\
&-\frac{23G''U''U'}{9\varphi U}-\frac{(G')^4}{6G^2}-\frac{5(G')^3}{6G\varphi}-\frac{(G')^2}{2\varphi^2}+\frac{5(G')^2(U'')^2}{36GU}-\frac{2G'(U'')^2}{9\varphi U}-\frac{(G')^2G''}{3G}\\
&-\frac{G'G''}{3\varphi}+\frac{2(G')^2U''}{3U}\Bigg)+s\Bigg(\frac{25(U')^6}{18G^2\varphi^4U^3}+\frac{167(U')^6}{54G\varphi^2U^4}-\frac{9(U')^6}{U^5}-\frac{13G'(U')^5}{12G^2\varphi^2U^3}\\
&+\frac{53(U')^5}{6G\varphi^3U^3}-\frac{25(U')^5}{2\varphi U^4}-\frac{5(G')^2(U')^4}{12G^3\varphi^2U^2}+\frac{11G'(U')^4}{6G^2\varphi^3U^2}+\frac{20G'(U')^4}{9G\varphi U^3}-\frac{175U''(U')^4}{54G\varphi^2U^3}\\
&+\frac{13U''(U')^4}{2U^4}+\frac{151(U')^4}{54G\varphi^4U^2}-\frac{1735(U')^4}{81\varphi^2U^3}+\frac{7(G')^2(U')^3}{2G^2\varphi U^2}+\frac{172G'(U')^3}{27G\varphi^2U^2}+\frac{17G'(U')^3}{6U^3}\\
&+\frac{4G''(U')^3}{9G\varphi U^2}-\frac{G'U''(U')^3}{6G^2\varphi^2U^2}-\frac{U''(U')^3}{G\varphi^3U^2}+\frac{23U''(U')^3}{\varphi U^3}-\frac{161(U')^3}{18\varphi^3U^2}-\frac{(G')^3(U')^2}{2G^3\varphi U}\\
&+\frac{4(G')^2(U')^2}{9G^2\varphi^2U}+\frac{2(G')^2(U')^2}{3GU^2}-\frac{119(U'')^2(U')^2}{108G\varphi^2U^2}+\frac{3(U'')^2(U')^2}{U^3}+\frac{8G'(U')^2}{3G\varphi^3U}\\
&+\frac{289G'(U')^2}{27\varphi U^2}-\frac{G'G''(U')^2}{2G^2\varphi U}-\frac{G''(U')^2}{3U^2}-\frac{3G'U''(U')^2}{G\varphi U^2}+\frac{373U''(U')^2}{54\varphi^2U^2}-\frac{(U')^2}{3\varphi^4U}\\
&-\frac{(G')^3U'}{4G^2U}+\frac{5(G')^2U'}{18G\varphi U}+\frac{4(U'')^2U'}{\varphi U^2}+\frac{9G'U'}{2\varphi^2U}-\frac{G'G''U'}{2GU}-\frac{2G''U'}{3\varphi U}-\frac{(G')^2U''U'}{G^2\varphi U}\\
&-\frac{40G'U''U'}{9G\varphi^2U}-\frac{9G'U''U'}{2U^2}-\frac{13G''U''U'}{9G\varphi U}+\frac{U''U'}{2\varphi^3U}+\frac{(G')^4}{12G^3}-\frac{(G')^3}{3G^2\varphi}-\frac{25(G')^2}{36U}\\
&-\frac{(G')^2}{2G\varphi^2}+\frac{(G'')^2}{3G}+\frac{2G'(U'')^2}{9G\varphi U}-\frac{(U'')^2}{6\varphi^2U}+\frac{(G')^2G''}{6G^2}+\frac{5G'G''}{6G\varphi}-\frac{25G'U''}{9\varphi U}-\frac{2G''U''}{3U}\Bigg)\\
&-\frac{25(U')^4}{108G^2\varphi^4U^2}-\frac{128(U')^4}{81G\varphi^2U^3}+\frac{9(U')^4}{4U^4}+\frac{13G'(U')^3}{36G^2\varphi^2U^2}-\frac{55(U')^3}{18G\varphi^3U^2}+\frac{7(U')^3}{2\varphi U^3}\\
&+\frac{5(G')^2(U')^2}{36G^3\varphi^2U}-\frac{G'(U')^2}{18G^2\varphi^3U}+\frac{53G'(U')^2}{27G\varphi U^2}+\frac{59U''(U')^2}{54G\varphi^2U^2}-\frac{2U''(U')^2}{3U^3}+\frac{(U')^2}{3G\varphi^4U}\\
&+\frac{8(U')^2}{\varphi^2U^2}+\frac{19G(U')^2}{2U^3}-\frac{G'U'}{2G\varphi^2U}-\frac{13G'U'}{2U^2}-\frac{2G'U''U'}{9G^2\varphi^2U}-\frac{U''U'}{2G\varphi^3U}-\frac{22U''U'}{3\varphi U^2}\\
&+\frac{(G')^3}{6G^3\varphi}+\frac{7(G')^2}{36GU}+\frac{(U'')^2}{6G\varphi^2U}-\frac{2(U'')^2}{3U^2}-\frac{G'}{\varphi U}+\frac{G'G''}{6G^2\varphi}+\frac{G''}{U}+\frac{G'U''}{9G\varphi U}
\end{aligned}
\tag{G.5}
$$

The coefficient α_{17} belonging to the structure $\Phi^{a\ \mu}_{;\mu}\Phi_{a;\nu}^{\ \ \nu}$:

$$
\begin{aligned}
\alpha_{17} = s&\left(-\frac{(U')^4}{4GU^2\varphi^2}-\frac{G'U'}{2U}+\frac{(G')^2}{4G}-\frac{(U')^2}{12U\varphi^2}-\frac{3(U')^4}{4U^3}\right)\\
&+\frac{(U')^2}{12GU\varphi^2}-\frac{U'}{U\varphi}+\frac{(U')^2}{4U^2}
\end{aligned}
\tag{G.6}
$$

The coefficient α_{18} belonging to the structure $(\Phi^{c\ \mu}_{;\mu} n_c)^2$:

$$\alpha_{18} = s^2\left(\frac{9G'(U')^3}{4U^2} + \frac{(G')^2}{8} + \frac{81(U')^6}{8U^4}\right) + s\left(\frac{(U')^4}{4GU^2\varphi^2} - \frac{(G')^2}{4G} + \frac{(U')^2}{12U\varphi^2} - \frac{27(U')^4}{4U^3}\right) - \frac{(U')^2}{12GU\varphi^2} + \frac{U'}{U\varphi} + \frac{13(U')^2}{2U^2} - \frac{U''}{U} + \frac{(N-1)(G')^2}{8G^2} \tag{G.7}$$

The coefficient α_{19} belonging to the structure $(\Phi^a_{,\mu}\Phi^{;\mu}_a)(\Phi^{b\ \nu}_{;\nu} n_b)$:

$$\begin{aligned}\alpha_{19} = s^2\Big(& -\frac{(U')^7}{4G\varphi^2U^4} - \frac{27(U')^7}{8U^5} + \frac{27(U')^6}{4\varphi U^4} - \frac{3G'(U')^5}{4G\varphi U^3} - \frac{3U''(U')^5}{4G\varphi^2U^3} - \frac{(U')^5}{12\varphi^2U^3} \\ & + \frac{15G'(U')^4}{8U^3} + \frac{3(G')^2(U')^3}{8GU^2} - \frac{7G'(U')^3}{4\varphi U^2} - \frac{9G''(U')^3}{4U^2} + \frac{3G'U''(U')^3}{4G\varphi U^2} \\ & - \frac{U''(U')^3}{4\varphi^2U^2} - \frac{3(G')^2(U')^2}{4G\varphi U} + \frac{G'U''U'}{4\varphi U} + \frac{(G')^3}{8G} - \frac{(G')^2}{2\varphi} - \frac{G'G''}{4}\Big) \\ & + s\Big(\frac{(U')^5}{6G\varphi^2U^3} + \frac{15(U')^5}{8U^4} + \frac{5G'(U')^4}{8G^2\varphi^2U^2} - \frac{2(U')^4}{\varphi U^3} + \frac{3G'(U')^3}{4G\varphi U^2} + \frac{U''(U')^3}{2G\varphi^2U^2} \\ & + \frac{3U''(U')^3}{2U^3} + \frac{37(U')^3}{36\varphi^2U^2} + \frac{3(G')^2(U')^2}{4G^2\varphi U} + \frac{41G'(U')^2}{24G\varphi^2U} - \frac{3G'(U')^2}{2U^2} \\ & - \frac{3U''(U')^2}{\varphi U^2} + \frac{(G')^2U'}{4GU} + \frac{G'U'}{\varphi U} + \frac{G''U'}{2U} - \frac{G'U''U'}{4G\varphi U} + \frac{U''U'}{12\varphi^2U} - \frac{(G')^3}{8G^2} \\ & + \frac{G'U''}{2U}\Big) + \frac{53(U')^3}{36G\varphi^2U^2} - \frac{(U')^3}{4U^3} - \frac{5G'(U')^2}{24G^2\varphi^2U} + \frac{14(U')^2}{\varphi U^2} - \frac{G'U'}{2G\varphi U} \\ & - \frac{U''U'}{12G\varphi^2U} - \frac{U''U'}{2U^2} + \frac{U'}{\varphi^2U} + \frac{2GU'}{U^2} - \frac{(G')^2}{4G^2\varphi} - \frac{G'}{2U} - \frac{U''}{\varphi U} + \frac{(N-1)(G')^2}{2G^2\varphi}\end{aligned} \tag{G.8}$$

The coefficient α_{20} belonging to the structure $(\Phi^a_{,\mu}n_a\Phi^{b,\mu}n_b)(\Phi^{c\ \nu}_{;\nu}\, n_c)$:

$$\begin{aligned}\alpha_{20} = s^3\Big(&\frac{81(U')^9}{8U^6} - \frac{81U''(U')^7}{2U^5} - \frac{45G'(U')^6}{8U^4} + \frac{81(U'')^2(U')^5}{2U^4} + \frac{9G'U''(U')^4}{U^3} \\ & + \frac{3(G')^2(U')^3}{8U^2} + \frac{9G'(U'')^2(U')^2}{2U^2} + \frac{3(G')^2U''U'}{2U} + \frac{(G')^3}{8}\Big) + s^2\Big(\frac{3(U')^7}{4G\varphi^2U^4} \\ & - \frac{45(U')^7}{8U^5} - \frac{45(U')^6}{4\varphi U^4} - \frac{G'(U')^5}{2G\varphi U^3} + \frac{9U''(U')^5}{2G\varphi^2U^3} + \frac{135U''(U')^5}{4U^4} + \frac{(U')^5}{4\varphi^2U^3} \\ & + \frac{5G'(U')^4}{4G\varphi^2U^2} - \frac{21G'(U')^4}{8U^3} + \frac{9U''(U')^4}{\varphi U^3} - \frac{3(G')^2(U')^3}{8GU^2} - \frac{63(U'')^2(U')^3}{2U^3} \\ & + \frac{4G'(U')^3}{3\varphi U^2} + \frac{9G''(U')^3}{4U^2} - \frac{13G'U''(U')^3}{4G\varphi U^2} + \frac{3U''(U')^3}{2\varphi^2U^2} + \frac{5G'(U')^2}{12\varphi^2U}\end{aligned}$$

$$
\begin{aligned}
&-\frac{15G'U''(U')^2}{4U^2}-\frac{(G')^2U'}{2U}+\frac{3(G')^2U''U'}{2GU}+\frac{23G'U''U'}{12\varphi U}+\frac{(G')^3}{8G}+\frac{3(G')^2}{4\varphi}\\
&-\frac{3G'(U'')^2}{2U}+\frac{G'G''}{4}\Bigg)+s\Bigg(\frac{7(U')^5}{4G\varphi^2U^3}+\frac{15(U')^5}{8U^4}-\frac{11G'(U')^4}{8G^2\varphi^2U^2}-\frac{(U')^4}{G\varphi^3U^2}\\
&+\frac{3(U')^4}{\varphi U^3}-\frac{25G'(U')^3}{12G\varphi U^2}-\frac{U''(U')^3}{G\varphi^2U^2}-\frac{12U''(U')^3}{U^3}-\frac{7(U')^3}{3\varphi^2U^2}-\frac{19G'(U')^2}{8G\varphi^2U}\\
&+\frac{3G'(U')^2}{U^2}-\frac{3U''(U')^2}{\varphi U^2}-\frac{(U')^2}{3\varphi^3U}+\frac{(G')^2U'}{2GU}+\frac{15(U'')^2U'}{2U^2}-\frac{G'U'}{4\varphi U}+\frac{13G'U''U'}{12G\varphi U}\\
&+\frac{U''U'}{6\varphi^2U}-\frac{(G')^3}{8G^2}-\frac{(G')^2}{2G\varphi}-\frac{G'G''}{2G}\Bigg)-\frac{13(U')^3}{6G\varphi^2U^2}-\frac{29(U')^3}{4U^3}+\frac{11G'(U')^2}{24G^2\varphi^2U}\\
&+\frac{(U')^2}{3G\varphi^3U}-\frac{33(U')^2}{2\varphi U^2}+\frac{G'U'}{4G\varphi U}-\frac{3U'}{\varphi^2U}-\frac{U'U''}{6G\varphi^2U}+\frac{33U'U''}{2U^2}+\frac{3U''}{\varphi U}-\frac{U'''}{U}\\
&-\frac{(N-1)(G')^2}{4G^2\varphi}-\frac{(N-1)(G')^3}{8G^3}+\frac{(N-1)G'G''}{4G^2}
\end{aligned}
\tag{G.9}
$$

The coefficient α_{21} belonging to the structure $\Phi^{a\ \ \mu}_{;\mu}\Phi^{;\nu}_{a}\Phi^{b}_{,\nu}n_b$:

$$
\begin{aligned}
\alpha_{21}=s^2\Bigg(&-\frac{(U')^7}{2G\varphi^2U^4}-\frac{9(U')^7}{4U^5}+\frac{9(U')^6}{2\varphi U^4}+\frac{5G'(U')^5}{4G\varphi U^3}-\frac{15U''(U')^5}{4G\varphi^2U^3}+\frac{9U''(U')^5}{2U^4}\\
&-\frac{(U')^5}{6\varphi^2U^3}-\frac{5G'(U')^4}{4G\varphi^2U^2}-\frac{3G'(U')^4}{4U^3}-\frac{9U''(U')^4}{\varphi U^3}+\frac{5G'(U')^3}{12\varphi U^2}+\frac{5G'U''(U')^3}{2G\varphi U^2}\\
&-\frac{5U''(U')^3}{4\varphi^2U^2}+\frac{3(G')^2(U')^2}{4G\varphi U}-\frac{5G'(U')^2}{12\varphi^2U}+\frac{3G'U''(U')^2}{U^2}+\frac{(G')^2U'}{4U}\\
&-\frac{3(G')^2U''U'}{2GU}-\frac{13G'U''U'}{6\varphi U}-\frac{(G')^3}{4G}-\frac{(G')^2}{4\varphi}\Bigg)+s\Bigg(-\frac{23(U')^5}{12G\varphi^2U^3}+\frac{3(U')^5}{4U^4}\\
&+\frac{3G'(U')^4}{4G^2\varphi^2U^2}+\frac{(U')^4}{G\varphi^3U^2}-\frac{(U')^4}{\varphi U^3}+\frac{4G'(U')^3}{3G\varphi U^2}+\frac{U''(U')^3}{2G\varphi^2U^2}+\frac{3U''(U')^3}{2U^3}\\
&+\frac{47(U')^3}{36\varphi^2U^2}-\frac{3(G')^2(U')^2}{4G^2\varphi U}+\frac{2G'(U')^2}{3G\varphi^2U}-\frac{G'(U')^2}{2U^2}+\frac{6U''(U')^2}{\varphi U^2}+\frac{(U')^2}{3\varphi^3U}\\
&-\frac{3(G')^2U'}{4GU}-\frac{3G'U'}{4\varphi U}-\frac{G''U'}{2U}-\frac{5G'U''U'}{6G\varphi U}-\frac{U''U'}{4\varphi^2U}+\frac{(G')^3}{4G^2}+\frac{(G')^2}{2G\varphi}\\
&+\frac{G'G''}{2G}\Bigg)+\frac{25(U')^3}{36G\varphi^2U^2}-\frac{G'(U')^2}{4G^2\varphi^2U}-\frac{(U')^2}{3G\varphi^3U}+\frac{5(U')^2}{2\varphi U^2}+\frac{G'U'}{4G\varphi U}\\
&+\frac{U''U'}{4G\varphi^2U}-\frac{U''U'}{U^2}+\frac{2U'}{\varphi^2U}-\frac{13GU'}{2U^2}+\frac{(G')^2}{4G^2\varphi}+\frac{G'}{U}-\frac{2U''}{\varphi U}
\end{aligned}
\tag{G.10}
$$

References

1. Amsler, C. et al. (Particle Data Group): Review of particle physics. Phys. Lett. B **667**, 1 (2008)
2. DeWitt, B.S.: Dynamical Theory of Groups and Fields. Blackie & Son, London (1965)
3. Parker, L.E., Toms, D.J.: Quantum Field Theory in Curved Spactime. Cambridge University Press, Cambridge (2009)
4. Poisson, E., Pound, A., Vega, I.: The motion of point particles in curved spacetime. Living Rev. Rel. **7**, 6 (2004). Cited on 19 December 2011. http://relativity.livingreviews.org/Articles/lrr-2011-7
5. Shapiro, I.L., Takata, H.: One-loop renormalization of the four-dimensional theory for quantum dilaton gravity. Phys. Rev. D **52**, 2162 (1995)
6. Sirlin, A., Zucchini, R.: Dependence of the Higgs coupling $\bar{h}_{\overline{\rm MS}}(M)$ on $m(\mathrm{H})$ and the possible onset of new physics. Nucl. Phys. B **266**, 389 (1986)
7. Synge, J.L.: Relativity: The General Theory. North-Holland, Amsterdam (1960)
8. Tarrach, R.: The pole mass in perturbative QCD. Nucl. Phys. B **183**, 384 (1981)